Geologic and Tectonic Development of the Caribbean Plate Boundary in Northern Central America

Edited by

Paul Mann
Institute for Geophysics
Jackson School of Geosciences
University of Texas at Austin
J.J. Pickle Research Campus, Building 196 (ROC)
10100 Burnet Road (R2200)
Austin, Texas 78758-4445
USA

THE
GEOLOGICAL
SOCIETY
OF AMERICA®

Special Paper 428

3300 Penrose Place, P.O. Box 9140 ▪ Boulder, Colorado 80301-9140 USA

2007

Copyright © 2007, The Geological Society of America, Inc. (GSA). All rights reserved. GSA grants permission to individual scientists to make unlimited photocopies of one or more items from this volume for noncommercial purposes advancing science or education, including classroom use. For permission to make photocopies of any item in this volume for other noncommercial, nonprofit purposes, contact the Geological Society of America. Written permission is required from GSA for all other forms of capture or reproduction of any item in the volume including, but not limited to, all types of electronic or digital scanning or other digital or manual transformation of articles or any portion thereof, such as abstracts, into computer-readable and/or transmittable form for personal or corporate use, either noncommercial or commercial, for-profit or otherwise. Send permission requests to GSA Copyright Permissions, 3300 Penrose Place, P.O. Box 9140, Boulder, Colorado 80301-9140, USA.

Copyright is not claimed on any material prepared wholly by government employees within the scope of their employment.

Published by The Geological Society of America, Inc.
3300 Penrose Place, P.O. Box 9140, Boulder, Colorado 80301-9140, USA
www.geosociety.org

Printed in U.S.A.

GSA Books Science Editor: Abhijit Basu

Library of Congress Cataloging-in-Publication Data

Geologic and tectonic development of the Caribbean plate boundary in northern Central America / edited by Paul Mann.
    p. cm. — (Special paper (Geological Society of America) ; 428)
  Includes bibliographical references.
  ISBN 978-0-8137-2428-7 (pbk.)
    1. Geology—Central America. 2. Geology—Caribbean Area. 3. Plate tectonics—Central America. 4. Plate tectonics—Caribbean Area. I. Mann, Paul, 1956–.

QE210.G462 2007
557.28—dc22

2007010711

Cover: Oblique view from the east of the Caribbean–North America plate boundary in northern Central America. The viewer is suspended above Honduras on the Caribbean plate looking westward to southern Mexico on the North America plate. The bay to the south is the Gulf of Honduras (Caribbean Sea), the Pacific Ocean is to the west, and the Gulf of Mexico is to the north.
    Actively rifted area of Chortis block on the Caribbean plate occupies left foreground of view. Topographic data with 90 meter resolution was collected by NASA's Shuttle Radar Topographic Mission in 2000 and obtained through the CGIAR Consortium for Spatial Information (http://srtm.csi.cgiar.org/).
    Image was prepared by Lisa Bingham, who used Fledermaus visualization software provided by IVS3D. Cover design by Janet Everett.

10 9 8 7 6 5 4 3 2 1

Dedication

This volume on the geology and tectonics of northern Central America is dedicated to Professor William R. Muehlberger, a pioneer in geologic and tectonic studies in that region during his 38-year career at The University of Texas at Austin.

William Rudolf "Bill" Muehlberger was born in New York City on September 26, 1923, and began undergraduate studies in Geology at Caltech in the fall of 1941. Bill served in the U.S. Marine Corps during World War II and Korea, but was never called to combat. He received his bachelor's degree and then his master's degree, in 1949, from Caltech and defended his Ph.D. dissertation, "Geology of the Soledad Basin, California," at Caltech in January 1954. Bill met his sweetheart, Sally Provine, in 1948 and they married in 1949. They had two children, Karen (1952) and Eric (1956). Bill began his teaching career at The University of Texas at Austin in 1954 and served as chairman from 1966 to 1970. During his career, Bill supervised 61 master's and 26 doctoral students and served on the committees of many others, mainly relating to tectonics, structural geology and petrology, in Texas, New Mexico, Colorado, the southern mid-continent basement, Vermont, the Canadian Rockies, the Gulf of Mexico, Mexico, Guatemala, Honduras, Turkey, and on the Moon. Work on basement rocks produced the "Basement Rock Map of the United States" at a scale of 1:2,500,000 published by the United States Geological Survey in 1968. Bill also edited an updated version of Phillip King's "Tectonic Map of North America," a 6×6 ft wall map published by the American Association of Petroleum Geologists in 1992. Bill's early interest in astronomy eventually led him to an affiliation with the National Aeronautics and Space Administration in which he provided geologic training for astronauts for their lunar missions.

Many of the papers in this volume had their roots in the thesis work of Bill's students who participated in a unique program of geologic mapping and research in Central America and Mexico. Bill credits Gabriel

---

Contributed by Pete Emmet, Paul Mann, Mark Gordon, Rick Finch, and Rob Rogers.
Photo courtesy of Joseph Jaworski (University of Texas at Austin).

Dengo for the suggestion that he undertake a geological field program in Honduras, which was essentially unmapped at that time. The early Honduran mapping, in particular, established a rigorous regional framework for the basement geology, stratigraphy, and tectonics of the region and resulted in the field-based mapping of the first four geological quadrangles in Honduras. A later focus on Mexico resulted in significant regional mapping in Chiapas, in the Sierra Madre Oriental, and in the Big Bend area in northern Chihuahua, Mexico. The geologic maps created by Bill and his students remain in wide use for a variety of purposes including urban planning, mineral and oil exploration, and academic research.

Bill Muehlberger has made a life and a career of field-based geological studies. His interest, experience, insights, and enthusiasm helped foster the careers of many young geoscientists in field-based geology, including the *Apollo* astronauts. As an advisor and mentor, he allowed his students the independence to develop their own projects, supported them financially and scientifically, and gave them his wise editorial and administrative direction when appropriate. As a scientist with a long and distinguished career, he has been rewarded with many scientific accolades from The Geological Society of America and other organizations. But for those of us who were fortunate to have worked with him, perhaps his greatest legacy is the impact and influence he has had on our lives and careers.

Contents

1. Overview of the tectonic history of northern Central America 1
   P. Mann

2. Present motion and deformation of the Caribbean plate: Constraints from new
   GPS geodetic measurements from Honduras and Nicaragua 21
   C. DeMets, G. Mattioli, P. Jansma, R.D. Rogers, C. Tenorio, and H.L. Turner

3. Transtensional deformation of the western Caribbean–North America plate
   boundary zone 37
   R.D. Rogers and P. Mann

4. Tectonic terranes of the Chortis block based on integration of regional aeromagnetic
   and geologic data 65
   R.D. Rogers, P. Mann, and P.A. Emmet

5. Cretaceous intra-arc rifting, sedimentation, and basin inversion in east-central Honduras 89
   R.D. Rogers, P. Mann, R.W. Scott, and L. Patino

6. Colon fold belt of Honduras: Evidence for Late Cretaceous collision between the continental
   Chortis block and intra-oceanic Caribbean arc 129
   R.D. Rogers, P. Mann, P.A. Emmet, and M.E. Venable

7. Petrogenesis of Central American Tertiary ignimbrites and associated Caribbean
   Sea tephra 151
   B.R. Jordan, H. Sigurdsson, S. Carey, S. Lundin, R.D. Rogers, B. Singer,
   and M. Barquero-Molina

This volume was published with the helpful support
of the ACS Petroleum Research Fund.

Overview of the tectonic history of northern Central America

Paul Mann*

Institute for Geophysics, Jackson School of Geosciences, University of Texas at Austin, J.J. Pickle Research Campus, Bldg. 196 (ROC), 10100 Burnet Road (R2200), Austin, Texas 78758-4445, USA

TECTONIC PROCESSES IN NORTHERN CENTRAL AMERICA

The geology of northern Central America provides a complex record of how tectonic and volcanic processes operate in tandem (Fig. 1). This volume draws together a multidisciplinary group of papers that look at different aspects of the three main processes that have affected northern Central America since the early Cretaceous: subduction, collision, and strike-slip faulting. Several factors that make this region favorable for integrated studies of the type assembled for this volume are (1) the outcrop record is extensive with a rock record spanning the period from the Precambrian to the Holocene; (2) there is a rich basinal record from which tectonic history can be extracted and compared to the history of the surrounding ocean basins (Figs. 2 and 3); (3) for volcanic arc studies, the outputs of the system are accessible for study in the onland arc; and (4) the oceanic inputs to arc systems are reasonably well-constrained due to the large, digital, and rapidly expanding database of offshore geological and geophysical data (e.g., Ranero et al., 2003) (Figs. 2 and 3).

While this area has many advantages for studies of both modern and ancient plate boundary processes, I believe that the regional Cenozoic tectonic development of northern Central America is more complex than is presently perceived, and that an improved Mesozoic-Cenozoic tectonic framework is an important step for understanding these complexities (Mann et al., 2007). For example, some of the longer term tectonic processes that have shaped the modern Cocos-Caribbean and Nazca-Caribbean subduction plate boundaries include (1) the detachment, translation, and rotation of the continental Chortis block of northern Central America, which forms the basement for the northern part of the modern and Miocene Central American volcanic arc; (2) slab break-off events affecting the subducting Cocos plate of the type described by Rogers et al. (2002) in northern Central America in

the Miocene and by Ferrari (2004) for Miocene arcs in southern Mexico; and (3) the influence of the North America–Caribbean strike-slip system on the northern part of the Central American arc (Rosencrantz et al., 1988; Mann, 1999) (Fig. 1).

Such significant changes in plate motions and forcing functions must be taken into account in any effort to understand the ancient and active processes of the Central American subduction system, or "factory," in northern Central America (Plank et al., 2002; Mann et al., 2007). Unfortunately, current plate reconstructions fail to account for all the constraints derived from published literature and ongoing efforts.

TECTONIC RECONSTRUCTIONS OF NORTHERN CENTRAL AMERICA

To elucidate the tectonic setting of Central America, and more specifically to provide the regional setting of the topical papers in this volume, I present a quantitative set of plate reconstructions of the Central America and Caribbean region for the period of 165–0 Ma (Fig. 4A–4H). These reconstructions are modified from Mann (1999), Rogers (2003), and Mann et al. (2007).

I use these reconstructions as a framework to discuss unresolved questions concerning the position and character of plate boundaries influencing the geology of northern Central America since the Late Jurassic. Knowledge of the tectonic setting through time is necessary to support any conclusions concerning the types of sedimentary basins present, the nature of regional deformation events, and the temporal variation of the subducted oceanic input and volcanic and plutonic output of the northern Central American "subduction factory" (Fig. 2).

The reconstructions include constraints compiled from current and recent onshore geologic work (e.g., Balzer, 1999; Johnston and Thorkelson, 1997; Plank et al., 2002) that documents and dates critical tectonic processes and changes in the Central American volcanic arc as well as extensive offshore geophysical observations, which now includes seismic reflection and refraction

---

*paulm@ig.utexas.edu

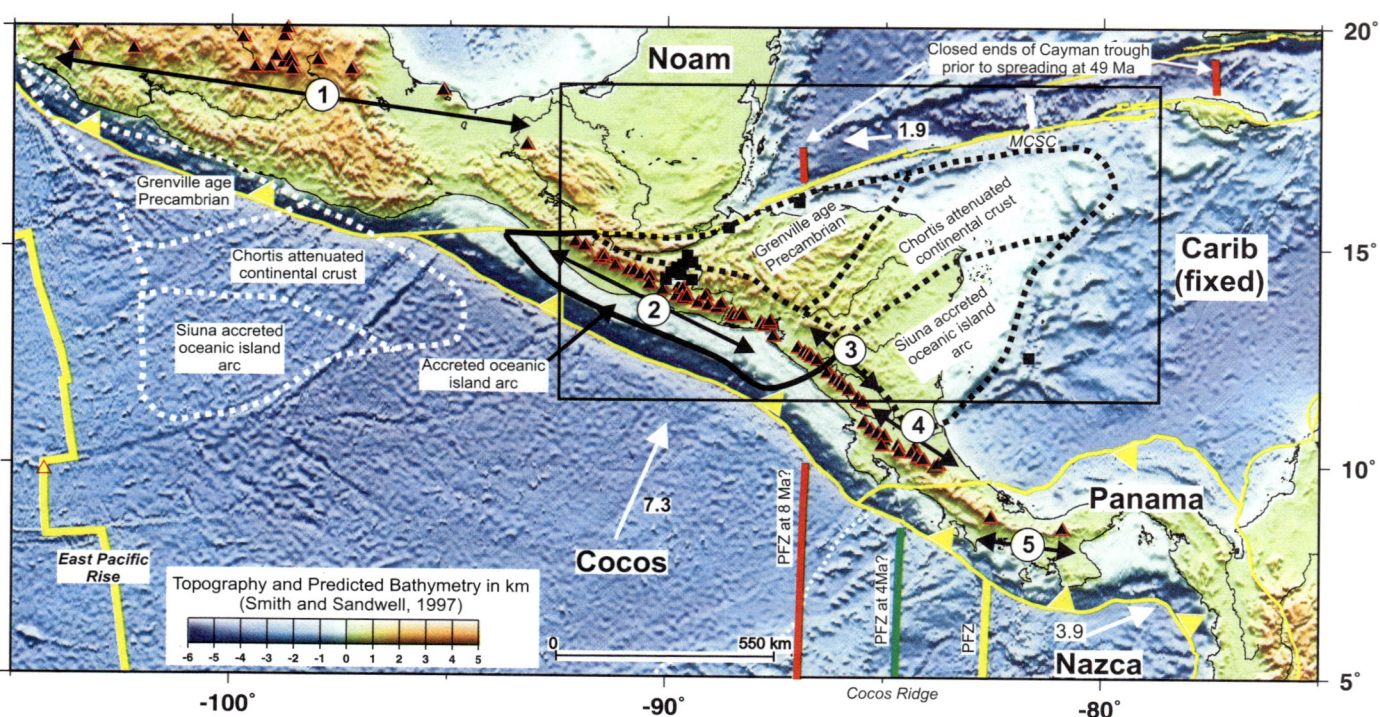

Figure 1. Tectonic and topographic setting of the Central American arc (volcanic arc segments 2–5), Trans-Mexican volcanic belt (segment 1), and the Middle American trench. Box shows study area discussed in this volume. Dark triangles represent Quaternary arc volcanoes; dark squares represent intraplate alkalic volcanoes. Plate motions in centimeters per year; data for Caribbean–North America plates are from DeMets et al. (2000), and data for Cocos-Caribbean and Nazca-Caribbean plates are from DeMets and Dixon (1999). Basement block types of overriding Caribbean plate include the following terranes described in the text: the Central Chortis terrane is underlain by Grenville age Precambrian crust detached from the southern margin of Mexico in the Late Cretaceous to early Tertiary (approximate reconstructed position of Chortis is shown); Eastern Chortis terrane is underlain by Jurassic metasedimentary rocks formed along the Mesozoic rifted margin of the Chortis block; the Siuna terrane is underlain by deformed rocks of oceanic origin that were accreted to the Eastern Chortis terrane during Late Cretaceous time (Venable, 1994; Walther et al., 2000; Rogers et al., Chapter 6, this volume); the Southern Chortis terrane is underlain by a Late Cretaceous island arc accreted to the Central Chortis terrane in Late Cretaceous time; the arc is inferred to represent a fragment of the Guerrero terrane of Mexico (Dickinson and Lawton, 2001); and the Northern Chortis magmatic zone is inferred to record a magmatic overprinting of parts of the Central and Eastern Chortis terranes. Age and position of the ends of the Cayman trough are from Leroy et al. (2000); positions of Cocos-Nazca-Caribbean triple junction are from McIntosh et al. (1993); white dashed outline is reconstructed Late Cretaceous position of the Chortis block from Rogers (2000). Carib—Caribbean; Noam—North America; MCSC—Mid-Cayman spreading center; PFZ—Panama fracture zone.

(McIntosh et al., 1993; Hinz et al., 1996; Christeson et al., 1999; von Huene et al., 2000; Walther et al., 2000; Ranero et al., 2003), gravity, magnetic (Barckhausen et al., 1998), and swath bathymetry data (von Huene et al., 2000) to constrain the tectonic history (Figs. 2 and 3).

METHODOLOGY FOR RECONSTRUCTIONS

Reconstructions presented in this overview are based on the following steps:

1. Compilation of all existing Cenozoic magnetic anomaly and fracture zone data from the Cayman trough (Caribbean Sea) and eastern Pacific Ocean; for the latter area, I use data from Udo Barckhausen and Hans Roeser of Bundesanstalt für Geowissentschaften und Rohstoffe (BGR), Germany (Fig. 2). They have an extensive anomaly and fracture zone database and have acquired new magnetic and bathymetric data in recent years (cf. Barckhausen et al., 2001). Much of these new offshore geophysical data are summarized in the volume edited by Bundschuh and Alvarado (2007).

2. Integration of main Cenozoic tectonic events in the surrounding ocean basins, which have potentially affected the subduction history of northern Central America, including (a) initiation and subsequent seafloor spreading in the Cayman trough (Fig. 3); (b) early rifting of the Farallon plate at 22.7 Ma; (c) super-fast spreading on the East Pacific Rise between 18 and 10 Ma; and (d) early history of the Cocos Ridge and its offset by the Panama fracture zone ca. 6 Ma (Fig. 2).

3. Integration of main Cenozoic tectonic events onshore in Central America that potentially affect the Central America convergent margin, including (a) eastward migration of the Chortis continental block to northern

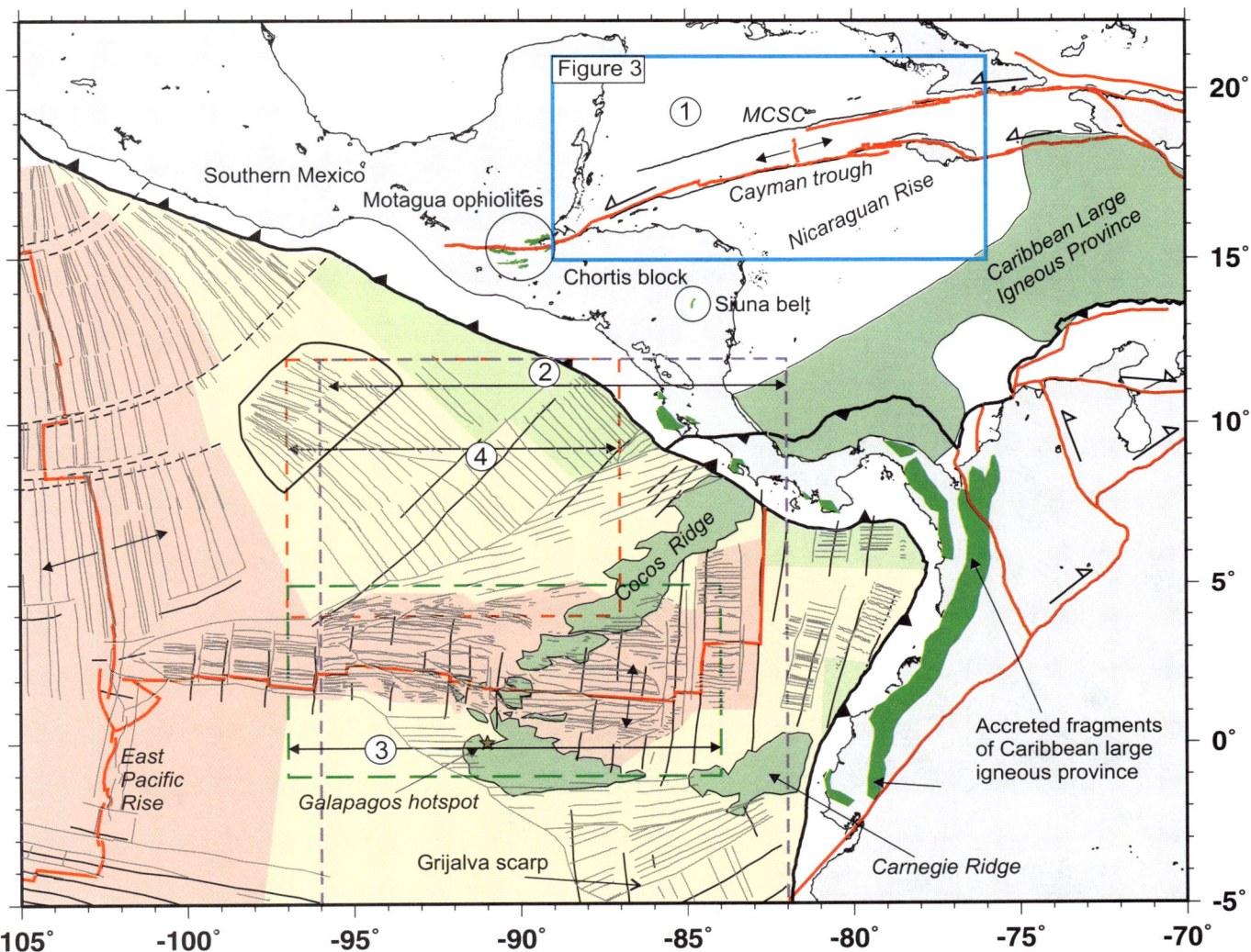

Figure 2. Marine magnetic anomalies and fracture zones that constrain tectonic reconstructions such as those shown in Figure 4 (ages of anomalies are keyed to colors as explained in the legend; all anomalies shown are from University of Texas Institute for Geophysics PLATES [2000] database): (1) Boxed area in solid blue line is area of anomaly and fracture zone picks by Leroy et al. (2000) and Rosencrantz (1994); (2) boxed area in dashed purple line shows anomalies and fracture zones of Barckhausen et al. (2001) for the Cocos plate; (3) boxed area in dashed green line shows anomalies and fracture zones from Wilson and Hey (1995); and (4) boxed area in red shows anomalies and fracture zones from Wilson (1996). Onland outcrops in green are either the obducted Cretaceous Caribbean large igneous province, including the Siuna belt, or obducted ophiolites unrelated to the large igneous province (Motagua ophiolites). The magnetic anomalies and fracture zones record the Cenozoic relative motions of all divergent plate pairs influencing the Central American subduction zone (Caribbean, Nazca, Cocos, North America, and South America). When incorporated into a plate model, these anomalies and fracture zones provide important constraints on the age and thickness of subducted crust, incidence angle of subduction, and rate of subduction for the Central American region. MCSC—Mid-Cayman Spreading Center.

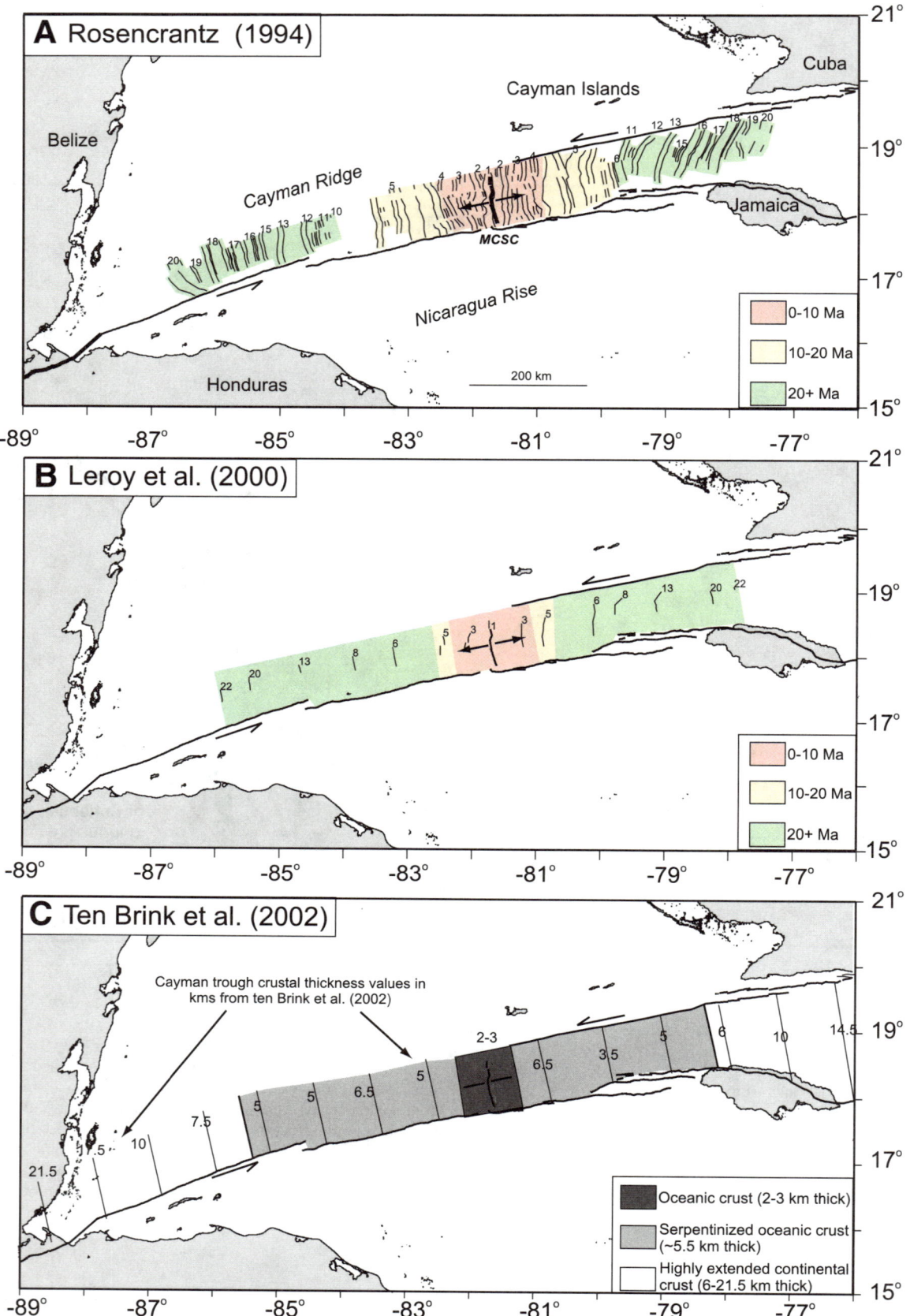

Figure 3. Marine magnetic anomaly interpretations of the Cayman trough. (A) Rosencrantz (1994) interpretation: numbers next to lineations identify magnetic anomalies. MCSC—Mid-Cayman Spreading Center. (B) Leroy et al. (2000) interpretation: numbers next to lineations identify magnetic anomalies. (C) ten Brink et al. (2002) interpretation showing larger area of thinned, continental crust. Numbers next to lineations represent their crustal thickness values.

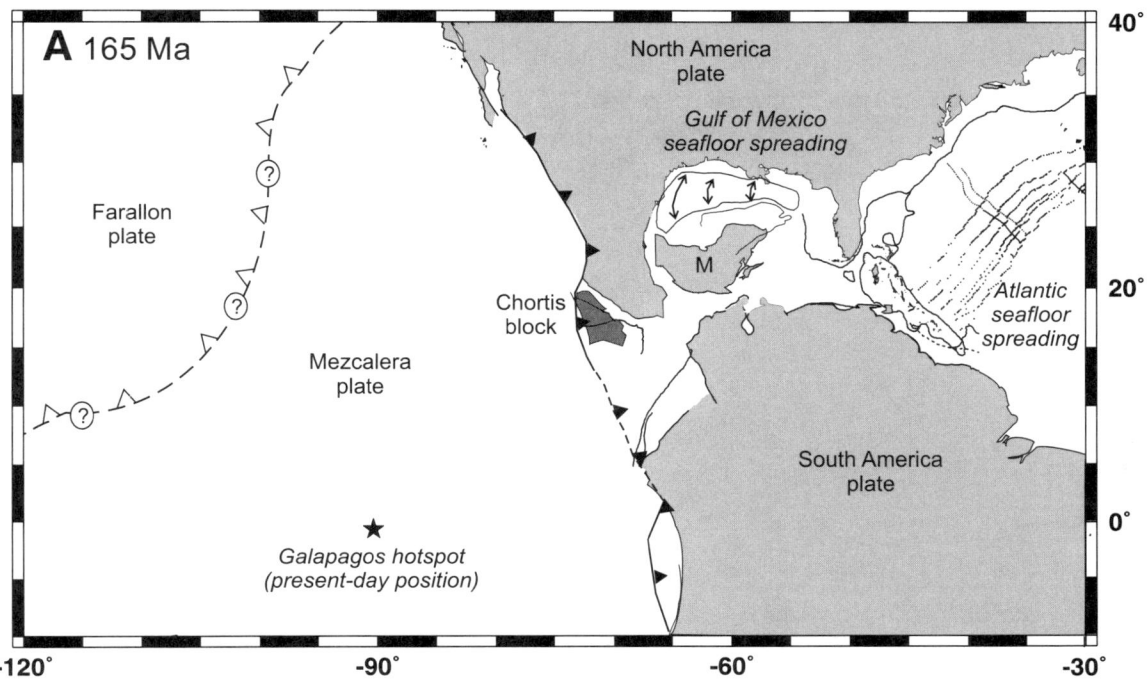

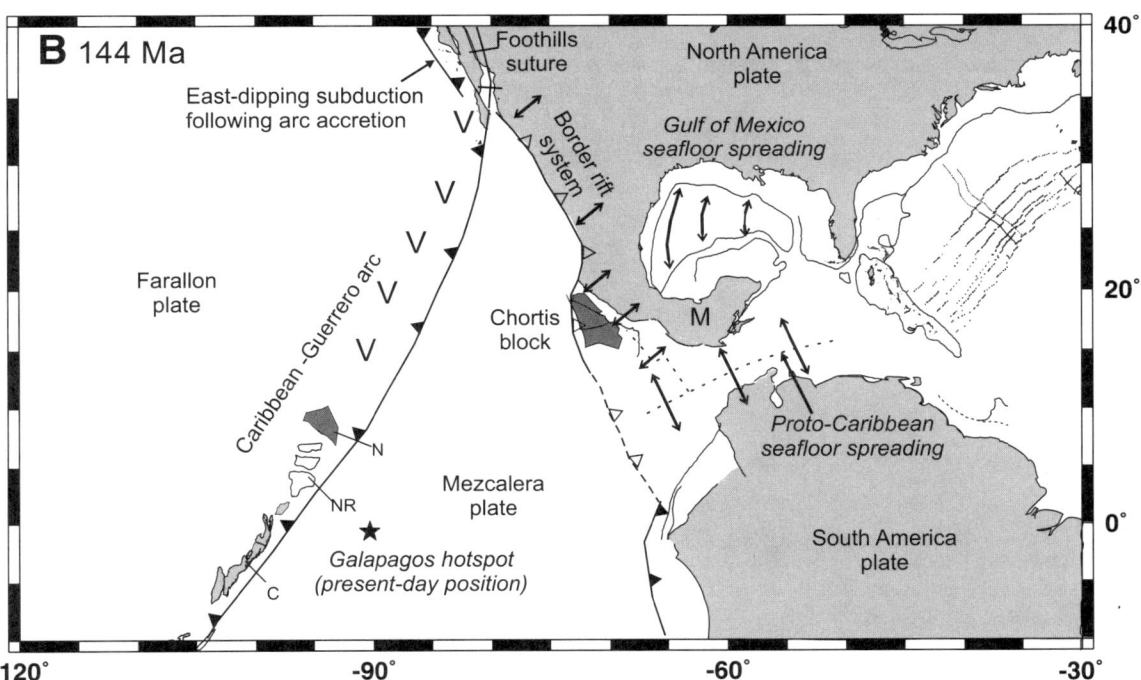

Figure 4. (*on this and following pages*) Reconstructions of the development of the Western Cordillera and Caribbean from Jurassic to present. (A) ca. 165 Ma; (B) ca. 144 Ma; (C) ca. 120 Ma; (D) ca. 90 Ma; (E) ca. 72 Ma; (F) ca. 49 Ma; (G) ca. 22 Ma; (H) present-day. See text for discussion. C—Cuba; CLIP—Caribbean large igneous province; CT—Cayman trough; G—Guerrero terrane; LA—Lesser Antilles; M—Maya block; N—Nicaragua; NR—Nicaraguan Rise; and Y—Yucatan basin. The countries of Costa Rica and Panama correspond to approximate area of the Chorotega block; the countries of Honduras, Nicaragua, and Guatemala correspond to the Chortis block.

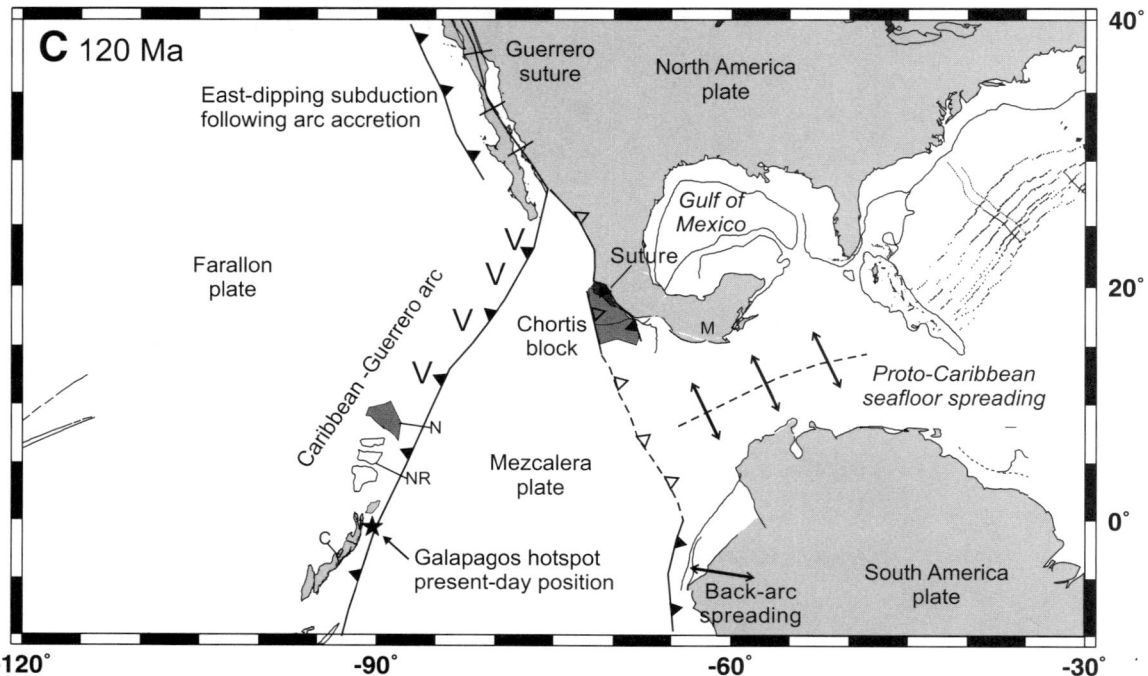
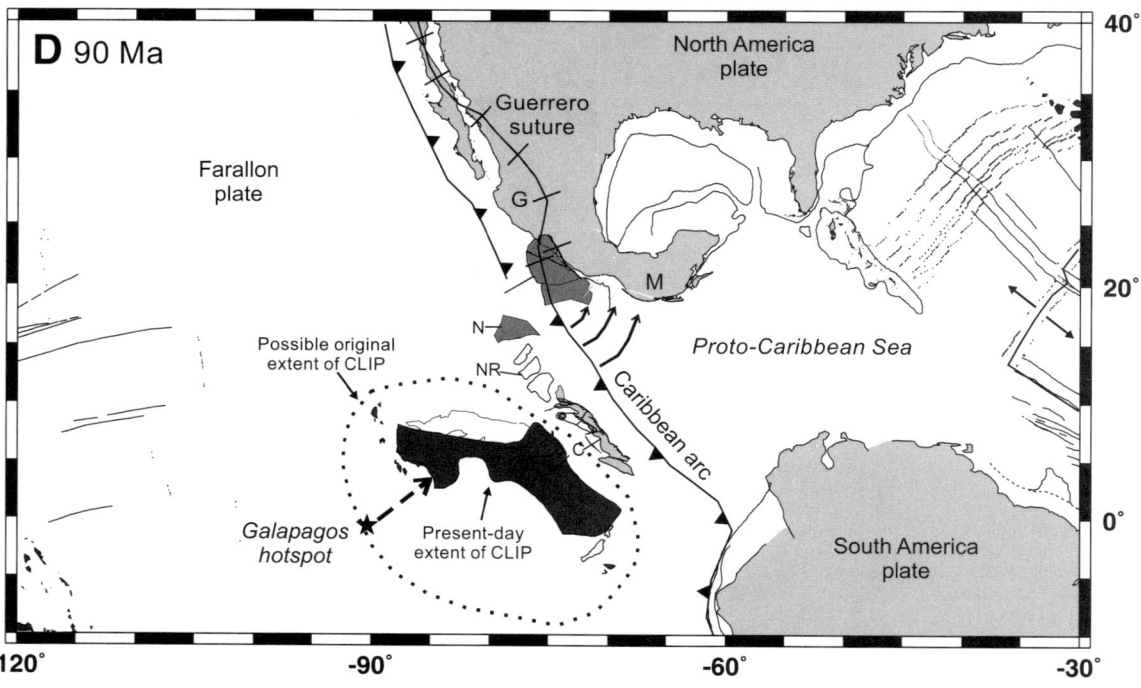

Figure 4. (*continued*)

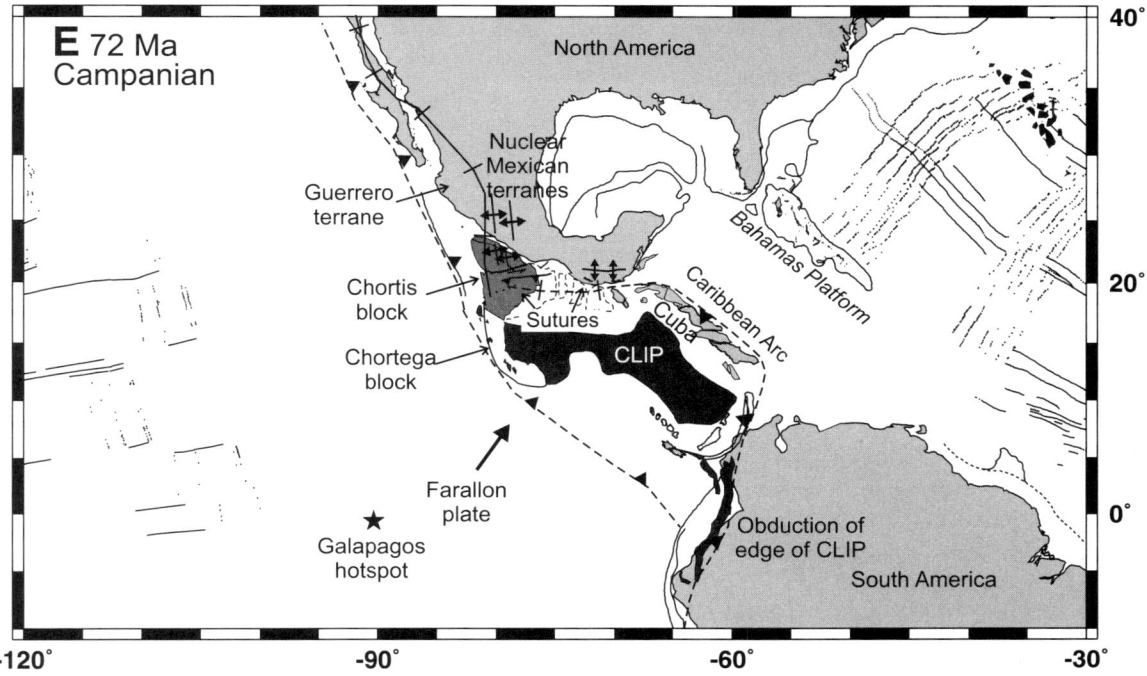
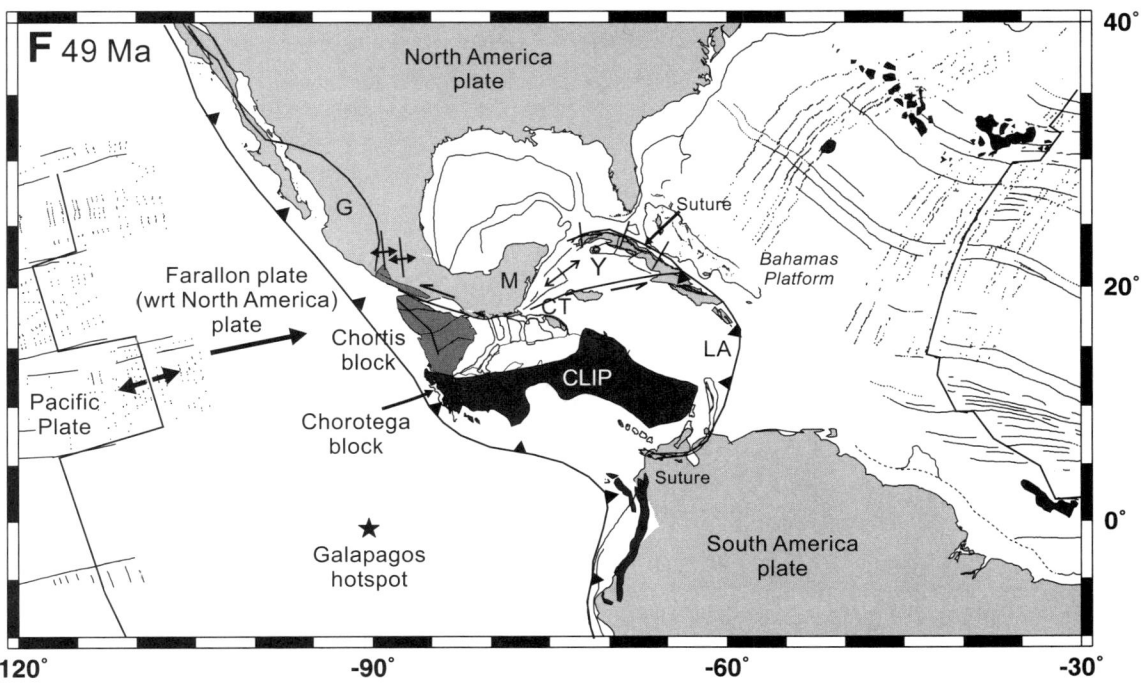

Figure 4. (*continued*)

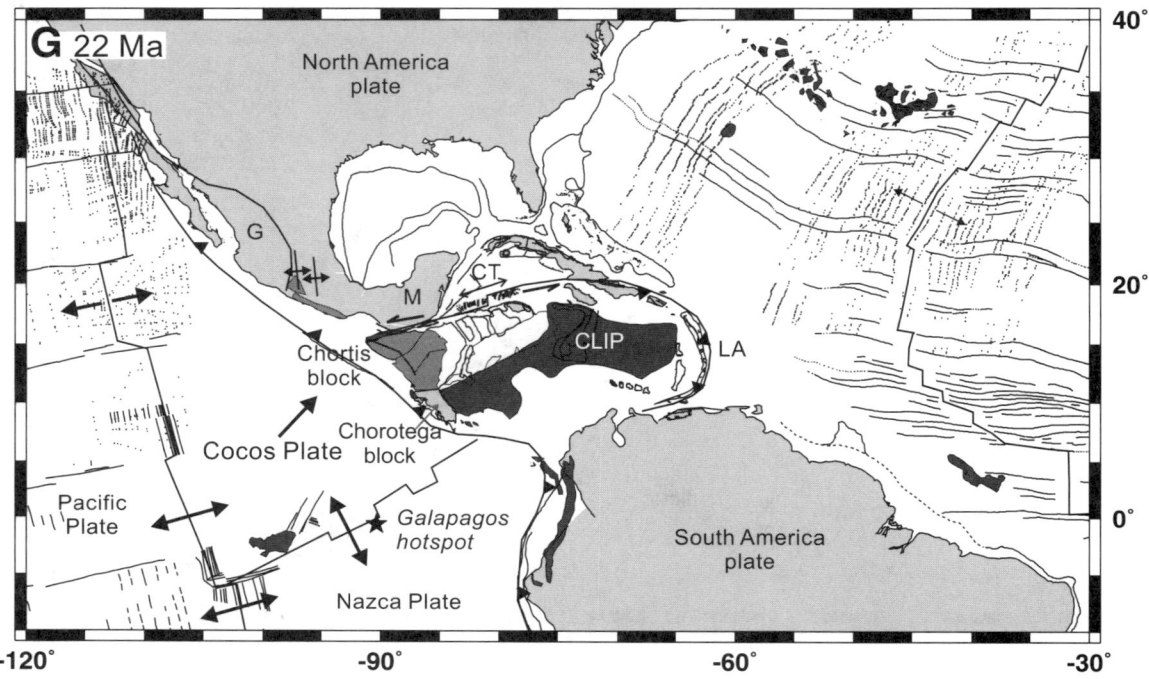
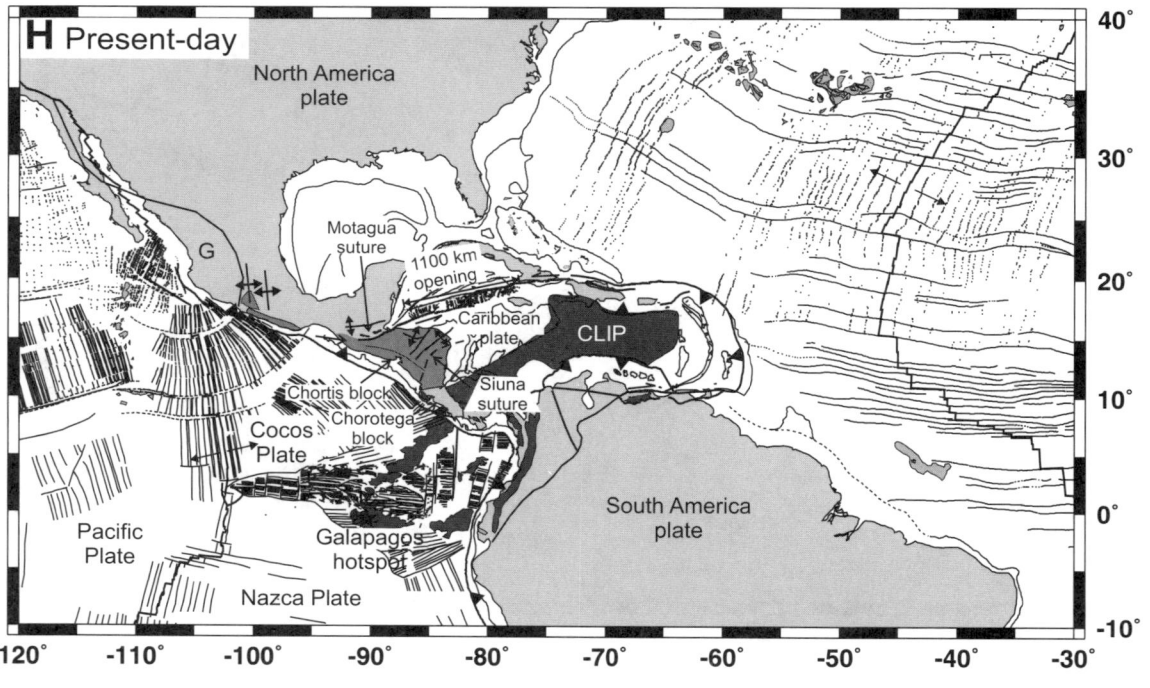

Figure 4. (*continued*)

Central America and progressive 35–28 Ma cessation of plutonic activity in southern Mexico (Rogers et al., Chapter 3, this volume; Schaff et al., 1995); and (b) NE to SW migration of the volcanic arc activity and opening of the Nicaraguan backarc basin from 10 to 0 Ma (Plank et al., 2002).

OVERVIEW OF THE TECTONIC HISTORY OF NORTHERN CENTRAL AMERICA

The geology of northern Central America exhibits complexities related to the mobility of the Caribbean plate and other small plates or microplates in the complexly deformed region between the much larger North and South American plates and along the convergent margin with eastern Pacific oceanic plates (Fig. 1). Much of the relative movements of these elements has been broadly outlined in previous reconstructions, such as those by Pindell and Barrett (1990), Pindell and Kennan (2001), and Meschede and Barckhausen (2000) (Fig. 2). However, I quantitatively reconstruct these motions by incorporating as many of the recent magnetic (Figs. 2 and 3), fracture zone, and geologic (Fig. 1) constraints as possible. These reconstructions show that many of the tectonic and geochemical complexities of the Cretaceous-Cenozoic geology of northern Central America can be related to plate tectonic parameters, such as subduction direction, arc-continent collision, oblique subduction, and slab break-off.

Some important tectonic elements of northern Central America that attest to its mobility, highlighted in our reconstructions of the region, include the following features.

1. Chortis Block

This area has been traditionally regarded as the Precambrian-Paleozoic continental nucleus of northern Central America (Fig. 1). Radiometric dating by Schaff et al. (1995) shows the diachronous nature of magmatism along the southern coast of Mexico that supports the proposed progressive, Cenozoic west to east translation and counter-clockwise rotation of the Chortis block from its late Cretaceous position proposed by Pindell and Dewey (1982) shown in Figure 1. This large-scale left-lateral offset is consistent with a minimum of 1100 km of left-lateral offset recorded on the narrow Cayman trough to the east (Rogers et al., Chapter 4, this volume) (Fig. 3).

Recent studies by Rogers (2003) and Rogers et al. (Chapter 4, this volume) show that the block is not homogeneous and can be divided into four tectonic terranes, shown on Figure 1: (1) the **Central Chortis terrane,** underlain by Grenville and Precambrian age continental crust detached from the southern margin of Mexico in the Late Cretaceous to Early Tertiary (approximate reconstructed position of Chortis is shown in Fig. 1); (2) the **Eastern Chortis terrane**, underlain by Jurassic metasedimentary rocks formed along the Mesozoic rifted margin of the Chortis block; (3) the **Siuna terrane**, underlain by deformed rocks of oceanic island origin accreted to the Eastern Chortis terrane during Late Cretaceous time (Venable, 1994; Walther et al., 2000; Rogers et al., Chapter 6, this volume); and (4) the **Southern Chortis terrane** is underlain by a Late Cretaceous island arc accreted to the Central Chortis terrane in Late Cretaceous time. The arc is inferred to represent a fragment of the Guerrero terrane of Mexico (Dickinson and Lawton, 2001); Northern Honduras is characterized by an elongate belt of magmatic overprinting of parts of the Central and Eastern Chortis terranes (Fig. 1).

In Figure 1, the boundaries between these varying basement types strike at right angles to the northern Central American volcanic arc and may impart important basement controls on the tectonic structures of the arc and types of magmas erupted along this Cocos-Caribbean subduction boundary. For example, the large Central American ignimbrite flare-up that erupted between 19 and 10 Ma and is aligned along the northwest-trending boundary of continental terranes of the Chortis block suggests that magma emplacement is controlled by this crustal structure (Rogers et al., 2002; Jordan et al., this volume). Movements along block boundaries within Chortis may also account for the changes in strike directions of the active arc segments in Nicaragua and Costa Rica that are numbered 2 and 3 on Figure 1. The interpreted basement boundaries also coincide fairly well with many of the offsets that segment the volcanic arc.

2. Cayman Trough

This 1100-km-long oceanic basin formed as a pull-apart basin at the 100-km-long Mid-Cayman spreading center between 49 Ma and present (Rosencrantz et al., 1988; Rosencrantz, 1994; Leroy et al., 2000) (Fig. 3). The marine magnetic anomalies and fracture zones produced by seafloor spreading provide a partial record of the motion between the North America and Caribbean plates and the Chortis block (Fig. 1), provided that the complexities of the Gonave microplate, at the southeastern edge of the Cayman trough in Jamaica and Hispaniola, are taken into account (Sykes et al., 1982; DeMets et al., this volume). Recent work by Müller et al. (1999) suggests that the Caribbean plate has remained fixed relative to Atlantic and Indian Ocean hotspots for much of the Cenozoic. By understanding the motion of the Caribbean plate, including the Chortis block, I should be able to improve the kinematic constraints of northern Central America.

KEY TECTONIC EVENTS AFFECTING CENTRAL AMERICA SHOWN IN THE PLATE RECONSTRUCTIONS

The reconstructions shown in Figures 4A–4H coincide with major tectonic events. Integration of main Cenozoic tectonic events in the surrounding ocean basins, which have potentially affected northern Central America, include the following events:
- Initiation of spreading of the Cayman trough ca. 49 Ma (diffuse rifting that preceded organized seafloor spreading at the Mid-Cayman spreading center began earlier [cf. Mann and Burke, 1990]) (Fig. 3).

- Slowdown or cessation of Cayman trough spreading at 26–20 Ma (Rosencrantz, 1994) (Fig. 3).
- Early rifting and breakup of the Farallon plate at 22.7 Ma and the original distribution of now-dispersed fragments that include the Cocos, Malpelo, and Carnegie ridges (Meschede and Barckhausen, 2000) (Fig. 2).
- Super-fast spreading on the East Pacific Rise between ca. 18 and 10 Ma (Wilson, 1996).

The main Cenozoic tectonic events in Central America include the following:

- Eastward migration relative to the Chortis continental block to northern Central America and progressive 35–28 Ma cessation of plutonic activity in southern Mexico (Schaff et al., 1995).
- NE to SW migration of the volcanic arc and opening of the Nicaraguan backarc basin from 10 to 0 Ma (Balzer, 1999; Plank et al., 2002).

The reconstructions also integrate the main Cenozoic tectonic events affecting the Caribbean plate as a whole, including relative stability of the Caribbean plate with respect to the Atlantic and Indian Ocean hotspots since 38 Ma (Müller et al., 1999) and the rapid north-south convergence of the North and South American plates from 38 to 10 Ma and their slower convergence from 10 to 0 Ma (Dixon and Mao, 1997).

PLATE RECONSTRUCTIONS OF CENTRAL AMERICA: 165–0 Ma

Plate Reconstructions of Central America and the Eastern Pacific

Several plate tectonic models for the Caribbean and Pacific Ocean appeared during the late 1980s and early 1990s. These works include Ross and Scotese (1988), Rosencrantz et al. (1988), Mayes et al. (1990), Atwater and Severinghaus (1990), and Pindell and Barrett (1990). Since these early efforts, new marine magnetic anomaly–fracture zone data and new interpretations of existing data have been published regarding these regions (Cande and Haxby 1991; Wilson and Hey, 1995; Wilson, 1996; DeMets and Traylen, 2000; Leroy et al., 2000; Barckhausen et al., 2001; Rogers, 2003) (Fig. 2). Some of these data were incorporated in more recent plate models by Meschede et al. (1998), Meschede and Barckhausen (ODP Leg 170 Scientific Results, 2000), Mann (1999), and Müller et al. (1999). Mann (1999) incorporated the North America–South America–Africa motions determined by Müller et al. (1999), using the PLATES (2000) software and database to expand the areal extent of his reconstructions. He did not incorporate the magnetic anomaly picks of Leroy et al. (2000) or Meschede and Barckhausen (2000), who worked independently.

Following Mann (1999), I present a series of plate reconstructions from the mid-Late Jurassic to the present-day that depict the evolution of Central America in the regional context of the southern Cordillera of western North America, the Caribbean plate, the Cayman trough of the Caribbean, and the oceanic plates of the Pacific Ocean (Figs. 4A–4H). An earlier but similar version of these reconstructions appears in Rogers (2003) and Mann et al. (2007).

The reconstructions were made in the mantle reference framework of Müller et al. (1999) and illustrate the westward migration of the North and South American plates relative to a Caribbean plate fixed in the mantle reference frame. The present position of the Galapagos hotspot provides a stationary point of reference of this framework and is shown in all reconstructions in Figure 4. It should be noted that the geologic evidence for the existence of the Galapagos hotspot can only be dated back to ca. 95 Ma (Hoernle et al., 2002).

The principal constraints on the plate motions shown on the reconstructions are published seafloor spreading anomalies and finite rotation poles by previous workers that were compiled by Rogers (2003) and Mann et al. (2007) (marine magnetic anomalies used are shown on Figs. 2 and 3). The plate circuit used for the reconstructions is Caribbean to North America (Rosencrantz, 1994); North America to Africa and South America to Africa (Müller et al., 1999); Africa to Antarctica (Royer et al., 1988); Antarctica to Nazca (Tebbens and Cande, 1997; Rosa and Molnar, 1988) and Antarctica to Pacific (Tebbens and Cande, 1997; Cande et al., 1995); and Nazca to Cocos (Wilson and Hey, 1995; Barckhausen et al., 2001). The nature and location of ancient plate boundaries shown in the reconstructions of Mexico and the Caribbean in Figures 4A–4H are based on geologic constraints summarized in the synthesis of Mexican geology by Dickinson and Lawton (2001) and in the syntheses of Caribbean and northern South American geology by Pindell and Barrett (1990), Pindell (1994), and Mann (1999).

165 Ma (Late Jurassic)

The reconstructions start in the Late Jurassic and resemble the reconstructions of Dickinson and Lawton (2001) for the earlier evolution of Mexican terranes (Fig. 4A). At this time, North America and South America are shown just prior to their separation to form the now-subducted proto-Caribbean seaway. Opening of the Gulf of Mexico during this period rotated the Maya (Yucatan) block south to Central America (Marton and Buffler, 1994). The continental terranes of the Chortis block were adjacent to the autochthonous Mexican terranes. The Cretaceous margins of Chortis were likely accreted from elements of the Guerrero-Caribbean arc formed at the leading edge of the Farallon plate that consumed the Mezcalera plate as it advanced from the west (Tardy et al., 1994; Moores, 1998). East-dipping subduction along the western margin of the Americas occurred across the Central America region (Fig. 4A).

144 Ma (Early Cretaceous)

At the start of the Cretaceous, North America and South America continued to separate and form the proto-Caribbean seaway (Pindell and Barrett, 1990). The oceanic crust of the proto-Caribbean was later consumed by the eastward and northeastward advance of the Guerrero-Caribbean arc shown in Figure 4B. Opening of the proto-Caribbean formed the Eastern Chortis terrane of attenuated continental crust (Pindell and Dewey, 1982). The opening event also rifted the Juarez terrane of Mexico (Dickinson and Lawton, 2001).

Rifting along the southeastern margin of North America formed the Arperos basin (Tardy et al., 1994). Widespread rifting along the western margin of North America at this time is attributed to trench rollback as subduction of the oceanic Mezcalera plate slowed during the approach of the Guerrero-Caribbean arc from the west (Dickinson and Lawton, 2001) (Fig. 4B).

I adopt the interpretation by Tardy et al. (1994), Burke (1988), and Moores (1998) that the Caribbean arc and the Guerrero arc are parts of the same segments of the same intra-Pacific ocean arc system that entered the Caribbean region during Cretaceous time. Diachronous collision of this arc with the eastern North America margin progressed from north (Sierran foothills of California) to south (Guerrero terrane of southern Mexico) (Dickinson and Lawton, 2001) (Fig. 4B). Following this arc-continent collision, eastward-dipping subduction of the Farallon plate stepped outboard (eastward) of the newly accreted Guerrero terrane.

An alternative view, which I do not support, proposes that the present-day area of the Caribbean was created during the period of 130–80 Ma but that this newly created area has remained relatively stationary with respect to North and South America and was not consumed by the Caribbean arc system (Frisch et al., 1992; cf. preface in Mann, 1995, for a review of salient points of both proposed concepts). In my view, the alternative viewpoint of Frisch et al. (1992) fails to explain the diachronous west to east timing of foreland basin subsidence and thrust deformation related to the diachronous collision between the Caribbean arc and the passive margins of North and South America (cf. Pindell and Barrett, 1990; Mann, 1999).

Rifting of the southeastern Mexican margin during the early Cretaceous detached part of the Oaxaca and Mixteca terranes from nuclear Mexico (Del Sur block of Dickinson and Lawton, 2001) to form the Chortis block. Rifting of Chortis may have occurred along the southward extension of the Arperos basin, as proposed by Pindell and Kennan (2001), or along a failed rift arm of the proto-Caribbean spreading center (Fig. 4A). During this time, the Chortis block underwent intrablock rifting and deposition of pre-Aptian, terrigenous siliciclastic rocks (Rogers, 2003; Rogers et al., Chapter 5, this volume).

120 Ma (Early Late Cretaceous)

Extension between North and South America continued and the Guerrero-Caribbean arc advanced eastward to subduct the proto-Caribbean oceanic basin (Fig. 4C). By this time, collision of the Guerrero terrane with the margin of eastern North America and closure of the Arperos basin were complete to the latitude of Baja (W. Dickinson, 2003, personal commun.).

I propose that prior to the diachronous closure of the Guerrero-Caribbean arc against southeastern Mexico, Chortis-Mexico convergence occurred along a short-lived, eastward-dipping subduction zone (Fig. 4C). This subduction produced intra-arc rifting and arc volcanism in the overriding Chortis block by 126 Ma (Drobe and Oliver, 1998; Rogers, 2003; Rogers et al., Chapter 5, this volume). Termination of this subduction cycle between Chortis and southeastern Mexico is recorded by a well-dated, 120 Ma subduction complex along the northern edge of the Chortis block, which is presently exposed on the southern margin of the Motagua Valley of Guatemala (Sisson et al., 2003; Harlow et al., 2004). Structural and stratigraphic continuity between Chortis and southeastern Mexico at this time is suggested by (1) the geochemical similarity between the volcanics erupted on the Chortis block and the Teloloapan volcanic rocks of Mexico; and (2) the similar Mesozoic stratigraphy and structural trends shared by both areas (Rogers et al., Chapters 4 and 5, this volume).

90 Ma (Late Cretaceous)

The Guerrero-Caribbean arc continued diachronous suturing along the eastern and southern thinned, continental edges of the Chortis block (Fig. 4D). The short-lived, middle Cretaceous volcanic arc, intra-arc basins, and associated mixed carbonate-clastic deposition on the Chortis block were terminated by a collisional event recorded by the deposition of clastic sedimentary rocks of late Cretaceous age in Honduras (Rogers, 2003; Rogers et al., Chapter 5, this volume). Collision-related shortening inverted intra-arc basins and created the four alignments of deformed Cretaceous sedimentary rocks seen in the present-day geology of Honduras (Rogers et al., Chapter 5, this volume). Strong shortening effects are also seen in southern Mexico at this time (Dickinson and Lawton, 2001).

By this time, the Guerrero-Caribbean arc had overridden the Galapagos hotspot, heralding a vigorous period of submarine oceanic plateau volcanism that began by at least 95 Ma and was widespread by 88 Ma (Sinton et al., 1997; Hoernle et al., 2002). These elements became amalgamated as the Chorotega block of southern Central America.

72 Ma (Late Cretaceous)

By the latest Cretaceous, the Caribbean arc, now adjacent to the thick, young, and buoyant Caribbean oceanic plateau, continued to migrate and collide to the northeast. Convergence between the arc and the southern rifted margin of Honduras (Eastern Chortis terrane) led to the obduction of Guerrero-Caribbean arc material (Siuna terrane of northern Nicaragua; Venable, 1994) (Fig. 4E) and to the formation of the Colon deformed belt (Rogers et al., Chapter 6, this volume; Fig. 1). Arc-Chortis convergence was expressed as left-lateral strike-slip motion along the Guayape fault system that developed at this time along the rifted structural grain of the southern Jurassic margin of Chortis (Eastern Chortis terrane). Shortening also occurred at this time between the Chortis and Maya blocks as recorded by the well-dated emplacement of ophiolites onto the Maya block (Donnelly et al., 1990; Sisson et al., 2003; Harlow et al., 2004). Eastward-dipping subduction developed between the Farallon plate and the eastern margin of the Caribbean oceanic plateau (Fig. 4E). The Caribbean arc detached from the Caribbean oceanic plateau, which was pinned on its northern and southern edges by collision with Chortis and northeastern South America. Continued north and eastward motion of the Caribbean arc by trench rollback detached Cuba from the Caribbean oceanic plateau and formed the Yucatan backarc basin south of Cuba and the Grenada backarc basin east of the present-day Lesser Antilles arc (Mann, 1999).

49 Ma (Eocene)

The northeastward migration of the Caribbean arc ended when part of the arc collided with the Bahamas carbonate platform (Fig. 4F). Collision transferred the Cuban area from the Caribbean plate to the North American plate as the strike-slip boundary moved southward. The Motagua-Cayman trough–Oriente fault zone developed to accommodate this new zone of left-lateral, strike-slip displacement between the North American and Caribbean plates (Rosencrantz et al., 1988; Mann, 1999).

Along the southeastern Mexican margin, eastward-dipping shallow subduction produced the Xolopa magmatic arc and the Northern Chortis magmatic zone (Schaff et al., 1995). The geometry of the Xolopa arc with respect to the Chortis-Farallon margin is highly oblique, similar to the present-day geometry of the Trans-Mexican volcanic belt with respect to the Cocos–North America margin (Fig. 4F). This geometry would suggest that the Chortis block occupied a forearc setting during oblique convergence of the Farallon plate relative to North America (Schaff et al., 1995). Southeast translation of the Chortis block was facilitated by extending the detachment zone parallel to the zone of Xolopa arc magmatism; this hot, weakened zone of arc was broken under oblique, Farallon–North America convergence, and as a result, the Chortis block was dislodged from Mexico and moved southeastward (Schaff et al., 1995). Magnetic and stratigraphic similarities suggest that a small remnant of the Chortis block was not dislodged in this manner and remained behind in Mexico (Teloloapan subterrane of Dickinson and Lawton, 2001) (Rogers, 2003; Rogers et al., Chapter 4, this volume).

22 Ma (Miocene)

Deformation arising from Farallon plate subduction to the northeast beneath North America and to the southeast beneath South America resulted in its breakup into the Cocos and Nazca plates at 23 Ma (Wortel and Cloetingh, 1981; Barckhausen et al., 2001) (Fig. 4G). The reorganization led to near-orthogonal subduction of the Cocos plate beneath the Chortis block (Wilson and Hey, 1995), and the abrupt termination of oblique subduction forces that previously drove the eastward motion of the Chortis block. A super-fast spreading period of this segment of the East Pacific Rise preceded the detachment of the Cocos slab beneath northern Central America. Rogers et al. (2002) and Jordan et al. (this volume) proposed that this slab detachment event produced large-scale topographic uplift in northern Central America. The development of the Nicaragua intra-arc depression, a backarc basin, was produced by a late Neogene phase of trenchward migration of the Central America volcanic arc and slab rollback (Plank et al., 2002) that was perhaps initiated by steepening of the Cocos slab following the break off of its subducted, downdip extension ca. 4–10 Ma (Rogers et al., 2002). Ferrari (2004) has proposed a similar process of slab break off for the arc in southern Mexico.

During this time, Central America became incorporated with the Caribbean plate and moved eastward relative to the North American plate. Several parallel, left-lateral strike-slip faults (Jocotan-Chamelecon, Polochic, and Motagua) developed in Guatemala along the Late Cretaceous Motagua Valley suture between the Chortis and Maya blocks. To the east, these faults connect to the Swan Islands fault zone of the Cayman trough. Internal deformation of the Chortis block and formation of transtensional rifts in the offshore Honduran borderlands resulted from divergence of the Caribbean plate motion vector from the azimuth of these plate boundary faults (Rogers, 2003; Rogers and Mann, Chapter 3, this volume).

0 Ma (Present-Day)

Presently, Central America is bounded by the Middle America trench and subduction system to the southwest and the strike-slip faults of the Motagua-Swan Islands to the north, and it is attached to the stable Caribbean plate to the east and southeast (Fig. 5G). GPS studies reported in this volume by DeMets et al. for northern Central America will improve constraints on block motions with this area.

DISCUSSION: UNRESOLVED TECTONIC PROBLEMS IN NORTHERN CENTRAL AMERICA

A challenge for future workers is to relate magmatic and deformational events recorded in the Precambrian to Holocene geologic record of northern Central America to changing plate boundary configurations of North America, Caribbean, and Cocos plates as predicted by the plate reconstructions shown in Figures 4A–4H. Part of the problem is that plate motion rates in this region have been very fast (e.g., 5–18 cm/yr range; Wilson, 1996). Future studies should seek consistency between onland and oceanic interpretations by working with data sets from both realms.

Precambrian Connections between North and South America

Fragments of Precambrian-age continental crust occur in central Honduras and in Guatemala (Fig. 1). Renne et al. (1989) proposed that these elements once formed a continuous belt

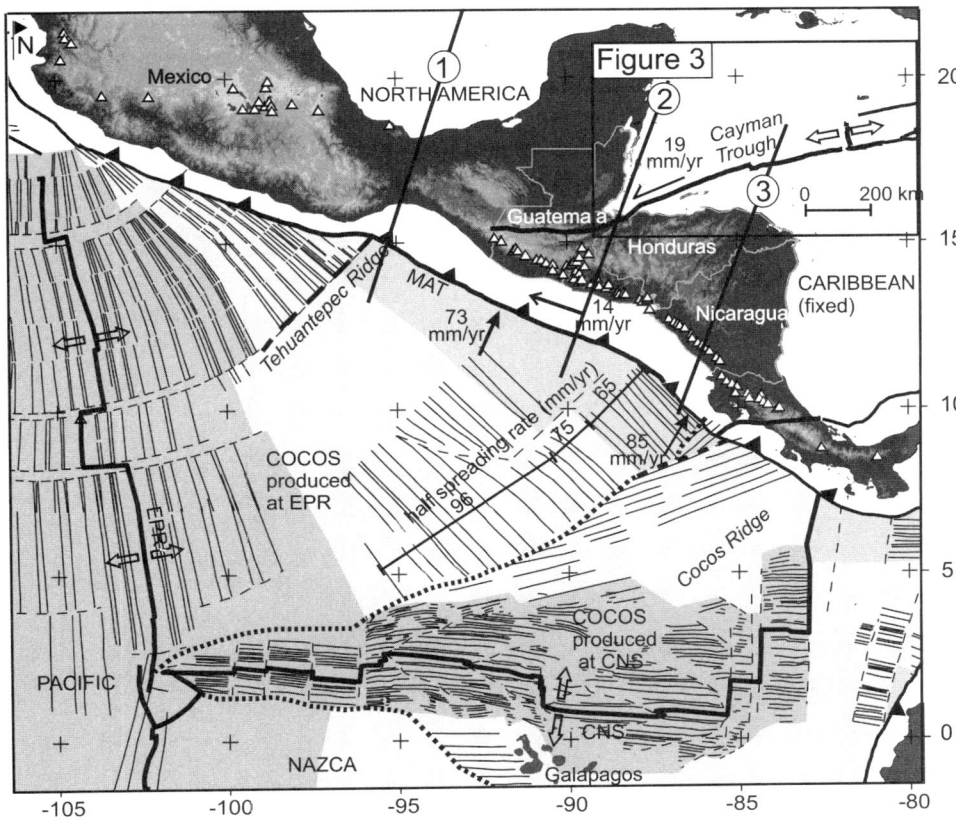

Figure 5. Present setting of Central America showing plates, Cocos crust produced at East Pacific Rise (EPR), and Cocos-Nazca spreading center (CNS), triple-junction trace (heavy dotted line), volcanoes (open triangles), Middle America Trench (MAT), and rates of relative plate motion (DeMets et al., 2000; DeMets, 2001). East Pacific Rise half spreading rates from Wilson (1996) and Barckhausen et al. (2001). Lines 1, 2, and 3 are locations of topographic and tomographic profiles in Figure 6.

linking the more continuous Grenville belts of North and South America. Centeño-García and Keppie (1999) suggest the fragments represent parts of multiple belts that originated as an isolated microcontinent in the eastern Pacific and were reorganized by plate motions since their formation. The Precambrian rocks of Central America represent a key, but largely unconstrained, element to the pre-Mesozoic configuration of the region. Since Precambrian rocks commonly have distinctive age provinces, composition, and geochemical signatures, the reconstruction of these blocks could be accomplished by working systematically and reassembling the blocks from the widely separated areas.

Slowing of Spreading in the Cayman Trough and Relationship to Events in Central America

Cayman trough history includes the initiation of oceanic spreading at 49 Ma and a slowdown in spreading rate from 26 to 20 Ma (Rosencrantz, 1994) (Fig. 3). The Cayman trough is a valuable, long-term recorder of motion between northern Central America on the Caribbean plate (Chortis block) and the North American plate (Fig. 3). Through closure conditions, these motions can be used to better-constrain motions of the Farallon plate and, after 22 Ma, the Cocos and Nazca plates beneath Central America. Cayman trough aeromagnetic data suggest that the motion was not steady but began quickly from ca. 49 Ma to 26 Ma, slowed down considerably between 26 and 20 Ma, and then maintained a slow but steady spreading rate between 20 and 0 Ma (Rosencrantz, 1994) (Fig. 3).

Slab Break-off of the Cocos Plate beneath Northern Central America

P-wave tomographic images of the mantle beneath northern Central America reveal a detached slab of the subducted Cocos plate (Figs. 6A–6C) (Rogers et al., 2002). Landscape features of the region of Honduras and Nicaragua above the detached slab are consistent with epeirogenic uplift produced by mantle upwelling following slab break-off between 10 and 4 Ma.

Following slab detachment, hot asthenospheric mantle flows inward to fill space vacated by the cold, more dense slab as it sinks into the mantle (cf. Wortel and Spakman, 2000). Thermomechanical modeling of the slab-detachment process demonstrates large (>500 °C) transitory heating of the base of the upper plate for several million years (cf. van de Zedde and Wortel, 2001). Asthenospheric upwelling can produce decompression-induced volcanism, and the geochemistry of the basaltic lavas from behind the volcanic front in Honduras and Guatemala is consistent with mantle upwelling (Walker et al., 2000). Slab detachment and uplift of the Central American plateau occurred between the end of the subduction-related ignimbrite flare-up at 10 Ma (Jordan

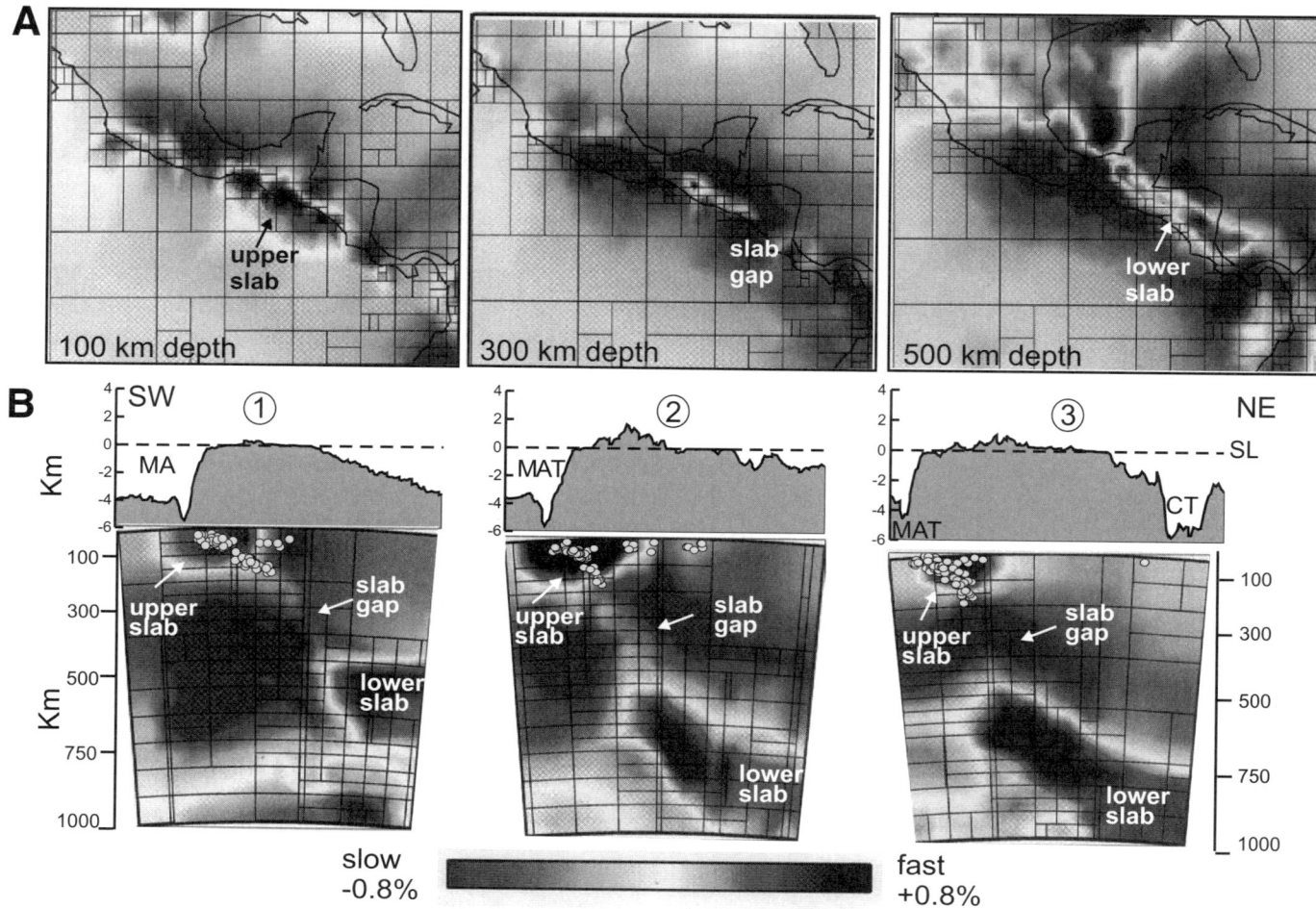

Figure 6. (A) Tomographic slices of the P-wave velocity of the mantle at depths of 100, 300, and 500 km beneath Central America. (B) Upper panels show cross sections of topography and bathymetry. Lower panels: tomographic profiles showing Cocos slab detached below northern Central America, upper Cocos slab continuous with subducted plate at Middle America Trench (MAT), and slab gap between 200 and 500 km. Shading indicates anomalies in seismic wave speed as a ±0.8% deviation from average mantle velocities. Darker shading indicates colder, subducted slab material of Cocos plate. Circles are earthquake hypocenters. Grid sizes on profiles correspond to quantity of ray-path data within that cell of model; smaller boxes indicate regions of increased data density. CT—Cayman trough; SL—sea level (modified from Rogers et al., 2002).

et al., this volume) and prior to 3.8 Ma, the time at which the tip of the Cocos slab was subducted (i.e., beginning of the "modern subduction" period in Central America) (Fig. 6).

In collisional settings, slab detachment occurs as the force of slab-pull near the trench is resisted by buoyant lithosphere, producing a tear in the downgoing slab (Wortel and Spakman, 2000). Rogers et al. (2002) suggest that this mechanism occurred along the non-collisional Middle America trench margin as a result of the decreasing age and increasing buoyancy of the incoming Cocos oceanic plate during the 19–10 Ma interval of super-fast spreading (Wilson, 1996) along the southernmost Cocos-Pacific segment of the East Pacific Rise. Although steady-state subduction at the Middle America Trench since 2.5 Ma has been inferred from the presence of cosmogenic isotopes in the modern arc lavas of Central America (Morris et al., 2002), tomographic observations showing slab break-off require highly variable rates of subduction and rates of slab melting during the Neogene.

Models for the Internal Deformation of the Chortis Block and the Eastern Caribbean Plate

Onshore extensional deformation of the western Caribbean plate has been explained as resulting from (1) counter-clockwise rotation of the Chortis block around the arcuate Motagua-Polochic plate boundary faults (Fig. 7A) (Burkart and Self, 1985; Gordon and Muehlberger, 1994) (Fig. 7A); and (2) as fault termination features along the North America–Caribbean margin (Guzmán-Speziale, 2001) (Fig. 7B). Offshore extensional deformation is generally attributed to a diffuse strike-slip plate margin south of the Swan Islands fault zone (Case et al., 1984). DeMets et al.

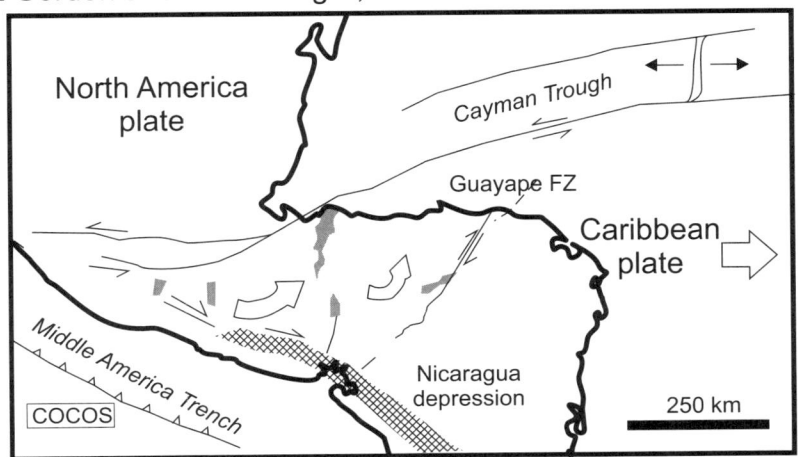
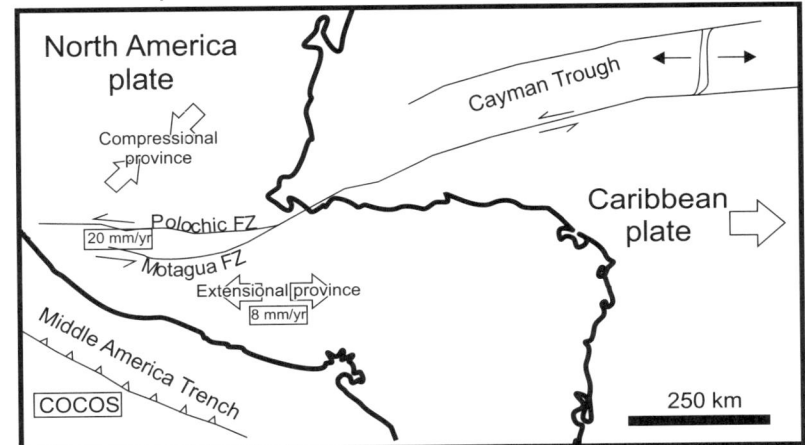
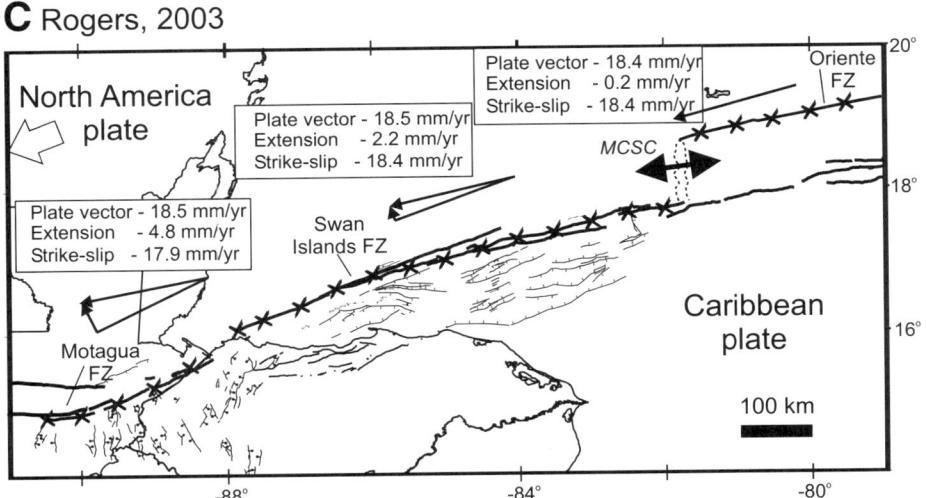

Figure 7. Internal deformation of the Chortis block, produced by: (A) counterclockwise rotation of Chortis block (Gordon and Muehlberger, 1994); and (B) fault termination (Guzmán-Speziale, 2001) structures north and south of the Motagua-Polochic strike-slip fault zone (FZ). (C) Transtension along the Motagua–Polochic–Swan Islands FZ. GPS-derived Caribbean plate velocity (DeMets et al., 2000) was calculated at 30 min increments (Xs on map) along the main North America–Caribbean plate boundary faults (Motagua–Swan Islands–Mid-Cayman spreading center–Oriente system). The three plate vectors shown are for points along the fault system at longitudes 89°W, 85°W, and 81°W and are decomposed to show the extensional and strike-slip component of the plate vector. The extensional component of motion is controlled by the angular divergence between the plate vector and the trend of the plate boundary fault. Note that the extensional component of the plate vector increases from 0.2 mm/yr at longitude 81°W near the Mid-Cayman spreading center (MCSC) to 4.8 mm/yr at longitude 89°W in the Motagua Valley of Guatemala and is consistent with the widening of plate margin deformation from east to west (Rogers, 2003; Rogers and Mann, this volume).

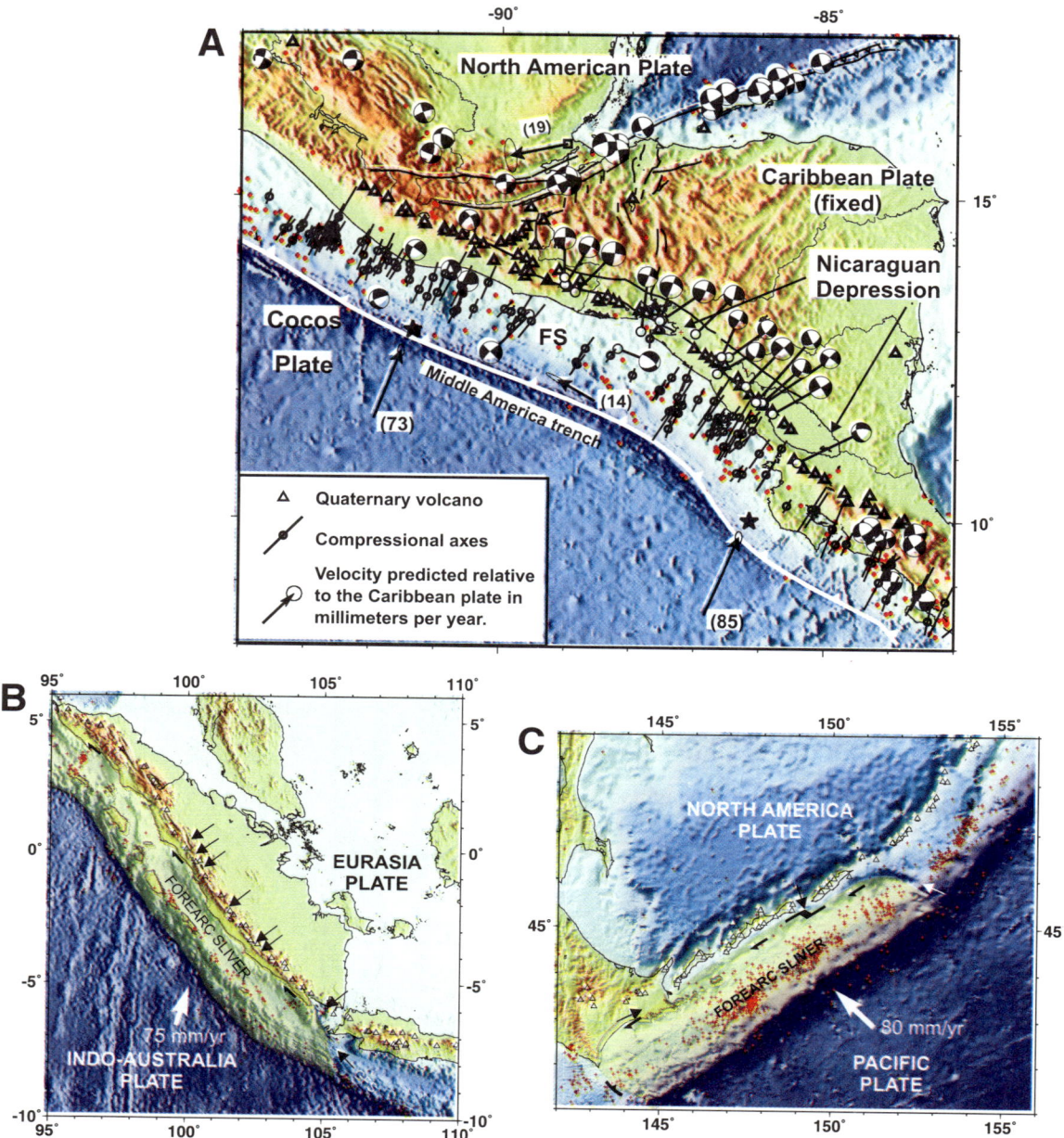

Figure 8. (A) The Central American forearc sliver (FS) is produced by the slight (~10°) oblique convergence of the Cocos plate at the Middle America trench, resulting in the northward migration of the forearc sliver (DeMets, 2001). Earthquake focal mechanisms show right-lateral strike-slip motion along the Central America arc. Motion vectors are shown in millimeters per year. For tectonic comparison, forearc sliver produced by oblique convergence of Indo-Australia plate and the Eurasia plate in Sumatra (B) and by convergence of Pacific plate and the North America plate in the Kurile Islands and northern Japan (C). In B black arrows highlight pull-apart basins at right-steps along the right-lateral Sumatra strike-slip fault zone that bounds the forearc sliver. In C small black and white arrows show trace of right-lateral strike-slip fault zone with one pull-apart segment, which bounds the Kurile forearc sliver.

(2000) relate the deformation along the northern margin of the Caribbean plate to the angle between the GPS-derived motion vector of the Caribbean plate and the azimuth of the plate boundary zone. Rogers (2003) and Rogers and Mann (this volume) refine the observations of Demets et al. (2000) for the eastern Caribbean plate and relate the pattern of active borderlands and oblique-slip faulting to the angular divergence of Caribbean motion and the local azimuth of the plate margin faults (Fig. 7C). This view of the western Caribbean deformation is also compatible with the fault termination model of Guzmán-Speziale (2001) (Fig. 7B).

Defining the Central American Forearc Sliver and the Origin of the Nicaraguan Depression

The Nicaraguan depression is a prominent Quaternary morphologic and structural depression aligned roughly with the belt of active Central American volcanoes and extending ~600 km from the northern Gulf of Fonseca in El Salvador and northern Nicaragua to the Caribbean Sea in Costa Rica (Fig. 8A). The Nicaraguan depression is an atypical backarc basin in that the depression commonly encompasses the entire active volcanic chain rather than occurring only in a backarc position. Two hypotheses have been proposed to explain the tectonic origin of this regional structure within the framework of the Central American volcanic arc. The first mechanism, supported by geochemical analyses and radiometric dating of volcanic rocks adjacent to the depression, supports the traditional two-dimensional view of arc-normal extension accompanying trenchward or southwestward migration of the arc from 24 Ma (Middle Miocene) to the present (Plank et al., 2002). Trenchward shifts in the position of the volcanic arc through time are related to a steepening or "rollback" in the dip of the subducted slab of the Cocos plate from ~50° at ca. 12 Ma to a current dip >65°. Increased slab dip was directly expressed by an increase in crustal extension of the volcanic arc with up to 20% pure shear over the past 12 Ma.

The second hypothesis states that the Nicaraguan depression has formed as a consequence of pull-apart extension at right-stepping stepovers along a major right-lateral fault system aligned with the active volcanic chain and parallel to the trend of the topographic and structural depression (Fig. 8A). The driving force for this deformation is slightly oblique subduction of the Cocos plate, which drives a forearc sliver along the right-lateral fault, formed along the thin, hot crust along the active volcanic axis (DeMets, 2001). This forearc sliver interpretation, which explains plate motions and strain partitioning in the obliquely subducting Sumatran and Kuril arcs, is consistent with the belt of damaging earthquakes along the Central American volcanic arc, which mostly exhibit right-lateral focal mechanisms (Figs. 8B and 8C). One controversy is whether the fault accommodating the strike-slip displacement of the forearc sliver is a single strike-slip fault zone parallel to the arc or is composed of "bookshelf faults" that accommodate strike-slip by motion on a series of oblique faults (La Femina et al., 2002).

FUTURE WORK

It is my hope that this review helps to stimulate increased work in Central America, particularly with regard to how the plate setting has evolved through time and how this plate evolution has changed the forcing functions along the northern Central American "subduction factory."

ACKNOWLEDGMENTS

Funding for field studies in Honduras related to this review was provided by the Petroleum Research Fund of the American Chemical Society (grant 33935-AC to Mann). The author acknowledges financial support for this publication provided by the University of Texas at Austin's Geology Foundation and Jackson School of Geosciences. Special thanks to Lisa Bingham for her help with text and maps. University of Texas Institute for Geophysics contribution no. 1869.

REFERENCES CITED

Atwater, T., and Severinghaus, J., 1990, Tectonic maps of the northeast Pacific, in Winterer, E.L., Hussong, D.M., and Decker, R.W., eds., The Eastern Pacific Ocean and Hawaii: Geological Society of America, Geology of North America, v. N, p. 15–20.

Balzer, V.G., 1999, The Late Miocene history of sediment subduction and recycling as recorded in the Nicaraguan volcanic arc [M.S. thesis]: Lawrence, Kansas, University of Kansas, 221 p.

Barckhausen, U., Roeser, H.A., and von Huene, R., 1998, Magnetic signature of upper plate structures and subducting seamounts at the convergent margin off Costa Rica: Journal of Geophysical Research, v. 103, p. 7079–7093, doi: 10.1029/98JB00163.

Barckhausen, U., Ranero, C.R., von Huene, R., Cande, S.C., and Roeser, H.A., 2001, Revised tectonic boundaries in the Cocos Plate off Costa Rica: Implications for the segmentation of the convergent margin and for plate tectonic models: Journal of Geophysical Research–Solid Earth, v. 106, p. 19,207–19,220.

Bundschuh, J., and Alvarado, G.E., eds., 2007, Central America: Geology, Resources and Hazards, v. 1: The Netherlands, Taylor and Francis/Balkema, Leiden, 1340 p, 20 plates.

Burkart, B., and Self, S., 1985, Extension and rotation of crustal blocks in northern Central America and effect on the volcanic arc: Geology, v. 13, p. 22–26, doi: 10.1130/0091-7613(1985)13<22:EAROCB>2.0.CO;2.

Burke, K., 1988, Tectonic evolution of the Caribbean: Annual Review of Earth and Planetary Sciences, v. 16, p. 201–230.

Cande, S.C., and Haxby, W.F., 1991, Eocene propagating rifts in the southwest Pacific and their conjugate features on the Nazca Plate: Journal of Geophysical Research, v. 96, B12, p. 19,609–19,622.

Cande, S., Raymond, C., Stock, J., and Haxby, W., 1995, Geophysics of the Pitman fracture zone and Pacific-Antarctic Plate motions during the Cenozoic: Science, v. 270, p. 947–953, doi: 10.1126/science.270.5238.947.

Case, J., Holcombe, T., and Martin, R., 1984, Map of geological provinces in the Caribbean region, in Bonini, W.E., Hargraves, R.B., and Shagam, R., eds., The Caribbean–South American Plate Boundary and Regional Tectonics: Geological Society of America Memoir 162, p. 1–30.

Centeño-García, E., and Keppie, J.D., 1999, Latest Paleozoic–early Mesozoic structures in the central Oaxaca Terrane of southern Mexico: Deformation near a triple junction: Tectonophysics, v. 301, no. 3–4, p. 231–242, doi: 10.1016/S0040-1951(98)00213-3.

Christeson, G.L., McIntosh, K.D., Shipley, T.H., Flueh, E.R., and Goedde, H., 1999, Structure of the Costa Rica convergent margin, offshore Nicoya Peninsula: Journal of Geophysical Research–Solid Earth, v. 104, no. B11, p. 25,443–25,468.

DeMets, C., 2001, A new estimate for present-day Cocos-Caribbean plate motion: Implications for slip along the Central American volcanic arc: Geophysical Research Letters, v. 28, p. 4043–4046, doi: 10.1029/2001GL013518.

DeMets, C., and Dixon, T.H., 1999, New kinematic models for Pacific–North America motion from 3 Ma to present: I, Evidence for steady motion and biases in the NUVEL-1A model: Geophysical Research Letters, v. 26, no. 13, p. 1921–1924, doi: 10.1029/1999GL900405.

DeMets, C., and Traylen, S., 2000, Motion of the Rivera plate since 10 Ma relative to the Pacific and North American plates and the mantle: Tectonophysics, v. 318, no. 1–4, p. 119–159, doi: 10.1016/S0040-1951(99)00309-1.

DeMets, C., Jansma, P.E., Mattioli, G.S., Dixon, T.H., Farina, F., Bilham, R., Calais, E., and Mann, P., 2000, GPS geodetic constraints on Caribbean-North America plate motion: Geophysical Research Letters, v. 27, no. 3, p. 437–440, doi: 10.1029/1999GL005436.

Dickinson, W., and Lawton, T., 2001, Carboniferous to Cretaceous assembly and fragmentation of Mexico: Geological Society of America Bulletin, v. 113, p. 1142–1160, doi: 10.1130/0016-7606(2001)113<1142:CTCAAF>2.0.CO;2.

Dixon, T.H., and Mao, A.L., 1997, A GPS estimate of relative motion between North and South America: Geophysical Research Letters, v. 24, no. 5, p. 535–538, doi: 10.1029/97GL00284.

Donnelly, T., Horne, G., Finch, R., and López-Ramos, E., 1990, Northern Central America: The Maya and Chortis blocks, in Dengo, G., and Case, J., eds., The Caribbean Region: Geological Society of America, Geology of North America, v. H, p. 37–76.

Drobe, J., and Oliver, D., 1998, U-Pb age constraints on Early Cretaceous volcanism and stratigraphy in central Honduras [abs]: Geological Society of America Abstracts with Programs, v. 30, no. 5, p. 12.

Ferrari, L., 2004, Slab detachment control on mafic volcanic pulse and mantle heterogeneity in central Mexico: Geology, v. 32, p. 77–80, doi: 10.1130/G19887.1.

Frisch, W., Meschede, M., and Sick, M., 1992, Origin of the Central American ophiolites; Evidence from paleomagnetic results: Geological Society of America Bulletin, v. 104, p. 1301–1314, doi: 10.1130/0016-7606(1992)104<1301:OOTCAO>2.3.CO;2.

Gordon, M., and Muehlberger, W., 1994, Rotation of the Chortis block causes dextral slip on the Guayape fault: Tectonics, v. 13, p. 858–872, doi: 10.1029/94TC00923.

Guzmán-Speziale, M., 2001, Active seismic deformation in the grabens of northern Central America and its relationship to the relative motion of the North America–Caribbean plate boundary: Tectonophysics, v. 337, p. 39–51, doi: 10.1016/S0040-1951(01)00110-X.

Harlow, G.E., Hemming, S.R., Avé Lallemant, H.G., Sisson, V.B., and Sorenson, S.S., 2004, Two high-pressure–low-temperature serpentinite-matrix mélange belts, Motagua fault zone, Guatemala: A record of Aptian and Maastrichtian collisions: Geology, v. 32, p. 17–20, doi: 10.1130/G19990.1.

Hinz, K., von Huene, R., Ranero, C.R., Flueh, E., Klaeschen, D., Leinbach, J., Ruoff, O., Bargeloh, H.O., Block, M., Fritsch, J., Kewitsch, P., Meyer, H., Schreckenberger, B., and Mrazek, I., 1996, Tectonic structure of the convergent Pacific margin offshore Costa Rica from multichannel seismic reflection data: Tectonics, v. 15, no. 1, p. 54–66, doi: 10.1029/95TC02355.

Hoernle, K., van den Bogaard, P., Irner, R., Lissinna, B., Hauff, F., Alvarado, G., and Garbe-Schoenberg, D., 2002, Missing history (16–71 Ma) of the Galapagos hotspot; implications for the tectonic and biological evolution of the Americas: Geology, v. 30, p. 795–798, doi: 10.1130/0091-7613(2002)030<0795:MHMOTG>2.0.CO;2.

Johnston, S.T., and Thorkelson, D.J., 1997, Cocos-Nazca slab window beneath Central America: Earth and Planetary Science Letters, v. 146, no. 3–4, p. 465–474, doi: 10.1016/S0012-821X(96)00242-7.

Jordan, B.R., Sigurdsson, H., Carey, S., Lundin, S., Rogers, R., and Barquero-Molina, M., 2007, this volume, Petrogenesis of Central American Tertiary ignimbrites and associated Caribbean Sea tephra, in Mann, P., ed., Geologic and Tectonic Development of the Caribbean Plate Boundary in Northern Central America: Geological Society of America Special Paper 428, doi: 10.1130/2007.2428(07)

La Femina, P., Dixon, T., and Strauch, W., 2002, Bookshelf faulting in Nicaragua: Geology, v. 30, p. 751–754, doi: 10.1130/0091-7613(2002)030<0751:BFIN>2.0.CO;2.

Leroy, S., Mauffret, A., Patriat, P., and de Lépinay, B.M., 2000, An alternative interpretation of the Cayman trough evolution from a reidentification of magnetic anomalies: Geophysical Journal International, v. 141, no. 3, p. 539–557, doi: 10.1046/j.1365-246x.2000.00059.x.

Mann, P., editor, 1995, Geologic and Tectonic Development of the Caribbean Plate Boundary in Southern Central America: Geological Society of America Special Paper 326, 349 p.

Mann, P., 1999, Caribbean sedimentary basins: Classification and tectonic setting from Jurassic to present, in Mann, P., ed., Caribbean Basins, Sedimentary Basins of the World: Amsterdam, The Netherlands, Elsevier Science B.V., v. 4, p. 3–31.

Mann, P., and Burke, K., 1990, Transverse intra-arc rifting: Paleogene Wagwater belt, Jamaica: Marine and Petroleum Geology, v. 7, p. 410–412, doi: 10.1016/0264-8172(90)90018-C.

Mann, P., Rogers, R., and Gahagan, L., 2007, Overview of plate tectonic history and its unresolved tectonic problems, in Bundschuh, J., and Alvarado, G.E., eds., Central America: Geology, Resources and Hazards, v. 1: Leiden, The Netherlands, Taylor and Francis/Balkema, p. 201–237.

Marton, G., and Buffler, R.T., 1994, Jurassic reconstruction of the Gulf of Mexico Basin: International Geology Review, v. 36, p. 545–586.

Mayes, C.L., Lawver, L.A., and Sandwell, D.T., 1990, Tectonic history and new isochron chart of the South Pacific: Journal of Geophysical Research, v. 95, no. 6, p. 8543–8567.

McIntosh, K., Silver, E., and Shipley, T., 1993, Evidence and mechanisms for forearc extension at the accretionary Costa Rica convergent margin: Tectonics, v. 12, p. 1380–1392.

Meschede, M., and Barckhausen, U., 2000, Plate tectonic evolution of the Cocos-Nazca spreading center, in Silver, E.A., Kimura, G., Blum, P., and Shipley, T.H., eds., Proceedings of the Ocean Drilling Project, Scientific Results: College Station, Texas, Ocean Drilling Program Leg 170, p. 1–10.

Meschede, M., Barckhausen, U., and Worm, H.-U., 1998, Extinct spreading on the Cocos Ridge, Terra Nova: The European Journal of Geosciences, v. 10, no. 4, p. 211–216.

Moores, E., 1998, Ophiolites, the Sierra Nevada, "Cordillera," and orogeny along the Pacific and Caribbean margins of North and South America: International Geology Review, v. 40, p. 40–54.

Morris, J., Valentine, R., and Harrison, T., 2002, ^{10}Be imaging of sediment accretion and subduction along the northeast Japan and Costa Rica convergent margins: Geology, v. 30, p. 59–62, doi: 10.1130/0091-7613(2002)030<0059:BIOSAA>2.0.CO;2.

Müller, R.D., Royer, J.-Y., Cande, S.C., Roest, W.R., and Maschenkov, S., 1999, New constraints on the Late Cretaceous/Tertiary plate tectonic evolution of the Caribbean, in Mann, P., ed., Caribbean Basins, Sedimentary Basins of the World series: Amsterdam, The Netherlands, Elsevier Science B.V., v. 4, p. 33–59.

Pindell, J., 1994, Evolution of the Gulf of Mexico and the Caribbean, in Donovan, S., and Jackson, T., eds., Caribbean geology: An introduction: Jamaica, University of the West Indies Publisher's Association, p. 13–39.

Pindell, J.L., and Barrett, S.F., 1990, Geologic evolution of the Caribbean: A plate-tectonic perspective, in Dengo, G., and Case, J.E., eds., The Caribbean Region: Geological Society of America, Geology of North America, v. H, p. 405–432.

Pindell, J., and Dewey, J.F., 1982, Permo-Triassic reconstruction of western Pangea and the evolution of the Gulf of Mexico/Caribbean region: Tectonics, v. 1, no. 2, p. 179–211.

Pindell, J., and Kennan, L., 2001, Kinematic evolution of the Gulf of Mexico and the Caribbean, in Fillon, D., ed., Transactions, 21st Bob Perkins Gulf Coast Section Society of Economic Paleontologists and Mineralogists (SEPM) Research Conference: Houston, Gulf Coast Section SEPM Foundation, p. 32.

Plank, T., Balzer, V., and Carr, M., 2002, Nicaraguan volcanoes record paleoceanographic changes accompanying closure of the Panama gateway: Geology, v. 30, p. 1087–1090, doi: 10.1130/0091-7613(2002)030<1087:NVRPCA>2.0.CO;2.

PLATES Project, 2000, Digital database: University of Texas Institute for Geophysics, http://www.ig.utexas.edu/research/projects/plates/ (Accessed May 2005).

Ranero, C., Morgan, P., McIntosh, K., and Reichert, C., 2003, Bending-related faulting and mantle serpentinization at the Middle America trench: Nature, v. 425, p. 367–373, doi: 10.1038/nature01961.

Renne, P.R., Mattinson, J.M., Hatten, C.W., Somin, M., Onstott, T.C., Millan, G., and Linares, E., 1989, ^{40}Ar/^{39}Ar and U-Pb evidence for late Proterozoic (Grenville-age) continental crust in north-central Cuba and regional tectonic implications: Precambrian Research, v. 42, p. 325–341, doi: 10.1016/0301-9268(89)90017-X.

Rogers, R., 2000, Slab break-off mechanism for the late Neogene uplift of Honduras, Northern Central America: Eos (Transactions, American Geophysical Union), v. 81, p. 1179.

Rogers, R., 2003, Jurassic-Recent tectonic and stratigraphic history of the Chortis block of Honduras and Nicaragua [Ph.D. dissertation]: University of Texas at Austin, 264 p.

Rogers, R., and Mann, P., 2007, this volume, Transtensional deformation of the western Caribbean–North America plate boundary zone, in Mann, P., ed., Geologic and Tectonic Development of the Caribbean Plate Boundary in Northern Central America: Geological Society of America Special Paper 428, doi: 10.1130/2007.2428(03).

Rogers, R., Karason, H., and van der Hilst, R., 2002, Epeirogenic uplift above a detached slab in northern Central America: Geology, v. 30, p. 1031–1034, doi: 10.1130/0091-7613(2002)030<1031:EUAADS>2.0.CO;2.

Rogers, R.D., Mann, P., Emmet, P.A., and Venable, M.A., 2007, this volume, Colon fold belt of Honduras: Evidence for late Cretaceous collision between the continental Chortis block and intraoceanic Caribbean arc, in Mann, P., ed., Geologic and Tectonic Development of the Caribbean Plate Boundary in Northern Central America: Geological Society of America Special Paper 428, doi: 10.1130/2007.2428(06).

Rogers, R.D., Mann, P., Scott, R.W., and Patino, L., 2007, this volume, Cretaceous intra-arc rifting, sedimentation and basin inversion in east-central Honduras, in Mann, P., ed., Geologic and Tectonic Development of the Caribbean Plate Boundary in Northern Central America: Geological Society of America Special Paper 428, doi: 10.1130/2007.2428(05).

Rogers, R.D., Mann, P., and Emmet, P.A., 2007, this volume, Tectonic terranes of the Chortis block based on integration of regional aeromagnetic and geologic data, in Mann, P., ed., Geologic and Tectonic Development of the Caribbean Plate Boundary in Northern Central America: Geological Society of America Special Paper 428, doi: 10.1130/2007.2428(04).

Rosa, J., and Molnar, P., 1988, Uncertainties in reconstructions of the Pacific, Farallon, Vancouver, and Kula plates and constraints on the rigidity of the Pacific and Farallon (and Vancouver) plates between 72 and 35 Ma: Journal of Geophysical Research, v. 93, p. 2997–3008.

Rosencrantz, E., 1994, Opening of the Cayman trough and the evolution of the northern Caribbean Plate boundary: Geological Society of America Abstracts with Programs, v. 27, no. 6, p. A-153.

Rosencrantz, E., Ross, M.I., and Sclater, J.G., 1988, Age and spreading history of the Cayman trough as determined from depth, heat flow, and magnetic anomalies: Journal of Geophysical Research, v. 93, p. 2141–2157.

Ross, M.I., and Scotese, C.R., 1988, A hierarchical tectonic model of the Gulf of Mexico and Caribbean region: Tectonophysics, v. 155, p. 139–168, doi: 10.1016/0040-1951(88)90263-6.

Royer, J.-Y., Patriat, P., Bergh, H., and Scotese, C., 1988, Evolution of the Southwest Indian Ridge from the Late Cretaceous (anomaly 34) to the Middle Eocene (anomaly 20): Tectonophysics, v. 155, p. 235–260, doi: 10.1016/0040-1951(88)90268-5.

Schaaf, P., Moran-Zenteño, D., Hernandez-Bernal, M., Solis-Pichardo, G., Tolson, G., and Kohler, H., 1995, Paleogene continental margin truncation in southwestern Mexico: Geochronological evidence: Tectonics, v. 14, p. 1339–1350, doi: 10.1029/95TC01928.

Sinton, C., Duncan, R., and Denyer, P., 1997, Nicoya Peninsula, Costa Rica: A single suite of Caribbean oceanic plateau magmas: Journal of Geophysical Research, v. 102, p. 15,507–15,520, doi: 10.1029/97JB00681.

Sisson, V., Harlow, G., Avé Lallemant, H., Hemming, S., and Sorensen, S., 2003, Two belts of jadeitite and other high-pressure rocks in serpentinites, Motagua fault zone, Guatemala: Geological Society of America Abstracts with Programs, v. 34, no. 4, p. 75.

Sykes, L.R., McCann, W.R., and Kafka, A.L., 1982, Motion of the Caribbean plate during last 7 million years and implications for earlier Cenozoic movements: Journal of Geophysical Research, v. 87, no B13, p. 10,656–10,676.

Tardy, M., Lapierre, H., Freydier, C., Coulon, C., Gill, J., Mercier de Lépinay, B., Beck, C., Martinez-R., Talavera-M., Ortiz-H., E., Stein, G., Bourdier, J., and Yta, M., 1994, The Guerrero suspect terrane (Western Mexico) and coeval arc terranes (the Greater Antilles and the Western Cordillera of Colombia): A late Mesozoic intra-oceanic arc accreted to cratonal America during the Cretaceous: Tectonophysics, v. 230, p. 49–73.

Tebbens, S., and Cande, S., 1997, Southeast Pacific tectonic evolution from early Oligocene to present: Journal of Geophysical Research, v. 102, p. 12,061–12,084, doi: 10.1029/96JB02582.

ten Brink, U.S., Coleman, D.F., and Dillon, W.P., 2002, The nature of the crust under Cayman trough from gravity: Marine and Petroleum Geology, v. 19, p. 971–987, doi: 10.1016/S0264-8172(02)00132-0.

van de Zedde, D.M.A., and Wortel, M.J.R., 2001, Shallow slab detachment as a transient source of heat at mid-lithospheric depths: Tectonics, v. 20, p. 868–882, doi: 10.1029/2001TC900018.

Venable, M., 1994, A geological tectonic, and metallogenetic evaluation of the Siuna terrane (Nicaragua) [Ph.D. dissertation]: Tucson, University of Arizona, 154 p.

von Huene, R., Ranero, C.R., Weinrebe, W., and Hinz, K., 2000, Quaternary convergent margin tectonics of Costa Rica, segmentation of the Cocos Plate, and Central American volcanism: Tectonics, v. 19, no. 2, p. 314–334, doi: 10.1029/1999TC001143.

Walker, J.A., Patino, L.C., Cameron, B.I., and Carr, M.J., 2000, Petrologic insights provided by compositional transects across the Central American arc: Southeastern Guatemala and Honduras: Journal of Geophysical Research, v. 105, p. 18,949–18,963, doi: 10.1029/2000JB900173.

Walther, C.H.E., Flueh, E.R., Ranero, C.R., von Huene, R., and Strauch, W., 2000, Crustal structure across the Pacific margin of Nicaragua: Evidence for ophiolitic basement and a shallow mantle sliver: Geophysical Journal International, v. 141, no. 3, p. 759–777, doi: 10.1046/j.1365-246x.2000.00134.x.

Wilson, D.S., 1996, Fastest known spreading on the Miocene Cocos-Pacific plate boundary: Geophysical Research Letters, v. 23, no. 21, p. 3003–3006, doi: 10.1029/96GL02893.

Wilson, D.S. and Hey, R.N., 1995, History of rift propagation and magnetization intensity for the Cocos-Nazca spreading center, Journal of Geophysical Research–Solid Earth, v. 100, no. B6, p. 10,041–10,056.

Wortel, R., and Cloetingh, S., 1981, On the origin of the Cocos-Nazca spreading center: Geology, v. 9, p. 425–430, doi: 10.1130/0091-7613(1981)9<425:OTOOTC>2.0.CO;2.

Wortel, R., and Spakman, W., 2000, Subduction and slab detachment in the Mediterranean-Carpathian region: Science, v. 290, p. 1910–1917, doi: 10.1126/science.290.5498.1910.

MANUSCRIPT ACCEPTED BY THE SOCIETY 22 DECEMBER 2006

Present motion and deformation of the Caribbean plate: Constraints from new GPS geodetic measurements from Honduras and Nicaragua

Charles DeMets*
Department of Geology and Geophysics, University of Wisconsin, Madison, Wisconsin 53706, USA

Glen Mattioli
Pamela Jansma
Department of Geology, University of Arkansas, Fayetteville, Arkansas 72701, USA

Robert D. Rogers
Department of Geology, California State University, 801 W. Monte Vista Ave., Turlock, California 95382, USA

Carlos Tenorio
Department of Physics, Universidad Nacional Autónoma de Honduras, Tegucigalpa, Honduras

Henry L. Turner
Department of Geology, University of Arkansas, Fayetteville, Arkansas 72701, USA

ABSTRACT

Velocities from six continuous and 14 campaign sites within the boundaries of the Caribbean plate, including eight new sites from previously unsampled areas of Honduras and Nicaragua at the western edge of the Caribbean plate, are described and tested for their consistency with Caribbean–North America plate motion and a rigid Caribbean plate model. Sites in central Honduras and Guatemala move 3–8 mm yr^{-1} westward with respect to the Caribbean plate interior, consistent with distributed east-to-west extension in Guatemala and the western two-thirds of Honduras. A site in southern Jamaica moves 8 ± 1 mm yr^{-1} westward relative to the Caribbean plate interior, indicating that most or all of Jamaica is unsuitable for estimating Caribbean plate motion. Two sites in southern Hispaniola also exhibit anomalous motions relative to the plate interior, consistent with a tectonic bias at those sites. An inversion of the velocities for 15 sites nominally located in the plate interior yields a well-constrained Caribbean plate angular velocity vector that predicts motion similar to previously published models. Data bootstrapping indicates that the solution is robust to better than 1 mm yr^{-1} with respect to both the site velocities that are used

*chuck@geology.wisc.edu

to estimate the plate angular velocity and the site velocity uncertainties. That velocities at seven of eight GPS sites in eastern Honduras and Nicaragua are consistent with the motions of sites elsewhere in the plate interior indicates that much or all of eastern Honduras and Nicaragua move with the plate interior within the 1–2 mm yr^{-1} resolution of our data. It further suggests that the morphologically prominent, but aseismic Guayape fault of eastern Honduras is inactive. Tests for possible east-to-west deformation across the Beata Ridge and Lower Nicaraguan Rise in the plate interior establish a 95% upper bound of ~2 mm yr^{-1} for any deformation across the two features, significantly slower than a published estimate of 9.0 ± 1.5 mm yr^{-1} during the past 23 Ma for deformation across the Beata Ridge.

Keywords: Caribbean plate, Central America, tectonics, plate motion.

1. INTRODUCTION

Over the past two decades, a fundamental objective of neotectonic research in the Caribbean region has been to determine the present motion of the Caribbean plate using Global Positioning System (GPS) technology (Dixon et al., 1991). Prior to the use of GPS for measuring present plate motions, Caribbean–North America plate velocities were estimated from conventional marine geophysical and seismologic observations. The predictions of the latter estimates varied widely, ranging from 11 ± 6 mm yr^{-1} of sinistral strike-slip motion (Jordan, 1975; Stein et al., 1988; DeMets et al., 1990, 1994) to 37 ± 10 mm yr^{-1} (95% uncertainty) of oblique convergence (Sykes et al., 1982) along much of the Caribbean–North America plate boundary. The wide range of predicted motions resulted from disagreements about which, if any, data constituted reliable measures of Caribbean plate motion, including whether earthquake slip vectors from the Middle America and Lesser Antilles trenches are systematically biased by strain partitioning (Sykes et al., 1982; Stein et al., 1988; DeMets, 1993, 2001; Deng and Sykes, 1995) and whether magnetic anomalies from the Cayman spreading center record the full Caribbean–North America rate (Sykes et al., 1982; Rosencrantz and Mann, 1991).

The first unambiguous geodetic determination of present-day Caribbean plate motion was reported by Dixon et al. (1998) from GPS measurements made at three sites during the early to mid-1990s (CRO1, ROJO, and SANA in Fig. 1). Relative to sites on the North America plate, all three stations moved 18–20 mm yr^{-1}, ~80% faster than predicted by the widely used NUVEL-1A model (DeMets et al., 1994). Subsequent geodetic measurements at additional sites in the eastern Caribbean confirmed this result (MacMillan and Ma, 1999; DeMets et al., 2000; DeMets, 2001; Sella et al., 2002) and further demonstrated that Caribbean–South America plate motion significantly exceeds that predicted by NUVEL-1A (Weber et al., 2001; Sella et al., 2002). It is thus now well established that the Caribbean plate moves significantly faster than predicted by NUVEL-1A.

All previous geodetic models of Caribbean plate motion have two significant, though unavoidable, drawbacks related to their underlying geodetic data. The first is that only one site velocity from the western half of the Caribbean plate was used by previous authors to constrain their estimates of Caribbean plate motion; namely, the velocity of campaign site SANA on San Andres Island, several hundred kilometers east of the Nicaraguan coast (Fig. 1). The lack of independent geodetic observations from the western Caribbean precludes any assessment of the accuracy and precision of published estimates of Caribbean plate motion in this region. Similarly, in the eastern Caribbean, three of the four GPS velocities that anchor all previously published estimates of Caribbean plate motion (sites BARB, CRO1, and ROJO in Fig. 1) are for sites that are close enough to seismically active plate boundary faults to raise concerns that their velocities might be biased by steady interseismic or long-term transient postseismic strain related to the earthquake cycle on those faults (e.g., Pollitz and Dixon, 1998). The sparse data from all areas of the Caribbean plate preclude the usual tests for velocity outliers due to factors such as GPS monument instability or localized tectonic effects.

The scientific motivations for the present analysis are twofold. First, field-based studies of the Caribbean plate boundaries require at minimum an accurately defined estimate of the motion of the plate interior in order to characterize complex deformation in those field areas. These include geodetic studies initiated by several groups in the late 1990s in large areas of Central America and ongoing GPS projects in Hispaniola (Calais et al., 2002), the Lesser Antilles (Jansma and Mattioli, 2005), Jamaica (DeMets and Wiggins-Grandison, 2007), and Venezuela (Weber et al., 2001).

The second motivation for this work is to determine whether the Caribbean plate undergoes significant internal deformation. A variety of geologic and seismic observations have been cited as evidence for such deformation, possibly driven by slow convergence of the South and North America plates across the Caribbean region (Dixon and Mao, 1997; Müller et al., 1999). Holcombe et al. (1990) describe evidence for young faulting and volcanism within seismically active rifts imaged by marine seismic reflection profiles from the Lower Nicaraguan Rise and propose that diffuse east-to-west rifting of the rise occurs in response to sinistral shear along its bounding escarpments. Heubeck and Mann (1991) suggest that the Caribbean plate consists of rigid subplates east and west of the Beata Ridge (BR in Fig. 1) coinciding with the Venezuelan and Colombian basins, and a subplate in western Central

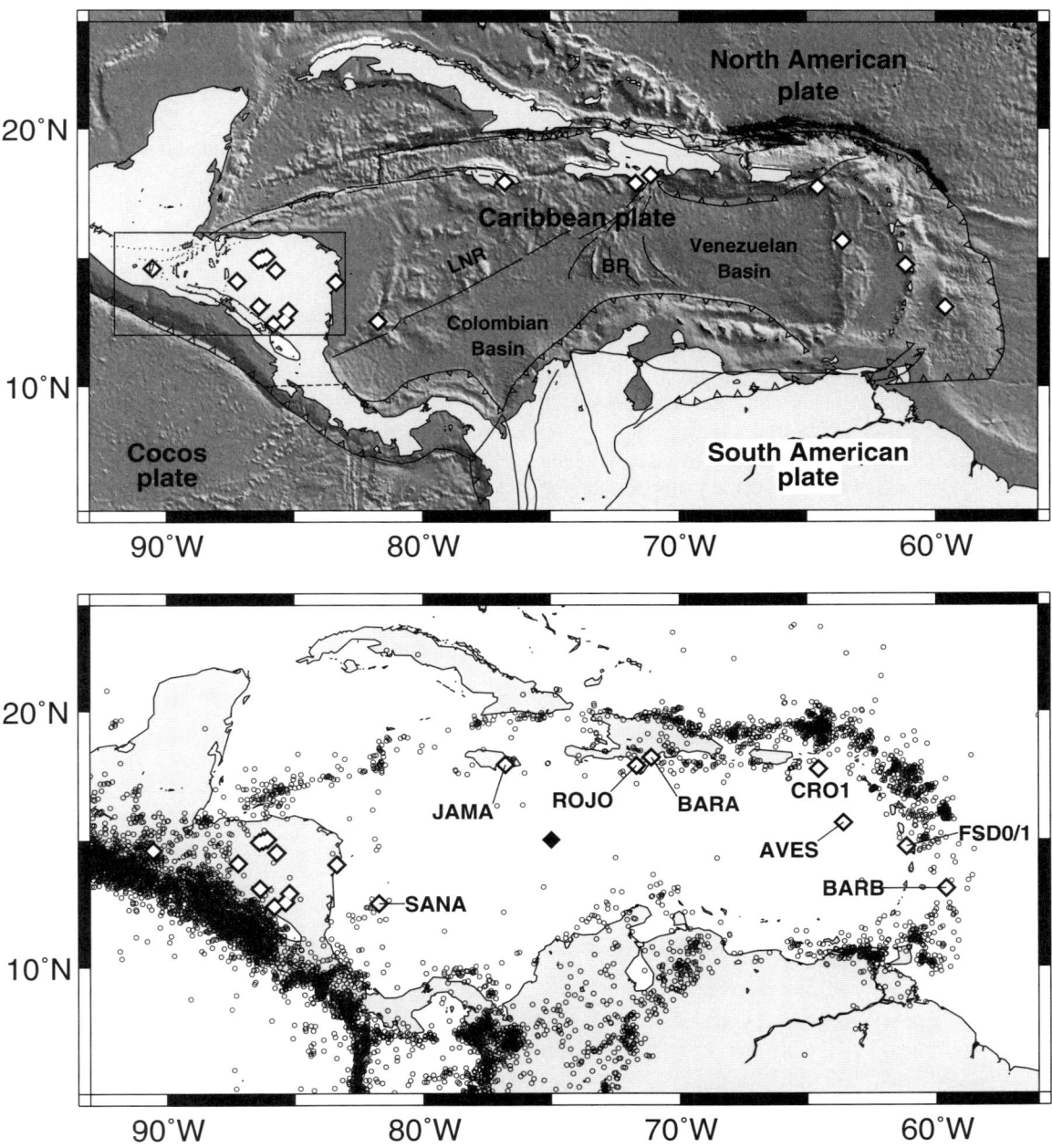

Figure 1. Seismotectonic setting of the Caribbean plate. BR in upper diagram is the Beata Ridge. Line LNR shows the general location and trend of the Lower Nicaraguan Rise, which extends ~1000 km northeastward from the Caribbean coast of Nicaragua toward southern Hispaniola. Open diamonds show locations of GPS sites whose velocities are employed in this study. Filled diamond shows 15°N, 75°W fiducial location employed for the analysis. Area enclosed in rectangle is displayed in Figure 6. All earthquakes above depths of 60 km and with surface- or body-wave magnitudes >3.5 for the period 1963 through 2004 are shown in the lower diagram. AVES, BARA, BARB, CRO1, FSD0/1, JAMA, ROJO, and SANA are site names discussed in text.

American west of the Honduras Depression. Consistent with this interpretation, Leroy and Mauffret (1996) interpret apparently reactivated reverse faults imaged in marine seismic profiles that cross the eastern flank of the Beata Ridge as evidence for contraction across the Beata Ridge and hence deformation within the Caribbean plate. Mauffret and Leroy (1999) further interpret compressional features along the Beata Ridge as evidence for NE-SW shortening between independently moving microplates flanking the Beata Ridge and estimate that the convergence rate across the Beata Ridge has averaged 9 ± 1.5 mm yr^{-1} for the past 23 Ma.

Here, we describe new GPS velocities for 12 sites from the western half of the Caribbean plate (Fig. 1) and use these in combination with the motions of eight additional sites from the central and eastern Caribbean plate to achieve the objectives described above. We first specify and interpret the velocities of all 20 Caribbean plate GPS sites relative to the North America plate in order to identify and exclude velocities that do not record motion of the plate interior. We then invert the velocities of 15 sites whose motions are mutually consistent to determine a best-fitting Caribbean plate angular velocity vector. We describe and interpret residual site velocities with respect to the predictions of the best-fitting angular velocity vector and employ formal data importances to determine the amount of information that the individual site velocities contribute to our best-fitting model. In light of evidence that a single station velocity (CRO1) supplies 40% of the model information, we employ data bootstrapping to determine how robust our estimate of Caribbean plate motion is with respect to the 15 site velocities and their estimated uncertainties. We then construct an alternative best estimate of Caribbean plate motion that more evenly distributes the information contributed by the site velocities and excludes two sites in the eastern Caribbean that exhibit evidence for small tectonic biases. We conclude by estimating for the first time a rigorous upper bound on possible east to west internal deformation of the Caribbean plate.

2. GPS DATA AND ANALYSIS

Table 1 summarizes information about the 20 campaign and continuous GPS measurements used in the analysis (Fig. 1). From Central America, we employ new data from 11 stations, eight of which are campaign sites that we installed and first occupied in 2000–2001 in aseismic, interior areas of Nicaragua and Honduras. Observations at these sites span intervals of 2.1 yr (PUEC) to 5.2 yr (CMP1). The other three Central American sites are the continuous stations ESTI, GUAT, and TEGU/TEG1, which were installed in 2000 and are operated by the U.S. National Oceanic and Atmospheric Administration (NOAA). We also update the velocity for site SANA, located on San Andres Island east of Nicaragua, using new data collected in 2003. From the central and eastern areas of the Caribbean plate, we use continuous data from sites in Jamaica (JAMA), southern Hispaniola (BARA), the Virgin Islands (CRO1), and Barbados (BARB), complemented by campaign data from sites AVES, ROJO, FSD0, and FSD1.

We exclude all sites on the Puerto Rico–Virgin Islands block, which moves westward at a rate of 2.6 ± 1.0 mm yr^{-1} relative to the Caribbean plate interior (Jansma and Mattioli, 2005), and also exclude all sites in Hispaniola that are located north of the Enriquillo fault, which accommodates significant long-term

TABLE 1. GPS SITE INFORMATION AND OCCUPATION HISTORY

Site name (country)	Lat (°N)	Long (°W)	Station days occupied per calendar year						Site velocity mm yr^{-1}	
			2000	2001	2002	2003	2004	2005	North	East
CMP1 (Honduras)	14.5092	85.7146	3					2	5.3 ± 2.0	10.6 ± 3.0
GLCO (Honduras)	15.0298	86.0699	2				3	2	4.7 ± 2.2	8.6 ± 3.3
MNTO (Honduras)	14.9168	86.3805	3				3		6.1 ± 2.1	11.1 ± 3.2
SFDP (Honduras)	14.9659	86.2449	3				3		5.4 ± 2.1	9.7 ± 3.5
PORT (Nicaragua)	12.5731	85.3671	3			4	4		3.4 ± 1.6	10.8 ± 6.2
PUEC (Nicaragua)	14.0421	83.3820		4		4			2.8 ± 4.4	6.4 ± 5.9
RIOB (Nicaragua)	12.9209	85.2206	4			4	4		5.6 ± 1.8	9.6 ± 7.8
TEUS (Nicaragua)	12.4098	85.8136	5		5		5		4.7 ± 2.4	10.0 ± 2.0
TEGU (Honduras)	14.0905	87.2056	245	282	333	236	366	64	3.7 ± 0.7	7.4 ± 0.8
GUAT (Guatemala)	14.5904	90.5202	156	333	342	314	326	261	2.1 ± 0.6	2.1 ± 0.8
ESTI (Nicaragua)	13.0996	86.3621	203	325	332	56			12.8 ± 1.1	10.9 ± 1.6
SANA (Colombia)	12.5238	81.7294	Occupied 1994 (5), 1996 (3), 1998 (6), 2000 (6), and 2003 (5)						6.9 ± 0.4	13.8 ± 1.2
JAMA* (Jamaica)	17.9390	76.7810	258	356	325	138			10.1 ± 0.7	1.6 ± 0.9
BARB (Barbados)	13.0879	59.6091	Semi-continuous from 1997–2001 (580 station days)						15.2 ± 1.0	10.8 ± 1.8
CRO1 (Virgin Isl.)	17.7569	64.5843	Continuous from Oct. 1995–present						12.5 ± 0.4	9.9 ± 0.6
BARA (Dom. Rep.)	18.2087	71.0982				322	325	245	7.6 ± 0.9	8.0 ± 0.9
ROJO (Dom. Rep.)	17.9040	71.6745	Occupied 1994 (9), 1995 (2), 1998 (3), and 2001 (2)						7.8 ± 1.5	11.1 ± 2.4
AVES (Venezuela)	15.6670	63.6183	Occupied 1994 (18) and 1998 (10)						13.3 ± 2.0	11.7 ± 2.9
FSD0 (Martinique)	14.7348	61.1467	Occupied 1994 (4), 1998 (5), and 1999 (4)						15.0 ± 2.0	12.4 ± 3.0
FSD1 (Martinique)	14.7349	61.1465	Occupied 1994 (5), 1998 (11), and 1999 (3)						14.9 ± 1.8	14.2 ± 2.7

Note: Site velocities are relative to ITRF2000. Uncertainties are standard errors.
*Site JAMA also has 88 days of data from 1999.

slip relative to the plate interior (Mann et al., 1995, 2002; Calais et al., 2002).

All GPS code-phase measurements employed for this analysis, including observations from 151 continuous stations that anchor our North America plate reference frame, were analyzed using GIPSY software from the Jet Propulsion Laboratory (JPL). We employed a standard point-positioning analysis strategy (Zumberge et al., 1997) combined with resolution of integer phase ambiguities. Daily GPS station coordinates were first estimated in a nonfiducial reference frame (Heflin et al., 1992) employing precise fiducial-free satellite orbits and clocks from JPL. The loosely constrained station coordinates were then transformed to ITRF2000 (Altamimi et al., 2002) using daily seven-parameter Helmert transformations supplied by JPL. We also estimated and removed daily and longer-period regionally correlated noise between sites using a technique described by Marquez-Azua and DeMets (2003). Daily repeatabilities in the north, east, and down components of the GPS site coordinates are 2–4 mm, 3–5 mm, and 8–10 mm, respectively. Uncertainties in the GPS site velocities are estimated using procedures described by Mao et al. (1999), with white and flicker noise estimated from individual GPS time series and a further assumed contribution of 1 mm per $\sqrt{yr}$ from random monument walk.

3. RESULTS

3.1. Plate-Wide Velocity Field Relative to the North America Plate

We begin by examining the velocities of all 20 GPS sites relative to the North America plate interior, which constitutes a natural geological reference frame for sites located along the Lesser Antilles trench and northern boundary of the Caribbean plate. The angular velocity vector that specifies motion of the North America plate relative to ITRF2000 is determined from an inversion of the velocities of 151 sites from the plate interior, based on our own analysis of continuous data from these stations (Table 2). We omitted the velocities of all sites within 2000 km of Hudson Bay, where glacial isostatic rebound measurably affects site motions (Park et al., 2002; Mazzotti et al., 2005; Calais et al., 2006). We also excluded sites west of the Rio Grande Rift to avoid biases from any slow deformation west of the rift. The weighted root-mean-square residual motions of the 151 North America plate sites relative to the best-fit model predictions average 0.8 mm yr^{-1} in both the north and east velocity components. Uncertainties in the velocities predicted by the well-constrained North America plate angular velocity vector are smaller than ± 0.1 mm yr^{-1} and ± 0.8° at locations in the Caribbean and are thus not a limiting factor in the analysis described below.

Figure 2 shows the 20 Caribbean GPS velocities after their transformation into the North America plate reference frame. Velocities range from 11 to 23 mm yr^{-1} and generally point toward N75°E ± 5° (Figs. 2 and 3). More than half of the velocities agree within their errors with the predictions of previous GPS-based models for Caribbean–North America plate motion (Dixon et al., 1998; DeMets, 2001; Sella et al., 2002). In particular, the four new campaign sites in eastern and central Honduras (CMP1, GLCO, MNTO, and SFDP) have an average weighted velocity of 19.3 ± 1.7 mm yr^{-1} toward N73.7°E ± 3° (see vector labeled "HND" in Fig. 2), and four of the five Nicaraguan sites (PORT, PUEC, RIOB, and TEUS) have an average weighted velocity of 17.8 ± 1.7 mm yr^{-1} toward N76.7°E ± 4° (see "NIC" in Fig. 2). Both averages are consistent within errors with the motion expected for sites that lie on the Caribbean plate interior (Fig. 3).

The motions of five sites depart significantly from their expected motions. Site ESTI in Nicaragua moves ~4 mm yr^{-1} faster than and 20° counterclockwise from the other Nicaraguan sites (Fig. 2). For security and logistical reasons, the GPS antenna at ESTI is mounted on a steel tower >7 m high (M. Chin, 2005, personal commun.). We thus suspect that monument instability may be the cause of the anomalous motion at ESTI and exclude this velocity from further analysis. Sites TEGU and GUAT, which are located ~150 km and ~500 km west of the cluster of four GPS sites in central eastern Honduras, move in the same direction as other nearby Central American sites (Figs. 2 and 3), but at rates that are 3 ± 1 mm yr^{-1} and 8 ± 1 mm yr^{-1} slower than predicted for sites on rigid Caribbean lithosphere. That the motions of TEGU and GUAT become progressively faster to the west is consistent with geologic (Manton, 1987) and seismic (Guzman-Speziale, 2001) evidence for significant extension across much of Honduras and Guatemala. The pattern of site velocities defined by the four campaign sites in eastern and central Honduras and continuous sites TEGU and GUAT strongly suggests that the western limit of stable areas of the Caribbean plate interior lies between TEGU and the four GPS sites in eastern-central Honduras.

TABLE 2. BEST-FITTING CARIBBEAN PLATE ANGULAR VELOCITY VECTOR INFORMATION

Plate pair	No. of sites	Angular velocity vector			Angular velocity vector covariances					
		λ (°N)	φ (°E)	ω (degrees/m.y.)	α_{xx}	α_{yy}	α_{zz}	α_{xy}	α_{xz}	α_{yz}
CA-ITRF2000	15	36.3	−98.5	0.255	0.350	3.017	0.520	−0.864	0.263	−0.859
NA-ITRF2000	151	−7.64	−86.21	0.196	0.011	0.182	0.107	0.015	−0.011	−0.125
CA-NA (15)	166	75.9	191.5	0.182	0.361	3.199	0.627	−0.849	0.252	−0.984
CA-ITRF2000	13	34.3	−96.8	0.270	0.132	0.831	0.175	−0.262	0.000	−0.184
CA-NA (13)	164	75.0	215.3	0.185	0.143	1.013	0.282	−0.247	−0.011	−0.309

Note: CA—Caribbean plate; NA—North America plate. First plate rotates counterclockwise around the pole relative to the second. Latitude, longitude, and angular rotation rate are specified by λ, φ, and ω, respectively. Elements of the symmetric 3 × 3 variance-covariance matrix are given in units of 10^{-8} radians^2 per m.y.^{2}. Variance-covariance matrix for the 13-site CA-ITRF2000 angular velocity vector is derived from bootstrapped solutions (see text).

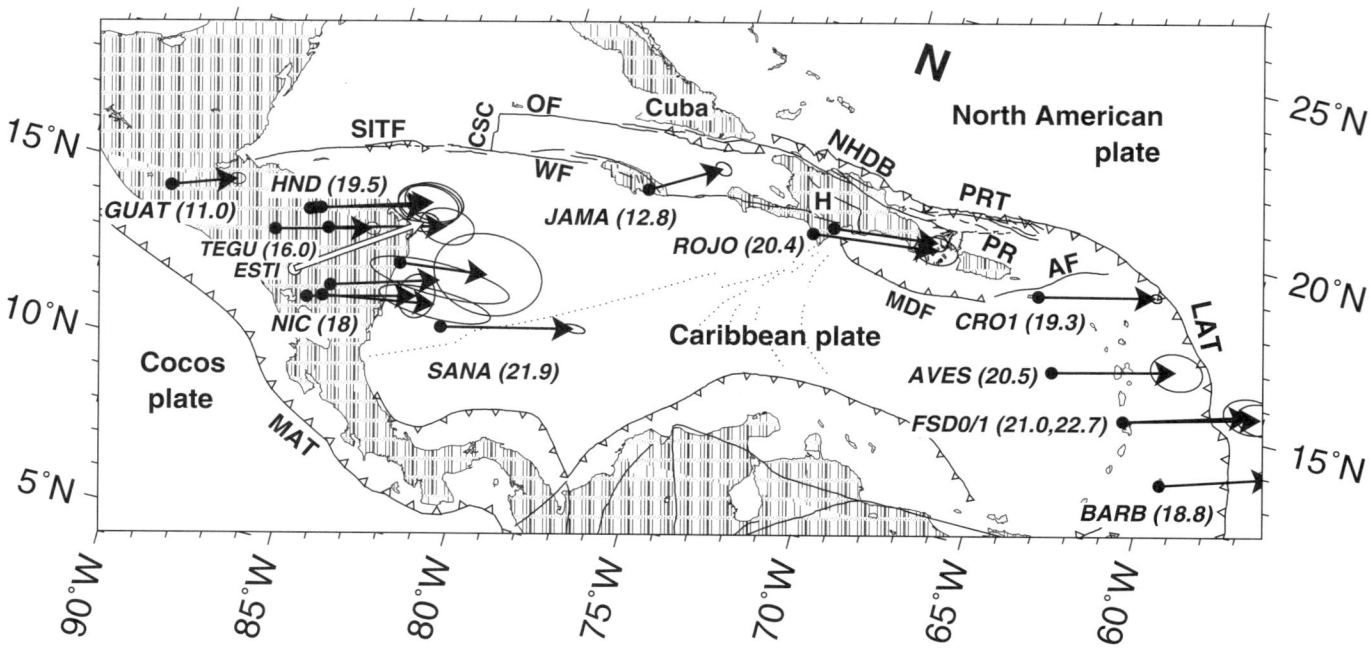

Figure 2. GPS site velocities relative to the North America plate on an oblique Mercator map projected about the best-fitting Caribbean–North America pole of rotation. Numerals by site names give the site rates in millimeters per year. Open arrow indicates motion of site ESTI, where the antenna is mounted on a tall steel tower of questionable stability. The location of GPS site BARA discussed in the text is shown in Figure 1. AF—Anegada fault; CSC—Cayman spreading center; H—Hispaniola; LAT—Lesser Antilles trench; MAT—Middle America trench; MDF—Muertos deformed belt; NHDB—Northern Hispaniola deformed belt; OF—Oriente fault; PR—Puerto Rico; PRT—Puerto Rico trench; SITF—Swan Islands transform fault; WF—Walton fault. "HND" and "NIC" refer to the weighted average velocities of GPS station clusters in the respective countries of Honduras and Nicaragua, as described in the text. AVES, BARB, CRO1, GUAT, ESTI, FSD0/1, JAMA, ROJO, SANA, and TEGU are site names discussed in the text.

Along the northern boundary of the Caribbean plate, the motion of site JAMA in southern Jamaica is 7 ± 1 mm yr^{-1} slower than and 17° counterclockwise from the velocity predicted for a Caribbean plate interior site (Figs. 2 and 3). Nineteen other GPS sites in Jamaica exhibit similar velocity deficits (DeMets and Wiggins-Grandison, 2007), indicating that most or all Jamaican GPS sites are unsuitable for estimating Caribbean plate motion.

In Hispaniola, site BARA also moves significantly slower (by 2 ± 1 mm yr^{-1}) than predicted for a plate interior site (Fig. 3). Its slip deficit with respect to the full Caribbean–North America rate is consistent with a GPS velocity gradient in Hispaniola that is interpreted by Dixon et al. (1998), Calais et al. (2002), and Mann et al. (2002) as evidence for elastic strain accumulation from locked plate boundary faults within and north of Hispaniola. Site BARA is thus located within the zone of interseismic elastic deformation associated with plate boundary faults in Hispaniola, making its velocity unsuitable for estimating the motion of the Caribbean plate interior.

We conclude that sites GUAT, TEGU, JAMA, and BARA are located in probable zones of distributed deformation and that monument instability may bias the velocity at site ESTI. Their velocities are thus not used below to constrain Caribbean plate motion.

3.2. Best-Fitting 15-Station Caribbean Plate Model and Residual Velocities

We next invert velocities of the 15 sites (Table 1) that appear to move with the Caribbean plate interior to define a best-fitting angular velocity vector for Caribbean plate motion relative to ITRF2000. The north and east velocity components for each site are weighted in the inversion by the reciprocal of their squared uncertainties (their variances), thereby ensuring that velocities for sites such as CRO1 with long, continuous time series contribute more to the solution than do sites that are infrequently occupied or that have shorter time series. The best-fitting angular velocity vector (Table 2) fits the 15 site velocities well, with respective weighted root-mean-square misfits of 0.8 mm yr^{-1} for the north velocity component and 1.3 mm yr^{-1} for the east velocity component. The misfits are comparable to site velocity misfits reported for other plates (Sella et al., 2002), but are smaller by ~25% than the assigned site velocity uncertainties. We therefore multiplied the angular velocity vector variances and covariances by reduced chi-square for the best-fitting solution (0.54) to ensure that the angular velocity uncertainties accurately reflect the dispersion of the site velocities with respect to the model predictions.

Figure 4 shows the site data importances, which constitute a formal measure of the amount of information that each site

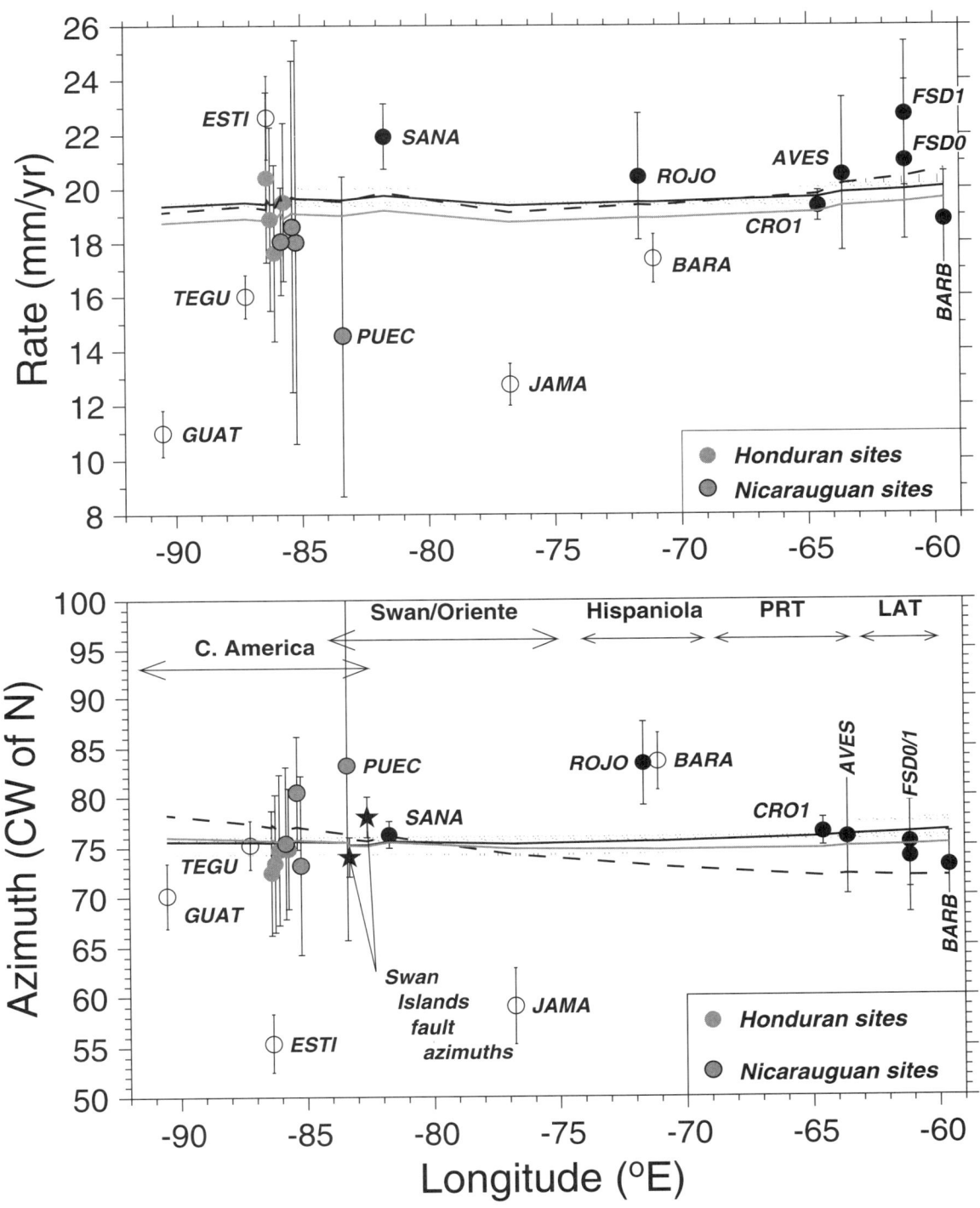

Figure 3. Rates (upper) and directions (lower) of GPS sites relative to North America plate. Solid and shaded circles show GPS velocities that are used to derive the best-fitting Caribbean–North America angular velocity, whose predictions are shown by the solid black line. Dashed and shaded lines show predictions of Caribbean–North America angular velocity vectors from DeMets (2001) and Sella et al. (2002), respectively. Open circles show the motions of sites that are not used to derive the best-fitting model due to obvious tectonic or other biases in their motions. Swan Islands transform fault azimuths in lower panel are from multibeam seafloor mapping (Rosencrantz and Mann, 1991). LAT—Lesser Antilles trench; PRT—Puerto Rico trench. AVES, BARA, BARB, CRO1, GUAT, ESTI, FSD0, FSD1, JAMA, PUEC, ROJO, SANA, and TEGU are site names discussed in the text.

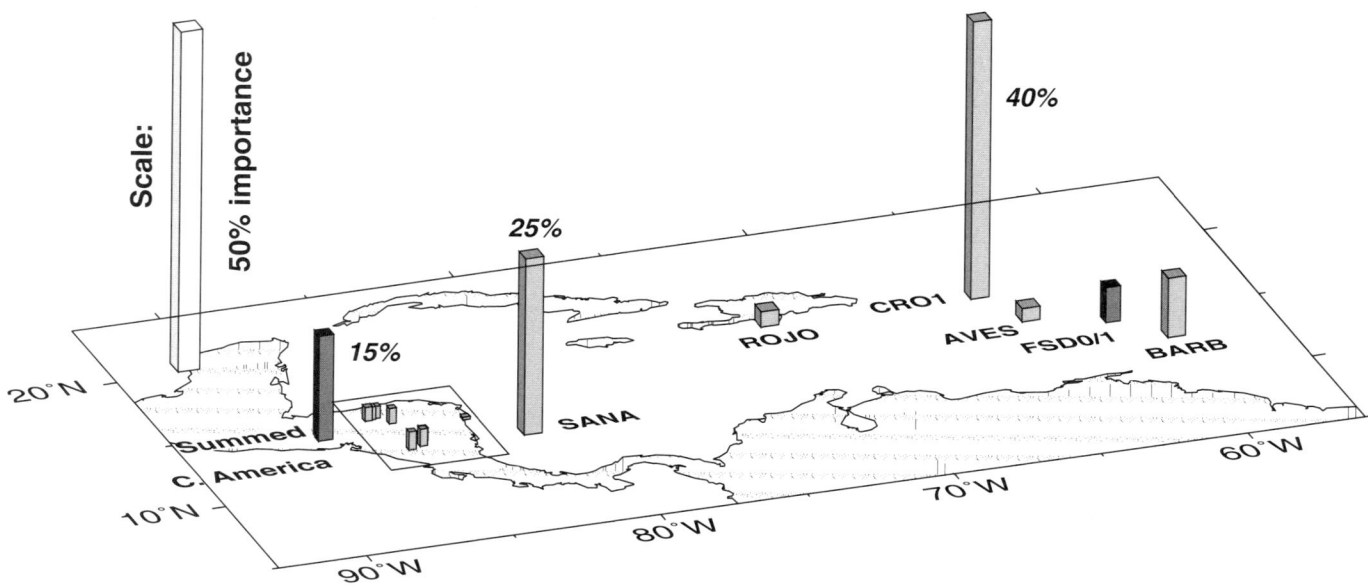

Figure 4. Data importances for velocities of Caribbean and Central American GPS sites that are used to derive the best-fitting, 13-station Caribbean-ITRF2000 angular velocity vector (Table 2). Darker gray-shaded bar shows the summed importances of the eight Central American sites within the boxed region. AVES, BARB, CRO1, FSD0/1, ROJO, and SANA are site names discussed in the text.

velocity contributes to the best-fitting angular velocity vector (Minster et al., 1974). For our application, the data importances are determined from the relative uncertainties of the individual site velocities and their geographic locations with respect to the other sites and the pole of rotation. They are thus a useful guide to the strengths and weaknesses of a data-poor model such as our own and can be used to develop strategies to overcome potential problems.

The best-fitting angular velocity vector derives much (40%) of its information from the velocity for site CRO1 (Fig. 4). This high importance is attributable to the small velocity uncertainty for CRO1, which has a 10-yr-long continuous time series. In contrast to CRO1, the other five sites in the eastern Caribbean have a cumulative importance of only 18%. Our own (and previous) estimates of Caribbean plate motion thus rely heavily on the velocity for CRO1 and, by implication, on the possibly incorrect assumption that site CRO1 accurately records motion of the undeforming Caribbean plate lithosphere. The good fit of the model to the velocity at CRO1 (Fig. 5), better than for any of the other three sites in the eastern part of the plate, results from the site velocity's high importance in the model and by itself cannot be taken as evidence that the site moves with the plate interior.

In the western region of the plate, the eight Honduran and Nicaraguan sites contribute a cumulative 15% of the model information, less than is contributed by site SANA (25%), which has been occupied more frequently and over a significantly longer period than have the eight Central American sites. Relative to previous solutions (e.g., DeMets et al., 2000; Sella et al., 2002), which use only the velocity at SANA to constrain plate motion west of Hispaniola, our model information is distributed more evenly between the western and eastern regions of the Caribbean plate, mainly because motion for the western half of the plate is now estimated from nine GPS velocities instead of just the velocity at site SANA. As a consequence, more meaningful tests of model robustness and plate rigidity are possible (described below).

Figures 5 and 6 show the residual velocities at all the sites with respect to motion predicted by our best-fitting, 15-station Caribbean-ITRF2000 angular velocity vector (Table 2). If the 15 site velocities and their estimated uncertainties accurately describe motion of the plate interior, the pattern of residual site velocities should be random in stable areas of the plate and show some systematic pattern in areas of diffuse or concentrated deformation. We next describe the residual site velocities, with particular attention to any evidence for tectonic or other systematic biases at individual sites.

Eastern and Central Caribbean

The velocities at sites AVES and CRO1 differ insignificantly (0.7 ± 2.8 mm yr^{-1} and 0.4 ± 0.6 mm yr^{-1}, respectively) from the velocities predicted by the best-fitting CA-ITRF2000 angular velocity vector (Fig. 5). The velocity for AVES contributes only 5% of the model information to the best-fitting angular velocity vector (Fig. 4). The small residual velocity at AVES thus constitutes useful evidence that the site is located in a stable area of the plate interior. Similarly, residual velocities at campaign sites FSD0 and FSD1 are both smaller than their estimated rate uncertainties of ± 3.5 mm yr^{-1}, indicating that neither moves relative to the plate interior at a rate that exceeds its estimated uncertainty.

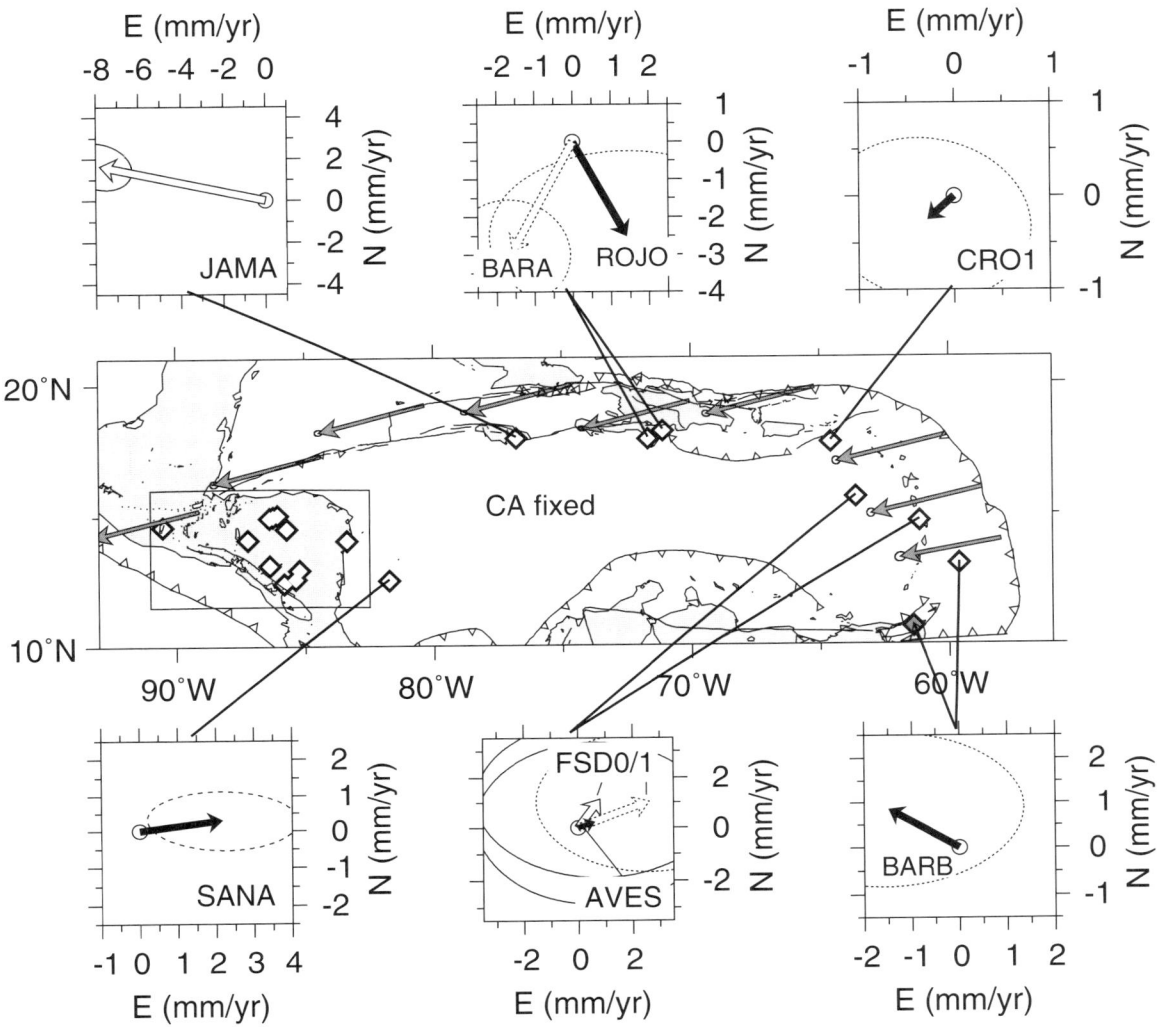

Figure 5. Velocities of circum-Caribbean GPS sites relative to the Caribbean plate (CA) interior after removing motion predicted by the best-fitting Caribbean-ITRF2000 angular velocity vector (Table 2). North America plate movement relative to the Caribbean plate (Table 2) is shown by shaded arrows for points along the plate boundary. Site velocity ellipses are two-dimensional, 1σ. Residual site velocities shown with open arrows were not used to derive the best-fitting CA-ITRF2000 angular velocity vector. Rectangle encloses area shown in Figure 6. AVES, BARA, BARB, CRO1, FSD0/1, JAMA, ROJO, and SANA are site names discussed in the text.

The residual velocity at site BARB points away from the Lesser Antilles trench toward the plate interior (Fig. 5). Although the misfit of 1.7 ± 1.7 mm yr^{-1} is only marginally significant, the geologic setting of this site makes it unlikely that it is part of the plate interior. Geological mapping indicates that Barbados, which is located at the crest of the extensive Lesser Antilles accretionary prism, formed by offscraping, back rotation, and shortening of marine sediments (Speed, 1983). The trench-normal component of the residual motion at BARB is consistent with active shortening of the accretionary wedge, possibly via permanent deformation within the wedge or by elastic strain from frictional locking of the subduction interface downdip from Barbados. Evidence for slow motion (~1 mm yr^{-1}) toward the plate interior at other sites in the Lesser Antilles (G. Mattioli, personal commun., 2006) suggests that the observed residual motion at site BARB is real and that surface deformation along the volcanic arc is probably influenced at a measurable level by elastic strain accumulation driven by frictional coupling across the Lesser Antilles subduction interface. Given the available observations, we exclude the velocity from BARB from the 13-station velocity model described in Section 3.4.

The residual velocities at sites BARA and ROJO in southern Hispaniola both have southward components (Fig. 5), increasing from 2 mm yr^{-1} at ROJO to 3 mm yr^{-1} at BARA. The southward motions of both sites toward the plate interior are consistent with a previously described gradient in the boundary-normal components of motion at other GPS sites from Hispaniola (Calais et al., 2002). Calais et al. interpret this gradient as an elastic response to

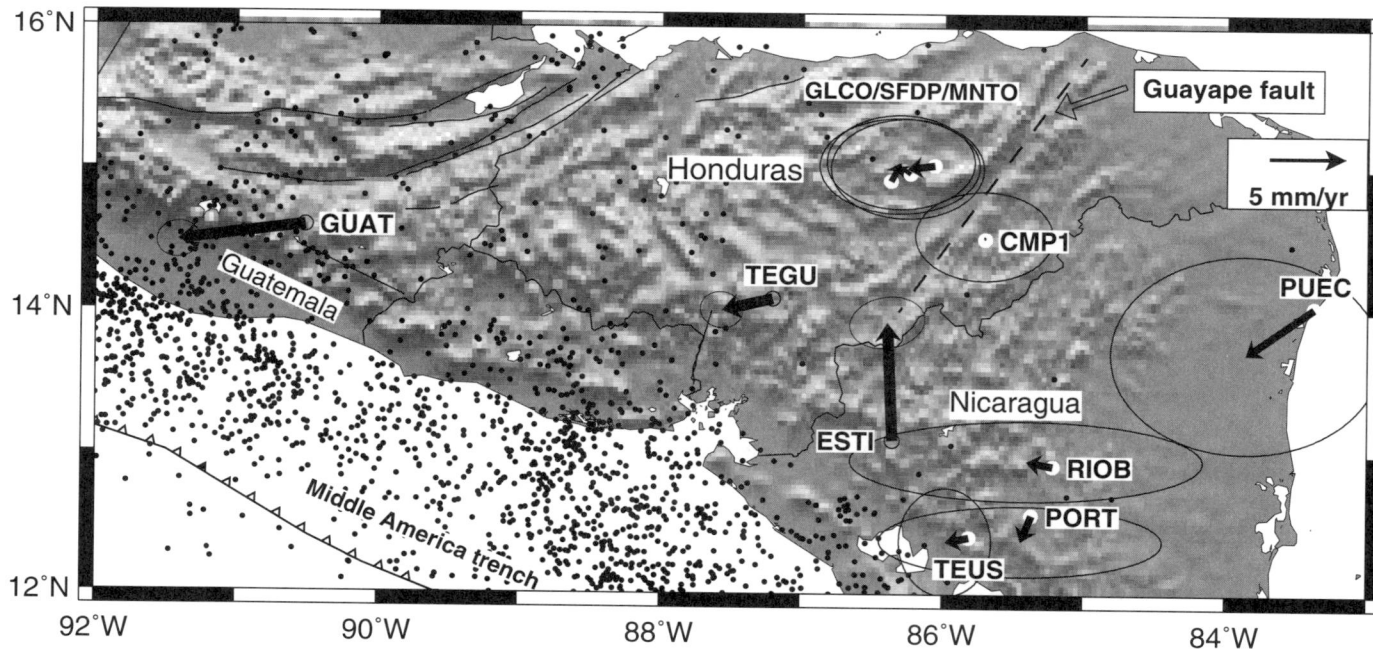

Figure 6. Residual velocities of Central American GPS sites relative to motion predicted by the best-fitting Caribbean-ITRF2000 angular velocity vector (Table 2). Site velocity ellipses are two-dimensional, 1σ. Open circles indicate epicenters of earthquakes located above a depth of 60 km with surface- or body-wave magnitudes greater than 3.5 for the period 1963 through 2004. CMP1, ESTI, GLCO/SFDP/MNTO, GUAT, PUEC, PORT, RIOB, TEGU, and TEUS are site names discussed in the text.

5 ± 2 mm yr^{-1} of dip-slip motion along the northern Hispaniola thrust fault (NHDP in Fig. 1), constituting roughly half of the 10 ± 1.5 mm yr^{-1} of boundary-normal convergence that should occur across Hispaniola if oblique convergence is fully partitioned into boundary-parallel and boundary-normal components (DeMets et al., 2000). Given the likelihood that the misfits at BARA and ROJO represent tectonic biases due to their proximity to active faults in Hispaniola, we exclude both site velocities from our 13-station solution that is described in Section 3.4.

Western Caribbean

For the nine sites from the western Caribbean whose velocities are used to estimate the best-fitting CA-ITRF2000 angular velocity vector, only the velocity for site SANA, which has the longest occupation history of any site in the western Caribbean, is significantly misfit (Fig. 5). It seems unlikely that the misfit is attributable to monument instability because the monument is embedded in bedrock. Elastic effects from major plate boundary faults are also an implausible explanation given that SANA is located >200 km from any major plate boundary faults. San Andres Island is, however, located adjacent to the seismically active San Andres Trough (Fig. 1), within a region of postulated distributed deformation in the Lower Nicaraguan Rise (Holcombe et al., 1990).

Residual velocities from the eight sites in Honduras and Nicaragua whose velocities are used to estimate Caribbean plate motion are smaller than their estimated standard errors (Fig. 6), with misfits for seven of the eight sites of only 0.2–1.9 mm yr^{-1}. The largest misfit occurs for site PUEC, which has shortest occupation history (2.1 yr) of the 20 sites we use. PUEC is located on the roof of a single-story concrete building, which may also contribute to the site's residual velocity. That all eight velocities from Honduras and Nicaragua are fit within their uncertainties despite their low data importances demonstrates that they are consistent with the higher importance velocities for sites in the eastern Caribbean (particularly CRO1). The three Honduran sites (GLCO, MNTO, and SFDP) that are clustered northwest of the Guayape fault (Fig. 6) exhibit no coherent pattern in their small residual velocities and no net motion relative to site CMP1, located immediately southeast of the Guayape fault. The Honduran velocities thus suggest there is little or no slip along the Guayape fault.

3.3 Model Sensitivity and Robustness

The 15-station best-fitting solution has several potential weaknesses that could degrade its accuracy. Prominent among these is its over-dependence on the velocity for site CRO1, which makes the solution sensitive to any systematic tectonic or other biases at that site. Of further concern are the possible tectonic biases in the velocities for sites BARB and ROJO. In light of these potential problems, we use two techniques to examine the robustness of our 15-station model predictions. We first examine the effects of the individual site velocities on the predictions of

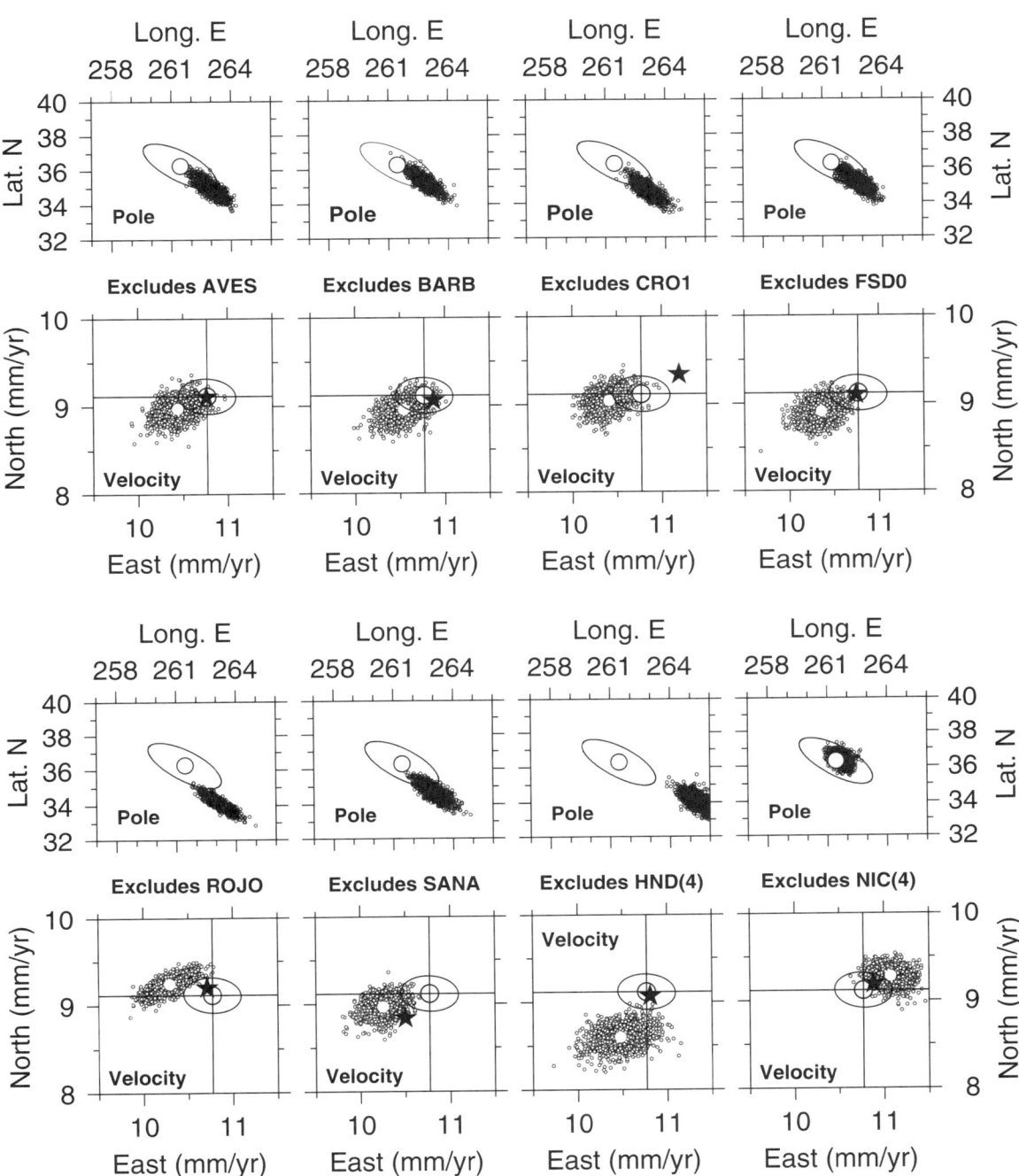

Figure 7. Outcome of data sensitivity and bootstrap analyses for the 15-station best-fitting Caribbean plate solution. Procedures are described in the text. HND and NIC represent the four Honduran site velocities and four Nicaraguan site velocities, respectively. Sub-panels labeled "Pole" show the 15-station, best-fitting pole (open circle), its two-dimensional, 1σ confidence ellipse, and bootstrap pole locations (dots) that are derived from bootstrapping of velocity subsets that exclude the site velocity or velocities that are specified beneath the sub-panel. Sub-panels with velocity estimates show only the velocity vector endpoints and exclude the vector origins (0, 0). Open circles and ellipses that are centered on the crosshairs are the velocities and two-dimensional, 1σ uncertainties at 15°N, 75°W derived from the 15-station, best-fitting angular velocity vector and its covariances. Stars show the velocities at 15°N, 75°W estimated from 14-station models that exclude one station velocity and weight the remaining 14 site velocities by their formal uncertainties from Table 1. The tiny circles in the velocity sub-panels show velocities at 15°N, 75°W predicted by the bootstrap poles. The open circle in the cloud of bootstrapped velocity predictions is the mean bootstrap velocity prediction. AVES, BARB, CRO1, FSD0, ROJO, and SANA are site names discussed in the text.

the best-fitting Caribbean plate angular velocity vector. We then examine whether our estimate of Caribbean plate motion changes significantly if we force a more even distribution of the kinematic information between the site velocities than is the case for the 15-station angular velocity vector that is described at the beginning of Section 3.2.

To examine the effect of the individual site velocities on the best-fitting model estimates, we systematically excluded the velocity of each of the 15 GPS sites, inverted the remaining 14 site velocities and their formal errors, and used the resulting best-fitting angular velocity vector to predict a velocity at 15°N, 75°W, near the center of the Caribbean plate. As is shown in Figure 7, the largest changes in the velocities predicted at this location are 0.4–0.5 mm yr^{-1} and 0.4° in direction and occur if we exclude either of the two highest importance site velocities, those for CRO1 or SANA. In contrast, omitting the velocities for the lower importance sites results in almost no change in the resulting model prediction, primarily because the low importance sites by definition have only a small influence on the best-fitting solution. Consequently, we conclude that the site motions estimated from the best-fitting 15-site angular velocity vector are robust at the level of 0.5 mm yr^{-1} and 0.4° with respect to the site velocities that are used to derive it.

Our assessment of the model robustness implicitly assumes that the formal site velocity uncertainties are approximately correct. If, however, one or more site velocities are affected by any tectonic or other systematic biases, such as monument instability, the formal velocity errors for those sites will underestimate the true errors and hence overweight those velocities in the inversion. Tectonic biases are of potential concern at four of the fifteen stations. Of particular concern is geodetic evidence for 1.9 ± 0.2 mm yr^{-1} of extension between CRO1 and a site in eastern Puerto Rico (Jansma and Mattioli, 2005), which may indicate there are active faults between or close to these two sites. For example, if faults in the Anegada basin (Masson and Scanlon, 1991), located adjacent to site CRO1, are actively accumulating interseismic elastic strain, this strain will bias the motion of site CRO1 relative to its motion with the rigid plate interior. Similarly, the poorly fit velocity at SANA on San Andres Island is also of concern given that the island is located adjacent to the San Andres Trough, where active seismicity, folding of young strata, and young volcanism are consistent with active extension (Holcombe et al., 1990). Finally, tectonic biases ≥1 mm yr^{-1} may exist in the velocities at BARB and ROJO, as described in Section 3.2.

Given the likelihood that one or more of the site velocities employed to derive our 15-station best-fitting angular velocity vector is influenced by nearby deformation, we explored a wider range of possible models for Caribbean plate motion by using data bootstrapping to expand the range of velocity weighting schemes (Efron and Tibshirani, 1986). Bootstrapping employs repeated random sampling of a parent data population to estimate alternative best-fitting models. For our application, we constructed 15 distinct "parent" velocity subsets as a starting point for the bootstrap analysis. Each of the parent velocity subsets excludes the velocity for one of the 15 GPS sites employed for our best-fitting model determination (Section 3.2), and hence establishes a basis for determining the influence of each of the 15 sites on the model.

From each of the 15 parent velocity subsets, we randomly selected 14 site velocities, saved the random data sample, and repeated the process so as to generate 1000 bootstrap data sets per parent velocity population. All of the site velocities in the parent data set are assigned identical uncertainties. As a consequence, the relative weights of the site velocities within any randomized sample are determined by the frequencies with which the velocities are randomly selected for that sample. Our bootstrapping procedure thus samples 1000 alternative data weighting schemes per parent velocity subset and is hence non-prejudicial with respect to the existence of tectonic or systematic biases in an individual site velocity. Our procedure implicitly assigns a weight of zero to the velocity of the site that is excluded from a given parent velocity subset, thereby allowing us to determine the influence of each of the 15 site velocities within the context of the bootstrap analysis.

Each bootstrap sample was inverted to derive its corresponding best-fitting angular velocity vector, giving rise to 15,000 individual bootstrap solutions. Figure 7 illustrates the results for the sites with the highest data importances in the 15-station best-fitting model. The average site velocities that are predicted by the bootstrap models never deviate by more than 0.6 mm yr^{-1} or 1.7° from the linear velocity predicted by our best-fitting 15-station model. The difference at CRO1, the most important site, is only 0.4 mm yr^{-1} and 0.8°. We conclude that the best-fitting 15-station model described in Section 3.2 is robust at a level of ± 0.6 mm yr^{-1}. Surprisingly, all but one of the averaged bootstrap velocities are slower than those predicted by the 15-station best-fitting model, thereby implying that slower plate motion is a robust characteristic of models that more evenly distribute the velocity information between the sites. The effect, however, is only a few tenths of a millimeter per year, too small to matter for most applications.

3.4. A 13-Station Bootstrap Model for Caribbean Plate Motion

Based on the possibility that the velocities for sites BARB and ROJO are biased by elastic effects associated with locking of nearby faults, we eliminate the velocities at these sites and employ bootstrapping of the velocities for the remaining 13 sites to estimate an alternative Caribbean plate angular velocity vector. In this modified data set, stations from the western Caribbean (nine) outnumber stations from the eastern Caribbean (four). Angular velocity vectors derived from bootstrapping of this more limited set of velocities will thus yield models that are more biased toward fitting the western Caribbean GPS site velocities than is the case for our 15-station analysis. The two solutions thus constitute approximate end-members that can be compared to further assess the robustness of the Caribbean plate angular velocity vector.

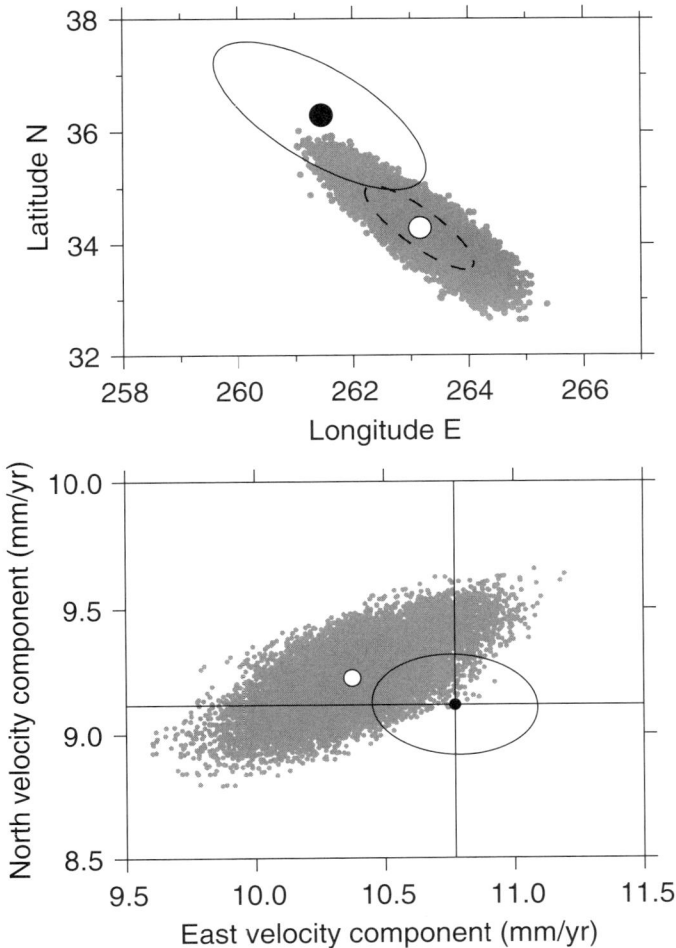

Figure 8. Upper: Thirteen-station bootstrap solution for Caribbean-ITRF2000 motion that omits velocities for sites BARB and ROJO. Open circle indicates mean location of bootstrap poles. Filled circle and ellipse show location of 15-station best-fitting pole and its two-dimensional, 1σ confidence ellipse. Individual bootstrap poles, shown with shaded circles, are derived by excluding one-by-one the velocities for sites AVES, CRO1, FSD0/1, SANA, and the eight Central American sites and inverting bootstrapped velocities for the remaining 12 sites. ITRF2000 is fixed. Lower: Velocities at central location on the Caribbean plate (15°N, 75°W) predicted by the CA-ITRF2000 13-station, bootstrap angular velocity vectors relative to the velocity predicted by the 15-station, best-fitting CA-ITRF2000 angular velocity vector (filled circle) and its formal two-dimensional, 1σ uncertainty. Open circle shows the linear velocity predicted by the mean bootstrapped angular velocity vector. The velocity origin (0, 0) is not shown.

Using the same bootstrapping procedure described in Section 3.3, 1000 velocity data sets were selected randomly from each of thirteen 12-station velocity subsets. Each of the 13,000 bootstrapped data sets was inverted to find its corresponding best-fitting angular velocity vector and the resulting solutions were averaged to determine a mean Caribbean-ITRF2000 angular velocity vector (Table 2). Variances and covariances that describe the uncertainties in the mean angular velocity vector were derived from the lengths and orientations of the three axes for the ellipsoid that encompasses 68.3% of the bootstrap solutions.

As is shown in Figure 8, bootstrapping the 13-station subset of velocities shifts the mean pole location to the southeast by 2.4 angular degrees and predicts motion that is 0.2 mm yr^{-1} slower and 1.4° degrees CCW from that predicted by the 15-station best-fitting model. The velocity difference slightly exceeds the prediction uncertainty of the 15-station, best-fitting Caribbean plate angular velocity vector (shown in Fig. 7) but is small in relationship to the tectonic signals being investigated around the margins of the Caribbean plate. Uncertainties in our determination of the Caribbean plate geodetic reference frame are thus not likely to constitute a limiting factor in studies of circum-Caribbean tectonics.

3.5. Testing for East-West Intraplate Deformation

We next employ the station velocities described in Section 3.4 to test for deformation within the Caribbean plate. The GPS stations available to us are mainly found at the far eastern and western ends of the Caribbean plate (Fig. 1) and are thus well located to test for the existence and magnitude of east-to-west intraplate deformation.

We begin by testing for deformation proposed by Mauffret and Leroy (1999), who estimate that the Beata Ridge has accommodated 9 ± 1.5 mm yr^{-1} of shortening since the early Miocene (23 Ma) based on their interpretation of marine seismic profiles from the eastern edge of the Beata Ridge. We tested for the proposed shortening by adding equal amounts of eastward motion to the velocities of sites located east of the Beata Ridge and westward motion to the velocities of sites west of the Beata Ridge. If any east-to-west shortening (or extension) occurs across one or more structures in the plate interior, then adding east-to-west deformation of the opposite sense to the velocities of sites that span the deforming zone will cancel some or all of the real deformation. The least-squares fit of a best-fitting angular velocity to GPS site velocities that are suitably corrected for any active shortening or extension should thus improve relative to the fit for the original, uncorrected GPS site velocities. Alternatively, if no deformation occurs, then imposing progressively larger east-west extension or shortening on the existing site velocities will yield progressively larger misfits for a best-fitting angular velocity vector. We use the F-ratio test to determine whether changes in the least-squares fit for different assumed amounts of east-west deformation are significant at a predefined confidence level. Our analysis thus establishes a rigorous upper bound on how much deformation could be occurring without rising above the detection threshold of the available GPS site velocities.

Figure 9 shows the least-squares fits to the 15 Caribbean site velocities for a series of models that impose progressively faster east-to-west extension or shortening across the plate interior. Relative to the least-squares fit to the original, unmodified GPS site velocities, the fit improves by ~5% for models that impose

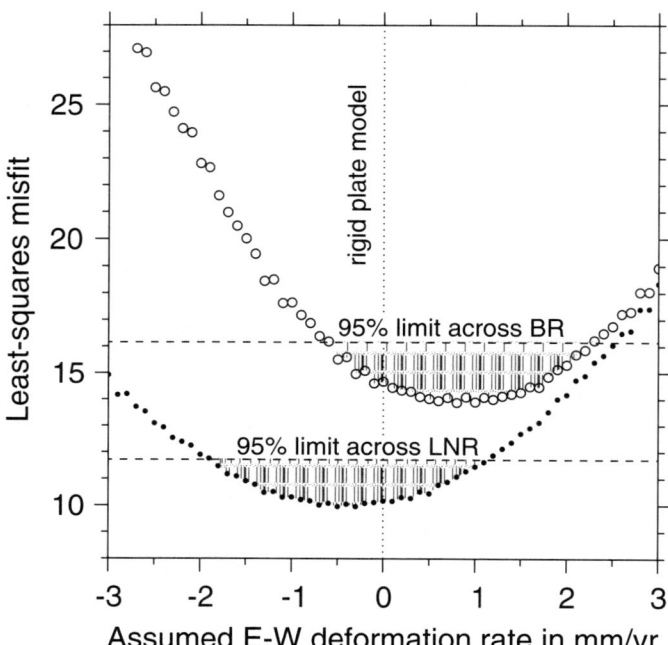

Figure 9. Comparison of least-squares misfits to GPS site velocities as a function of assumed rates of east-to-west extension (negative rates) or shortening (positive rates) across the Beata Ridge (BR) and Lower Nicaragua Rise (LNR). See text for description of how the assumed deformation is imposed on the raw GPS site velocities. Open circles show least-squares misfits for models that separate the 15 site velocities into groups east and west of the Beata Ridge; solid circles use the same velocities, but exclude site SANA located in the Lower Nicaragua Rise. Shaded region indicates the range of rates for assumed east-west plate deformation that do not increase the misfit at more than the 95% confidence level relative to the best-fitting model. The least-squares values associated with the 95% cutoff are determined using an F-ratio test for 1 versus 2*N-3-1 degrees of freedom, representing the number of north and east velocity components for n sites reduced by four adjustable parameters, three of which specify the best-fitting angular velocity and the fourth of which specifies the assumed east-west deformation rate.

as much as 1 mm yr^{-1} of shortening to the site velocities. The improvement in fit is not significant at any reasonable confidence level, indicating that shortening is not required. Overall, models that impose east-to-west plate shortening faster than ~2.5 mm yr^{-1} or extension faster than 0.5 mm yr^{-1} fit the data less well than for an assumed rigid plate at the 95% confidence level. Present-day shortening at rates as fast as the 9 ± 1.5 mm yr^{-1} rate is thus inconsistent with our GPS station velocities at high confidence levels.

We also tested the Holcombe et al. (1990) hypothesis that diffuse east-west extension of the Lower Nicaraguan Rise occurs in response to northeast-directed shear along its bounding escarpments. Doing so required only a minor modification to the above procedure; namely, we excluded the velocity for site SANA, which is located within the zone of diffuse deformation proposed by Holcombe et al. A model that corrects the GPS site motions for 0.5 mm yr^{-1} of east-west extension yields the best fit to the 14 remaining site velocities (Fig. 9), but fails to fit our station velocities significantly better than a simpler rigid plate model. The GPS velocities impose upper limits (at 95%) of 2 mm yr^{-1} of extension and 1 mm yr^{-1} of shortening for E-W oriented deformation across the Lower Nicaraguan Rise.

Our simple numerical experiments do not exclude more complex models of intraplate deformation in which, for example, deformations across the Beata Ridge and Lower Nicaragua Rise are nearly equal in magnitude, but opposite in sense, thereby canceling any integrated E-W directed deformation at the plate scale. The sparse distribution of sites in the plate interior unfortunately prevents us from testing more complex models.

4. DISCUSSION AND FUTURE WORK

The results reported in this paper have several useful implications. The eight new velocities from apparently stable areas of Honduras and Nicaragua allow for stronger tests of the model robustness than was previously possible. A variety of evidence indicates that our estimate of Caribbean plate motion is robust with respect to the 15 site velocities that are used to derive the best-fitting angular velocity vector and their estimated uncertainties. For example, removing any single site velocity and inverting the remaining velocities and their formal errors results in a maximum change of only 0.5 mm yr^{-1} and 0.4° in the motion estimated at a site in the plate interior (Fig. 7). Bootstrapping the velocities to sample a wider range of velocity weighting schemes confirms the apparent robustness of the solution, with estimated plate velocities that differ from the best-fitting solution by no more than 0.6 mm yr^{-1} and 1.7° (Fig. 7).

Two additional measures of the solution robustness reinforce this conclusion. The Swan Islands transform fault west of the Cayman spreading center (Fig. 2) is a narrow, seismically active fault that separates the Caribbean and North America plates. SeaMARC II multibeam mapping of this fault yields well-determined azimuths (Rosencrantz and Mann, 1991) that can be used to test the accuracy of geodetic estimates of Caribbean–North America plate motion. Our best-fitting Caribbean–North America plate angular velocity vector predicts slip directions that are only 2° clockwise and 2° counterclockwise from the measured azimuths at two locations where the fault has an easily interpreted trace (Fig. 3), insignificantly different within the uncertainties.

Our model predictions can also be compared to those of previous geodetic models, although such comparisons are less useful for validating model accuracy given that all published GPS-based models of Caribbean–North America plate motion are derived in part from significantly overlapping sets of GPS site velocities (e.g., Dixon et al., 1998; DeMets et al., 2000; DeMets, 2001; Sella et al., 2002). Nonetheless, the Caribbean–North America plate velocity predicted by our new angular velocity vector at the geographic center of the Caribbean plate, 19.7 ± 0.4 mm yr^{-1} toward N75.6°E ± 0.9° (1-σ), differs insignificantly from the 19.2 mm yr^{-1}, N74.9°E velocity predicted by the Sella et al. (2002)

model. Differences between the present model and that published by DeMets (2001) are also small (see dashed line in Fig. 3).

The new Central American GPS velocities have several important regional and local tectonic implications. Seven of the eight site velocities are misfit by <1.9 mm yr^{-1} by our best-fitting, 15-station model (Fig. 3), which indicates that significant areas of eastern Central America move with the Caribbean plate interior. The similarity of the motions of sites located on either side of the prominent but aseismic Guayape fault suggests that this fault is inactive. In contrast, progressively faster westward motion of 3 ± 1 mm yr^{-1} at TEGU and 8 ± 1 mm yr^{-1} (Fig. 6) at GUAT west of central Honduras indicates that distributed east-west extension occurs in areas of western Central America. The result is consistent with geologic and seismic observations of east-west extension across central and western Honduras and Guatemala (Manton, 1987; Guzman-Speziale, 2001).

Finally, simple, but rigorous numerical experiments with the GPS site velocities indicate that any east-to-west deformation across the Beata Ridge and Lower Nicaraguan Rise is unlikely to exceed 2 mm yr^{-1} and within the uncertainties is zero (Fig. 9). The kinematic evidence for insignificant east-to-west deformation agrees with results reported by Driscoll and Diebold (1998), who conclude that marine seismic data from the Beata Ridge do not require the occurrence of significant contraction across this structure since the Miocene. If such contraction has occurred, as suggested by Mauffret and Leroy (1999), our results suggest that it has now ceased.

Our results suggest useful strategies for future efforts to further improve estimates of Caribbean plate motion. One or more continuous GPS sites in eastern Honduras and/or Nicaragua would contribute geographically unique and well-constrained information about Caribbean plate motion. Long-term monitoring of the east-west length of a baseline between such sites and the existing continuous GPS station on St. Croix Island (CRO1) in the eastern Caribbean would provide a stronger basis for detecting any east-to-west intraplate deformation. Tectonic biases in the motions of sites in southern Hispaniola and Jamaica (Figs. 5–7) make these islands poor targets for monitoring the motion of the plate interior unless suitable corrections for elastic strain accumulation are undertaken. Future occupations of AVES and SANA are clearly warranted in light of their locations in the plate interior, and site velocities from the Lesser Antilles volcanic islands are needed to establish whether some islands move with the plate interior.

ACKNOWLEDGMENTS

We are grateful to colleagues at Instituto Nicaragüense de Estudios Territoriales in Nicaragua and Comisión Permanente de Contingencias in Honduras for their assistance in the field and J.B. De Chabalier at Institut de Physique du Globe de Paris for sharing data from the GPS sites in Martinique. We thank Miranda Chin at the U.S. National Oceanic and Atmospheric Administration for assistance in procuring information and data at CORS sites in Central America and Jamaica. We thank Roland Burgmann and Giovanni Sella for helpful reviews. This work was supported by funding from the National Science Foundation (EAR-0003550 [DeMets], EAR-0085432 [Mattioli and Jansma]), the U.S. National Aeronautics and Space Administration NASA (NAG5-6031), and the University of Arkansas.

REFERENCES CITED

Altamimi, Z., Sillard, P., and Boucher, C., 2002, ITRF2000: A new release of the International Terrestrial Reference Frame for earth science applications: Journal of Geophysical Research, v. 107, no. B10, 2214, doi: 10.1029/2001JB000561.

Calais, E., Mazabraud, Y., Mercier de Lepinay, B., Mann, P., Mattioli, G., and Jansma, P., 2002, Strain partitioning and fault slip rates in the northeastern Caribbean from GPS measurements: Geophysical Research Letters, v. 106, p. 1–9.

Calais, E., Han, J.Y., DeMets, C., and Nocquet, J.M., 2006, Deformation of the North American plate from a decade of continuous GPS measurements: Journal of Geophysical Research, v. 111, B06402, doi: 10.1029/2005JB004253.

DeMets, C., 1993, Earthquake slip vectors and estimates of present-day plate motions: Journal of Geophysical Research, v. 98, p. 6703–6714.

DeMets, C., 2001, A new estimate for present-day Cocos-Caribbean plate motion: Implications for slip along the Central American volcanic arc: Geophysical Research Letters, v. 28, p. 4043–4046, doi: 10.1029/2001GL013518.

DeMets, C., and Wiggins-Grandison, M., 2007, Deformation in Jamaica and motion of the Gonâve microplate from GPS and seismic data: Geophysical Journal International, v. 168, p. 362–378, doi: 10.111/j.1365-246X.2006.03236.x.

DeMets, C., Gordon, R.G., Argus, D.F., and Stein, S., 1990, Current plate motions: Geophysical Journal International, v. 101, p. 425–478.

DeMets, C., Gordon, R.G., Argus, D.F., and Stein, S., 1994, Effect of recent revisions to the geomagnetic reversal timescale on estimates of current plate motions: Geophysical Research Letters, v. 21, p. 2191–2194, doi: 10.1029/94GL02118.

DeMets, C., Jansma, P.E., Mattioli, G.S., Dixon, T.H., Farina, F., Bilham, R., Calais, E., and Mann, P., 2000, GPS geodetic constraints on Caribbean–North America plate motion: Geophysical Research Letters, v. 27, p. 437–440, doi: 10.1029/1999GL005436.

Deng, J., and Sykes, L.R., 1995, Determination of Euler pole for contemporary relative motion of Caribbean and North American plates using slip vectors of interplate earthquakes: Tectonics, v. 14, p. 39–53, doi: 10.1029/94TC02547.

Dixon, T.H., and Mao, A., 1997, A GPS estimate of relative motion between North and South America: Geophysical Research Letters, v. 24, p. 535–538, doi: 10.1029/97GL00284.

Dixon, T., Gonzalez, G., Katsigris, E., and Lichten, S., 1991, First epoch geodetic measurements with the Global Positioning System across the northern Caribbean plate boundary zone: Journal of Geophysical Research, v. 96, p. 2397–2415.

Dixon, T.H., Farina, F., DeMets, C., Jansma, P., Mann, P., and Calais, E., 1998, Relative motion between the Caribbean and North American plates and related boundary zone deformation from a decade of GPS observations: Journal of Geophysical Research, v. 103, p. 15,157–15,182, doi: 10.1029/97JB03575.

Driscoll, N.W., and Diebold, J.B., 1998, Deformation of the Caribbean region: One plate or two?: Geology, v. 26, p. 1043–1046, doi: 10.1130/0091-7613(1998)026<1043:DOTCRO>2.3.CO;2.

Efron, B., and Tibshirani, R., 1986, Bootstrap method for standard errors, confidence intervals and other measures of statistical accuracy: Statistical Science, v. 1, p. 54–77.

Guzman-Speziale, M., 2001, Active seismic deformation in the grabens of northern Central America and its relationship to the relative motion of the North America–Caribbean plate boundary: Tectonophysics, v. 337, p. 39–51, doi: 10.1016/S0040-1951(01)00110-X.

Heflin, M., Bertiger, W., Blewitt, G., Freedman, A., Hurst, K., Lichten, S., Lindqwister, U., Vigue, Y., Webb, F., Yunck, T., and Zumberge, J., 1992, Global geodesy using GPS without fiducial sites: Geophysical Research Letters, v. 19, p. 131–134.

Heubeck, C., and Mann, P., 1991, Geologic evaluation of plate kinematic models for the North American-Caribbean plate boundary zone: Tectonophysics, v. 191, p. 1–26, doi: 10.1016/0040-1951(91)90230-P.

Holcombe, T.L., Ladd, J.W., Westbrook, G., Edgar, N.T., and Bowland, C.L., 1990, Caribbean marine geology: Ridges and basins of the plate interior, in Dengo, G., and Case, J.E., The Caribbean Region: Boulder, Colorado, Geological Society of America, Geology of North America, v. H, p. 231–260.

Jansma, P.E., and Mattioli, G.S., 2005, GPS results from Puerto Rico and the Virgin Islands: Constraints on tectonic setting and rates of active faulting, in Mann, P., ed., Active Tectonics and Seismic Hazards of Puerto Rico, the Virgin Islands, and Offshore Areas: Geological Society of America Special Paper 385, p. 13–30.

Jordan, T.H., 1975, The present-day motions of the Caribbean plate: Journal of Geophysical Research, v. 80, p. 4433–4439.

Leroy, S., and Mauffret, A., 1996, Intraplate deformation in the Caribbean region: Journal of Geodynamics, v. 21, p. 113–122, doi: 10.1016/0264-3707(95)00037-2.

MacMillan, D.S., and Ma, C., 1999, VLBI measurements of Caribbean and South American motion: Geophysical Research Letters, v. 26, p. 919–922, doi: 10.1029/1999GL900139.

Mann, P., Taylor, F.W., Edwards, R.L., and Ku, T., 1995, Actively evolving microplate formation by oblique collision and sideways motion along strike-slip faults: An example from the northeastern Caribbean plate margin: Tectonophysics, v. 246, p. 1–69, doi: 10.1016/0040-1951(94)00268-E.

Mann, P., Calais, E., Ruegg, J.-C., DeMets, C., Jansma, P., and Mattioli, G.S., 2002, Oblique collision in the northeastern Caribbean from GPS measurements and geological observations: Tectonics, v. 37, doi: 10.1029/2001TC001304.

Manton, W.I., 1987, Tectonic interpretation of the morphology of Honduras: Tectonics, v. 6, p. 633–651.

Mao, A., Harrison, C.G.A., and Dixon, T.H., 1999, Noise in GPS coordinate time series: Journal of Geophysical Research, v. 104, p. 2797–2816, doi: 10.1029/1998JB900033.

Marquez-Azua, B., and DeMets, C., 2003, Crustal velocity field of Mexico from continuous GPS measurements, 1993 to June, 2001: Implications for the neotectonics of Mexico: Journal of Geophysical Research, v. 108, no. B9, doi: 10.1029/2002JB002241.

Masson, D.G., and K.M. Scanlon, 1991, The neotectonics setting of Puerto Rico: Geological Society of America Bulletin, v. 103, p. 144–154.

Mauffret, A., and Leroy, S., 1999. Neogene intraplate deformation of the Caribbean plate at the Beata Ridge, in Mann, P., ed., Caribbean Basins, Sedimentary Basins of the World: Amsterdam, Elsevier Science B.V., v. 4, p. 627–669.

Mazzotti, S., James, T.S., Henton, J., and Adams, J., 2005, GPS crustal strain, postglacial rebound, and seismic hazard in eastern North America: The Saint Lawrence valley example: Journal of Geophysical Research, v. 110, no. B11301, doi: 10.1029/2004JB003590.

Minster, J.B., Jordan, T.H., Molnar, P., and Haines, E., 1974, Numerical modeling of instantaneous plate tectonics: Geophysical Journal of the Royal Astronomical Society, v. 36, p. 541–576.

Müller, R.D., Royer, J.-Y., Cande, S.C., Roest, W.R., and Maschenkov, S., 1999, New constraints on the Late Cretaceous/Tertiary plate tectonic evolution of the Caribbean, in Mann, P., Caribbean Basins, Sedimentary Basins of the World series: Amsterdam, Elsevier Science B. V., v. 4, p. 33–59.

Park, K., Nerem, R.S., Davis, J.L., Schenewerk, M.S., Milne, G.A., and Mitrovica, J.X., 2002, Investigation of glacial isostatic adjustment in the northeast U.S. using GPS measurements: Geophysical Research Letters, v. 29, no. 11, 1509, doi: 10.1029/2001GL013782.

Pollitz, F.F., and Dixon, T.H., 1998, GPS measurements across the northern Caribbean plate boundary zone: Impact of postseismic relaxation following historic earthquakes: Geophysical Research Letters, v. 25, p. 2233–2236, doi: 10.1029/98GL00645.

Rosencrantz, E., and Mann, P., 1991, SeaMARC II mapping of transform faults in the Cayman Trough, Caribbean Sea: Geology, v. 19, p. 690–693, doi: 10.1130/0091-7613(1991)019<0690:SIMOTF>2.3.CO;2.

Sella, G.F., Dixon, T.H., and Mao, A., 2002, REVEL: A model for recent plate velocities from space geodesy: Journal of Geophysical Research, v. 107, B4, doi: 10.1029/2000JB000033.

Speed, R.C., 1983, Structure of the accretionary complex of Brabados; 1, Chalky Mount: The Geological Society of America Bulletin, v. 94, p. 92–116, doi: 10.1130/0016-7606(1983)94<92:SOTACO>2.0.CO;2.

Stein, S., DeMets, C., Gordon, R.G., Brodholt, J., Engeln, J.F., Wiens, D.A., Argus, D., Lundgren, P., Stein, C., and Woods, D., 1988, A test of alternative Caribbean plate relative motion models: Journal of Geophysical Research, v. 93, p. 3041–3050.

Sykes, L.R., McCann, W.R., and Kafka, A.L., 1982, Motion of Caribbean plate during the last 7 million years and implications for earlier Cenozoic movements: Journal of Geophysical Research, v. 87, p. 10,656–10,676.

Weber, J.C., Dixon, T.H., DeMets, C., Ambeh, W.B., Jansma, P., Mattioli, G., Saleh, J., Sella, G., Bilham, R., and Perez, O., 2001, GPS estimate of relative motion between the Caribbean and South American plates, and geologic implications for Trinidad and Venezuela: Geology, v. 29, p. 75–78, doi: 10.1130/0091-7613(2001)029<0075:GEORMB>2.0.CO;2.

Zumberge, J.F., Heflin, M.B., Jefferson, D.C., Watkins, M.M., and Webb, F.H., 1997, Precise point positioning for the efficient and robust analysis of GPS data from large networks: Journal of Geophysical Research, v. 102, p. 5005–5017, doi: 10.1029/96JB03860.

MANUSCRIPT ACCEPTED BY THE SOCIETY 22 DECEMBER 2006

The Geological Society of America
Special Paper 428
2007

Transtensional deformation of the western Caribbean–North America plate boundary zone

Robert D. Rogers*
Paul Mann

Institute for Geophysics, Jackson School of Geosciences, University of Texas at Austin, J.J. Pickle Research Campus, Bldg. 196 (ROC), 10100 Burnet Road (R2200), Austin, Texas 78758-4445, USA

ABSTRACT

Divergence, expressed as the angle between the plate motion vector and the azimuth of a plate margin fault, has been proposed to explain development of contrasting styles of transtensional deformation along transform margins. We present the western North America–Caribbean plate margin as a test of this hypothesis. Here, geologic, earthquake, marine geophysical, and remote sensing data show two distinct structural styles: (1) east-west extension along north-trending rifts normal to the plate margin in the western study area (western Honduras and southern Guatemala); and (2) NNW-SSE transtension along rifts subparallel to the plate margin in the eastern study area (northern Honduras and offshore Honduran borderlands region). Orientations of rifts in each area coincide with the angle of divergence between the GPS-derived plate motion vector and the azimuth of the plate boundary fault, such that the western zone of east-west extension has an angle >10°, while the eastern zone of NNW-SSE extension occurs when the angle of divergence is between 5° and 10°. A narrow transition area in north-central Honduras separates the plate boundary–normal rifts of western Honduras from the plate boundary–parallel rifts to the east.

Faults of the offshore Honduran borderlands extend onshore into the Nombre de Dios range and Aguan Valley of northern Honduras where tectonic geomorphology studies show pervasive oblique-slip faulting with active left-lateral river offsets and active uplift of stream reaches. Offshore, exploration seismic data tied to wells in the Honduran borderlands reveal active submarine faults bounding asymmetric half-grabens filled by middle Miocene clastic wedges with continued clastic deposition into Pliocene-Pleistocene. The north-trending rifts of western Honduras form discontinuous half-grabens that cut late Miocene ignimbrite strata. Plate reconstructions indicate the north-trending rifts of western Honduras developed in response to increased interplate divergence as the western margin of the Caribbean plate shifted from the Jocotan fault to the Polochic fault during the middle Miocene.

Keywords: Honduran borderlands, oblique divergence, geomorphology, Caribbean plate.

---

*Present address: Department of Geology, California State University Stanislaus, 801 West Monte Vista Avenue, Turlock, California 95382, USA; rrogers@geology.csustan.edu.

Rogers, R.D., and Mann, P., 2007, Transtensional deformation of the western Caribbean–North America plate boundary zone, *in* Mann, P., ed., Geologic and tectonic development of the Caribbean plate boundary in northern Central America: Geological Society of America Special Paper 428, p. 37–64, doi: 10.1130/2007.2428(03). For permission to copy, contact editing@geosociety.org. ©2007 The Geological Society of America. All rights reserved.

INTRODUCTION

Strain partitioning along plate margins produces complex but characteristic deformation patterns along plate boundary faults. Most previous studies of strain partitioning have come from transpressive regions where convergent plate motion is transformed into components of strike-slip and thrust faulting in either ancient (cf. Teyssier et al., 1995; Jones and Tanner, 1995; Claypool et al., 2002) or active settings (Marshall et al., 2000; Calais et al., 2002; Mann et al., 2002; Meckel et al., 2003). In contrast, there are few examples documented from active transtensional settings where obliquely divergent plate motion is partitioned into coexisting strike-slip and normal components of deformation (Ben-Avraham and Zoback, 1992; Ben-Avraham, 1992).

Key questions for understanding strain partitioning in transtensional settings include (1) how much strain is accommodated by strike-slip faulting versus normal faulting along each plate boundary segment; (2) what controls different styles of coexisting strike-slip and normal faulting along a plate boundary; (3) what controls the transition from one style of transtensional strain partitioning to another and the locations where the transitions occur; and (4) how are successive styles of transtension superimposed as a plate boundary evolves?

We present the northwestern margin of the Caribbean plate as an example of strain partitioning in a transtensional setting and examine in detail the resulting deformational patterns as observed onshore in Honduras and Guatemala and beneath the adjacent Caribbean Sea (Honduran borderlands). This region is ideal for such a study for several reasons:

1. GPS-based geodesy from DeMets et al. (2000) and DeMets et al. (this volume) places firm constraints on the variation of interplate motion and deformational styles across a 3100-km-long segment of the North America–Caribbean plate boundary (Fig. 1).
2. Plate boundary-related normal faults of a variety of orientations and ages are present in northern Central America and in adjacent areas of the Honduran borderlands (Fig. 2).
3. Plate margin faults have known variations in position and orientation that can be tied to quantitative plate reconstructions based on the opening history of the Cayman trough and surrounding plate pairs.

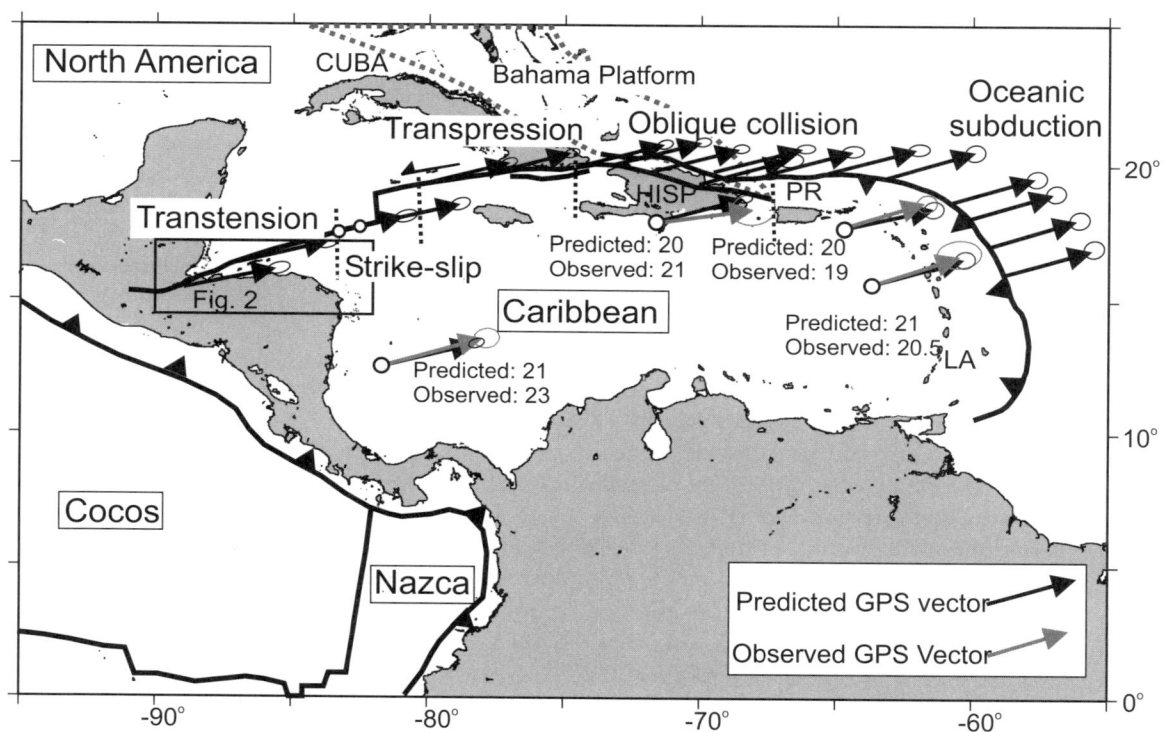

Figure 1. Plate-scale tectonic setting of study area in northern Honduras and northern Guatemala (boxed area shown in detail in Fig. 2). Caribbean–North America velocity predictions of DeMets et al. (2000) (black arrows) based on GPS velocities at four sites in the stable interior of the Caribbean plate (gray arrows) and two fault strike measurements in the strike-slip segment of the North America–Caribbean plate boundary northeast of Honduras (open circles). PR—Puerto Rico; LA—Lesser Antilles; HISP—Hispaniola.

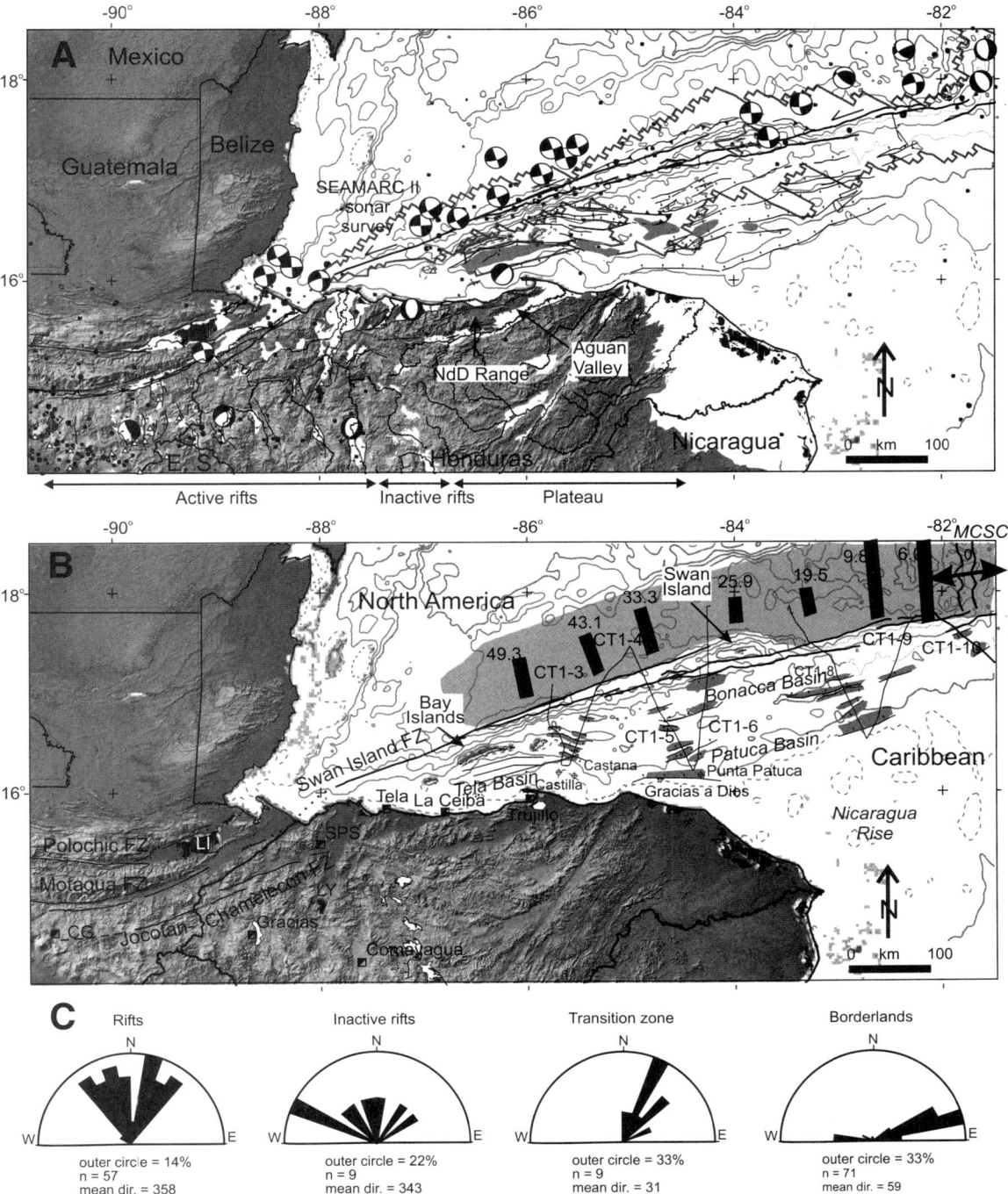

Figure 2. (A) Neotectonic features of the western North America–Caribbean plate boundary region of northern Central America. All onland and offshore faults shown have morphologic features indicating recent fault motion. Small black dots are earthquake epicenters compiled from the U.S. National Earthquake Information (NEIC) database, and earthquake focal mechanisms are compiled from the Harvard database. Bathymetric contours are 1000 meters except for the 100-meter line near shore. Gray shaded areas of Honduran borderlands represent actively filling turbidite basins from Pinet (1976). Gray outline is footprint of SeaMARC II sidescan sonar mosaic (Rosencrantz and Mann, 1991). White fill onshore is Quaternary alluvium. NdD—Nombre de Dios; ES—El Salvador. (B) Compilation of basement features and long-lived basins in northern Central America. Dark gray shaded areas of Honduran borderlands represent areas of Mio-Pliocene fault-bounded wedges of basinal sedimentary rocks. Gray shading offshore is ocean crust formed at Mid-Cayman spreading center; black bars are spreading anomalies with ages in millions of years (Leroy et al., 2000). UTIG multi-channel seismic lines shown include lines CT1-3 through CT1-10. Locations of single-channel seismic lines from UTIG SeaMARC II cruise in the Tela basin are also shown. CG—Guatemala City, Guatemala; SPS—San Pedro Sula, Honduras; LI—Lago Isabel, Guatemala; LY—Lago Yojoa, Honduras. (C) Rose diagrams show fault orientations from each region of Neogene faulting: (1) western rifts; (2) inactive rifts; (3) transition zone; and (4) borderland faults.

STRAIN PARTITIONING OF THE NORTHWESTERN CARIBBEAN PLATE

GPS-determined motion of the Caribbean plate relative to North America predicts significant along-strike variations in the style of deformation along the 3100-km-long plate boundary (DeMets et al., 2000) (Fig. 1). The plate boundary extends from the Motagua Valley of Guatemala to the Lesser Antilles arc and has a GPS-derived rate of mainly left-lateral displacement of 21 mm/yr. DeMets et al. (2000) note that GPS-based interplate velocities are consistent with the along-strike transition in structural styles from (1) transtension in northwestern Central America and the western Cayman trough where divergence between the Caribbean plate vector and the trend of the plate boundary faults predicts oblique opening over a wide area; (2) pure strike-slip faulting in the central Cayman trough where the Caribbean plate vector and the trend of the plate boundary faults are parallel with strike-slip faults (Rosencrantz and Mann, 1991); (3) transpression and oblique underthrusting in the eastern Cayman trough and southern Cuba where there is convergence between plate vector and plate bounding fault (Calais and Mercier de Lépinay, 1993); (4) even greater amounts of transpression in the Hispaniola region, particularly in the zone of contact between the Caribbean plate and Bahamas Platform (Mann et al., 2002); and (5) oblique subduction of oceanic crust beneath the northeastern edge of the Caribbean plate in Puerto Rico and the Virgin Islands (Jansma et al., 2000). DeMets et al. (2000) propose a correlation between the direction of plate motion and the degree of transtension with a divergence angle between fault and plate vector of ~5° marking the threshold between predominantly strike-slip structures and predominately transtensional structures.

REGIONAL GEOLOGY AND ACTIVE TECTONICS

The western transtensional region of the North America–Caribbean margin is a diffuse plate margin with three main groups of active faults: (1) linear faults mapped along the Motagua–Swan Islands strike-slip fault zone (Plafker, 1976; Rosencrantz and Mann, 1991); (2) more discontinuous, north-trending normal faults associated with rift structures of western Honduras and Guatemala; and (3) a broad, 125–150-km-wide zone of submarine faults occupying the offshore region of shelfal to abyssal depths (200–2000 m) known as the Honduran borderlands (Pinet, 1971; Case and Holcombe, 1980). The north-trending faults of western Honduras and southwestern Guatemala bound rifts at right angles to the trend of the arcuate, strike-slip Motagua fault plate boundary (Fig. 2). These north-trending normal faults have been interpreted by previous workers as intraplate deformation and block rotations about an arcuate, convex-southward left-lateral strike-slip fault system (Plafker, 1976; Burkart and Self, 1985) and as fault termination structures related to the termination of left-lateral slip of the Motagua fault zone (Langer and Bollinger, 1979; Guzman-Speziale, 2001).

In the Honduran borderlands and the Nombre de Dios range of northern Honduras, normal faults bound elongate rifts and intervening basement blocks that are subparallel to the main left-lateral strike-slip plate boundary fault (Swan Islands fault zone) (Fig. 2). The westward extent of the borderland faults disappear at the longitude of the north-trending Yojoa-Sula rift basin at 88° west (Fig. 2A).

The Motagua fault offsets a staircase of Quaternary river terraces in the Motagua Valley of Guatemala that yield long-term, left-lateral slip rates of 0.45–1.88 cm/yr (Schwartz et al., 1979). There is no quantitative estimate of the cumulative lateral offset on the Motagua fault since alluvium of the Motagua Valley obscures many of the adjoining and presumably offset rock units. The Motagua fault is directly along strike of the offshore Swan Islands fault to the east and is flanked by the active, left-lateral Polochic fault zone to the north (Burkart, 1983; Burkart, 1994) and the inactive Jocotan-Chamelecon fault zone to the south (Ritchie, 1976). The Jocotan-Chamelecon fault is inferred to be inactive because its trace has become fragmented by east-west opening of more recent, north-oriented rifts (Ritchie, 1976; Plafker, 1976). Burkart (1983) documents 130 km of left-lateral displacement on drainages along the Polochic fault that occurred between 10 and 3 Ma. Burkart (1994) proposes a chronology of plate boundary faulting that started along the Jocotan fault between 20 and 10 Ma and switched to the Polochic fault at 10 Ma, with the Motagua fault active since 3 Ma. A 230-km-long, left-lateral surface rupture averaging 1.08 m of horizontal and 0.3 m of vertical displacement occurred on 4 February 1976 along the Motagua fault of Guatemala (Plafker, 1976; Kanamori and Stewart, 1978). This M7.5 event also splayed to the south and activated ruptures along north-south oriented rift faults in Guatemala, where focal mechanisms indicated dominantly normal displacements (Langer and Bollinger, 1979).

The Swan Islands fault zone forms a prominent semicontinuous fault scarp on the seafloor (Rosencrantz and Mann, 1991). Its location is consistent with the high level of earthquake activity shown by the left-lateral earthquake focal mechanisms in the area (Fig. 2A). The fault juxtaposes oceanic crust of the Cayman trough generated at the Mid-Cayman spreading center over the past 49 m.y. with continental crust of the Honduran borderlands to the south (Fig. 2B). The Swan Islands fault steps right at the Swan Islands restraining bend, a major right-step in the left-lateral fault trace associated with topographic uplift of the Swan Islands of Honduras and active convergent deformation of the seafloor observed on sidescan and seismic reflection profiles (Mann et al., 1991). Mann et al. (1991) and Leroy et al. (2000) propose that the right step and restraining bend in the Swan Islands fault zone formed when the Mid-Cayman spreading center lengthened in a southward direction by 25 km between 19.5 and 25.9 Ma (Fig. 2B). The Swan Islands fault zone extends eastward, where part of the Caribbean–North America interplate motion is transformed into seafloor spreading along the 100-km-long Mid-Cayman spreading center (Leroy et al., 2000) and the other part of the interplate motion continues along strike-slip faults extending along the southern edge of the Cayman trough to the island of Jamaica (Rosencrantz and Mann, 1991).

Earthquake epicenters with M > 4.0 from the National Earthquake Information Center (NEIC) database and focal mechanisms for earthquakes from the Harvard University catalog of Centroid Moment Tensors with 4.0 < M < 7.1 are compiled on a topographic basemap in Figure 2A. The greatest concentration of epicenters aligns in a belt parallel to the Swan Islands–Motagua fault zone, a continuous zone of subaerial and submarine active, left-lateral faulting that accommodates much of the present-day North America–Caribbean plate motion (Molnar and Sykes, 1969; Rosencrantz and Mann, 1991; Deng and Sykes, 1995). Earthquake focal mechanisms along this fault are dominantly left-lateral (Deng and Sykes, 1995; Van Dusen and Doser, 2000; Cáceres et al., 2005) (Fig. 2A). Earthquake events in offshore areas of the Honduran borderlands and Nicaragua Rise are widely scattered and infrequent, indicating low levels of activity in these more intra-plate areas and/or poor seismograph coverage (Fig. 2A).

Earthquake epicenters with focal mechanisms that are indicative of east-west extension are spatially associated with faults bounding north-trending rifts of western Honduras and Guatemala. Here, 12 rifts are characterized by steep, fault-bounded, intermontane valleys filled by late Miocene to Quaternary age sediments and lakes (Manton, 1987; Gordon and Muehlberger, 1994) (Fig. 2A). Guzman-Speziale (2001) used earthquake focal mechanisms in the rift area to calculate an east-west rate of extension of 8 mm/yr across all the rifts and proposed that the rifts terminate strike-slip displacement along the Motagua-Polochic fault zone.

OBJECTIVES OF THIS PAPER

Our objectives are to document active deformation south of the plate margin and variation in structural style of rifts using compilations of previous studies, tectonic geomorphology studies of onshore areas, and marine geophysical studies of offshore areas in the Honduran borderlands. We relate the orientation of faulting to Caribbean–North America plate motion partitioned into along-strike components of boundary-parallel strike-slip and boundary-normal extension for the western 1000 km of the margin. Finally we examine the implication of the relation of deformation styles to plate kinematics in context of the evolution of the margins since the early Miocene. The first objective—documenting the active deformation—is presented first for the onshore areas and then for the offshore areas.

ONSHORE TECTONIC GEOMORPHOLOGY AND STRUCTURE

Morphologic Indicators of Active Tectonics

In order to relate active deformation to plate margin kinematics, we first distinguish regions that are actively deforming. Onshore, we expand on three morphologic provinces defined by Rogers et al. (2002) by including interpretation of fault scarps from previous geologic mapping and interpretation of LANDSAT imagery shown in Figure 2, defining the eastern extent of rifting and its boundaries with adjacent, more tectonically stable morphologic provinces. Previously, some workers assumed that rifting is confined to western Honduras and that eastern Honduras is part of the stable Caribbean plate (e.g., Plafker, 1976; Burkart and Self, 1985), while others (Manton, 1987; Gordon and Muehlberger, 1994; Avé Lallemant and Gordon, 1999) extend the zone of active deformation into eastern Honduras.

Morphologic Provinces of Honduras

Tectonic geomorphology and hypsometric analysis reveal four main morphologic provinces within Honduras, which are outlined in white and numbered 1–4 on Figure 3A (see Rogers et al. [2002] for description of methods). Three zones originally defined by Rogers et al. (2002) are the western rifts province (zone 1), the plateau province (zone 2) and the eastern province (zone 3). The plateau province is subdivided into a region of inactive rifts (zone 2a) and the undeformed core plateau (zone 2b). Based on this study, the north coast (zone 4) forms a fourth morphologic region not described by Rogers et al. (2002).

The western rifts morphological province exhibits basin and range topography (zone 1, Fig. 3A) and is underlain by pre-Jurassic basement rocks, folded Cretaceous rocks, and overlying Miocene ignimbrite deposits in the western area (Rogers et al., 2002; Jordan et al., this volume). The variable elevation in this region reflects the extreme topographic relief related to the formation of half-grabens and elevated rift shoulders (Fig. 3B). The Central American plateau originally extended across the area of zone 1, but the plateau area has been disrupted by the formation of north-trending rifts. The longitudinal profiles of the five largest rivers in the western rifts have low gradient reaches on rift valley floors but abruptly steepen when they cross onto the uplifted rift shoulders (Fig. 3C).

To the east of the active rifting, geologic mapping reveals inactive rifts formed by north-striking normal faults cutting across a terrain of pre-Jurassic basement rocks, folded Cretaceous sedimentary rocks, and Miocene volcanic deposits (King, 1972, 1973; Rogers and O'Conner, 1993; Markey, 1995) (zone 2a, Fig. 3A). Cumulative hypsometry from this area shows a hypsometric pattern similar to that observed in zone 2b, the stable core of the Central American plateau. Rivers in this zone have dominantly straight longitudinal profiles along their entire lengths river (Fig. 3C). This pattern is similar to river profiles from the plateau zone.

The plateau province (zone 2b, Fig. 3A) represents the core of the moderately dissected Central American plateau defined by Rogers et al. (2002). The rocks underlying this morphologic zone consist of pre-Jurassic metamorphic basement rocks and folded Cretaceous rocks (Rogers et al., Chapter 5, this volume). High-level erosion surfaces characterize the relatively smooth upper surface of the plateau (Helbig, 1959; Manton, 1987). Cumulative hypsometry reveals that 50% of this zone is

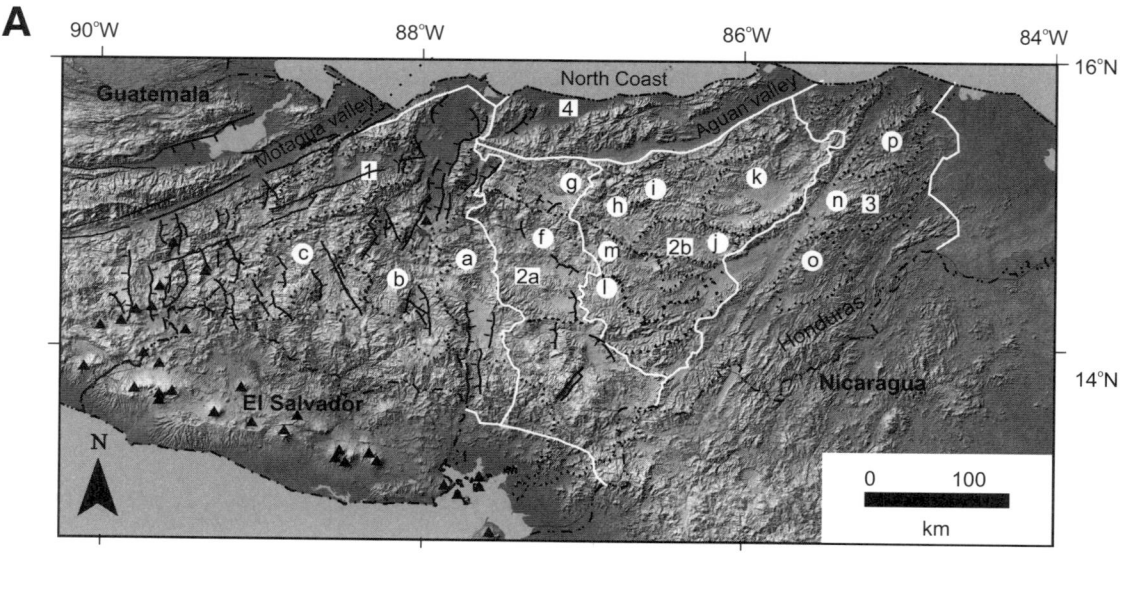

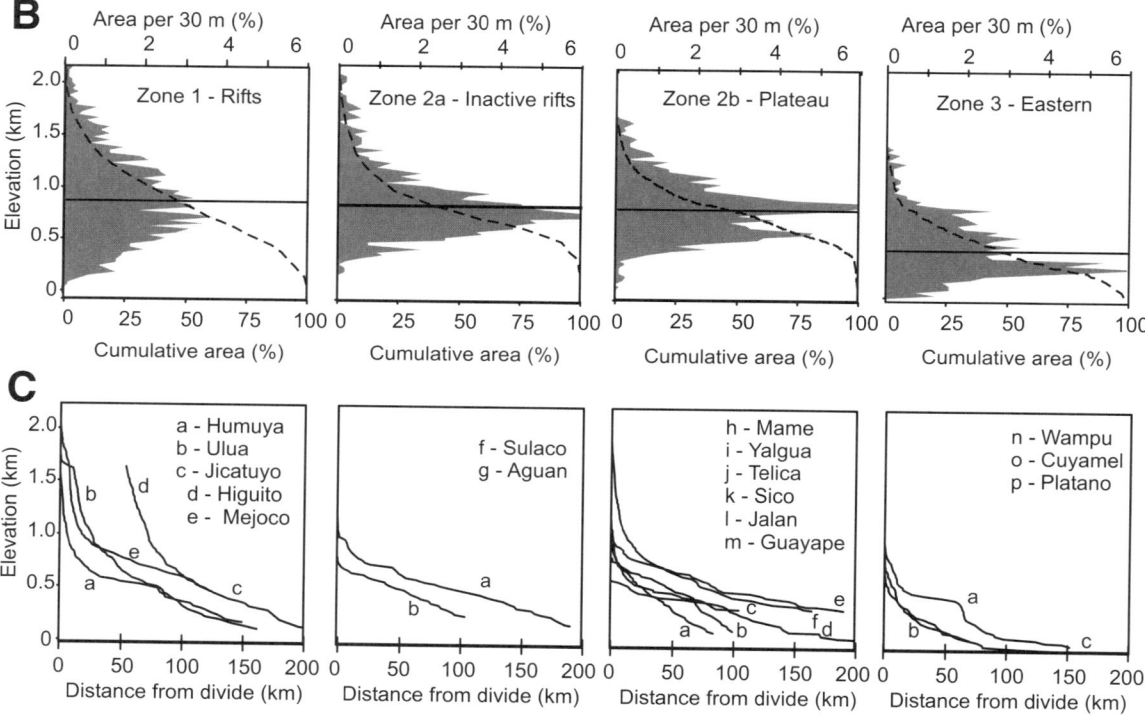

Figure 3. (A) Digital elevation model of northern Central America showing four regions of Honduras (H) outlined by white lines: Zone 1—active, north-trending rifts in western Honduras; Zone 2a—inactive rifts in central Honduras; Zone 2b—Honduran plateau; Zone 3—eastern mountains of Honduras; Zone 4—North Coast Province. Normal fault scarps from Figure 2 are shown as black lines. Watershed boundaries are shown by dotted, black lines. Solid triangles are Quaternary volcanoes. Letters defined in C. (B) Hypsometry (histogram of elevation versus area) of zones 1 through 3 in Honduras. Differential (shaded) and cumulative (dashed line) hypsometries have same vertical (elevation) axis but different horizontal axes. Solid line shows mean elevation. (C) Longitudinal profiles of trunk rivers for watersheds noted by letters in 3A. Rios Higuito and Mejoco are tributaries to Río Jicatuyo in zone 1.

concentrated between 700 and 1000 m above sea level (masl) (Fig. 3B). The plateau is preserved because (1) it is located in an inland area removed from the Caribbean Sea and Pacific Ocean and subjected to lower rainfall amounts (1–1.5 m/yr) than surrounding areas to the west (>2 m/yr) and east (>3 m/yr); and (2) it is a tectonically stable area located on the Caribbean plate and therefore not subjected to active faulting. Six rivers in this zone have straight longitudinal profiles along their entire lengths (Fig. 3C) similar to those in the inactive rift zone. These rivers entrenched into bedrock following the uplift of the Central American plateau.

The basement rocks underlying the eastern lowlands of Honduras consist of pre-Mesozoic metamorphic basement rocks and folded Cretaceous rocks (Fig. 3A, zone 3) (Rogers et al., Chapter 6, this volume). This zone represents the deeply dissected eastern edge of the Central American plateau defined by Rogers et al. (2002). The longitudinal profiles of rivers in this zone display the classic concave profile of graded rivers in tectonically stable regions, adjusted to this higher rainfall area adjacent to the Caribbean Sea. The stepped profile of the Río Wampu (Fig. 3C, profile a) is interpreted as a special case related to the breaching of its divide (the stepped area) and piracy of tributaries west of the divide.

The North Coast's Sierra Nombre de Dios and adjacent Aguan Valley (zone 4, Fig. 3A) is one of the most seismically active (Fig. 2A) and highest relief (maximum elevation: 2643 masl) areas of Honduras. In addition to these suggestions of tectonic activity, LANDSAT imagery displays the late Quaternary strike-slip and normal faulting crosscutting the region (Manton, 1987; Manton and Manton, 1999; Avé Lallemant and Gordon, 1999). In this region, deformation trends are subparallel to the plate bounding Swan Island transform and perpendicular to the north-trending rifts of the western rift (Fig. 2).

Tectonic Geomorphology of Western Rifts

Structural features aligned perpendicular to the plate margin include the north-trending normal faults and associated rifts of western Honduras and southeastern Guatemala (Plafker, 1976; Mann and Burke, 1984; Manton, 1987; Gordon and Muehlberger, 1994; Guzman-Speziale, 2001). The rifts maintain a trend almost perpendicular to the arcuate Motagua fault zone and extend from the Guatemala City graben in the west to western Honduras (Fig. 2A). These rifts are generally half-grabens, most with active eastern bounding faults. Two rifts occur in the 30-km-wide zone bounded by the Motagua and Jocotan faults and three rifts occur in the triangular wedge defined by the Middle America volcanic arc to the west and the Jocotan fault and its western extension to the north (Plafker, 1976).

Further east, the rifts assume a more northerly orientation and are wider, with larger Quaternary-filled floors. The discontinuous full grabens in this area comprise the "Honduras depression" described by Muehlberger (1976). Quaternary basaltic volcanism occurs in the rift near Lake Yojoa (Wadge and Wooden, 1982). Volcanic flows from recent cores north of Lake Yojoa have been dated using K-Ar as 0.5 Ma (ENEE, 1987).

None of the rifts have been studied in detail and the rifts are mainly known by 1:50,000-scale quadrangle mapping (Kozuch, 1991) incorporated into maps shown in Figures 2 and 3A. Geologic mapping reveals no evidence for Quaternary motion along the faults bounding the easternmost rifts (King, 1972, 1973; Markey, 1995; Rogers, 1995) (Fig. 2B). Gordon (1994) carried out fault striation studies on older, Tertiary rocks in western Honduras and found a complex set of extension axes that are likely related to the pre–10.5 Ma opening history of the rifts.

Because all active rifts are sharply defined topographically and appear to be currently subsiding, older, pre-Quaternary rocks characterizing their development are not exposed. Vertical motion on the faults bounding the rifts is as much as 2 km (Everett, 1970), and rifting appears to have been incipient during the deposition of middle Miocene ignimbrites (Dupré, 1970). Gordon and Muehlberger (1994) conclude that the rifting in Honduras began after 10.5 Ma, which is consistent with Hemphillian-age (9.0–6.7 Ma) mammalian fauna found in rift valley-fill deposits (Webb and Perrigo, 1984).

Tectonic Geomorphology of the North Coast

The Nombre de Dios range forms a heavily faulted dome with its core underlain by sheared high-grade gneiss and schist, felsic intrusions, and locally mafic volcanic strata of pre-Cenozoic age (Fig. 4B). Manton (1987) observes that the dominant east-northeast trend of recent faults is parallel to the trend of faults and folds in the older rocks. For a domal uplift of this large size and high elevation, younger, flanking ridges of sedimentary rocks are limited. This suggests that most of the clastic, erosional products of the uplift have been transported directly offshore into basins of the Honduran borderlands (see the section Geologic and Bathymetric Setting of the Honduran Borderlines).

The LANDSAT image in Figure 4A and the geologic map in Figure 4B display the five major faults of the Nombre de Dios range that form prominent topographic lineaments and are discussed in this section (Table 1). All faults are linear, have east-northeast strikes, and are parallel to offshore faults known from marine geophysical mapping of the offshore Honduran borderlands and Swan Islands fault zone and compiled on the map in Figure 4B.

Highly deformed marine turbidites of Miocene age are overlain by a south-dipping valley-fill conglomerate south of the city of Trujillo, indicating that the Nombre de Dios range and adjacent Aguan valley to the south may have once formed part of the submarine Honduran borderland province in pre-Miocene time (Manton and Manton, 1999). A low relief area apparent on topographic maps and imagery on the footwall of the La Esperanza fault appears to be a marine terrace and is cut by lineaments at the eastern end of the Aguan Valley, suggesting recent subaerial emergence of the terrace and continued activity of the La Esperanza fault (Fig. 4).

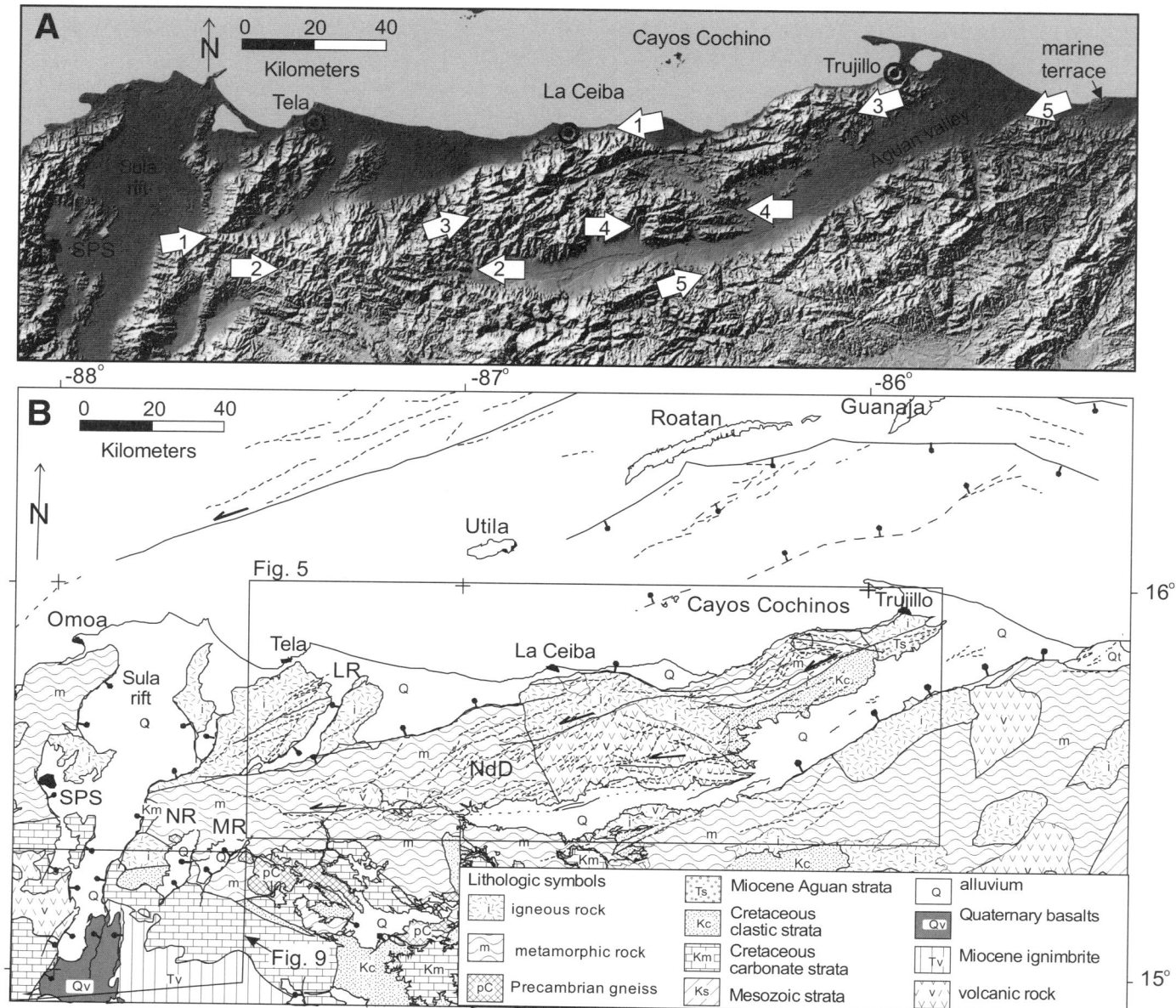

Figure 4. (A) LANDSAT image showing topographic lineaments of major faults of northern Honduras (numbered arrows are at endpoints of the following fault lineaments): 1—La Ceiba fault; 2—Aguan fault; 3—Río Viejo fault; 4—Lepaca fault; and 5—La Esperanza fault. (B) Mapped geology (Kozuch, 1991; Manton and Manton, 1999) and faults for approximately the same region as the LANDSAT image. Mapped faults are solid lines; lineaments are indicated by dashed lines. Offshore faults are taken from marine geophysical data compiled in Figure 2. Barbs indicate faults with vertical throw that form major topographic mountain fronts. Note parallelism between offshore and onshore fault trends. LR—Leon rift; MR—Morazon rift; NdD—Nombre de Dios range; NR—Negrito rift; O—Omoa; SPS—San Pedro Sula.

The Nombre de Dios range is a highly asymmetric uplift, with its drainage divide south of range midline (Fig. 5A). This asymmetry appears to be related to greater precipitation on the Caribbean side and a rainshadow on the landward (Aguan Valley) side. The north-flowing Río Cangrajal and Río Papaloteca have penetrated farthest into the Nombre de Dios range and have captured tributaries draining the south flank. This southward penetration is evidenced by remnants of a stranded paleo-divide north of and parallel with the modern divide. The modern divide is characterized by a low-relief surface interpreted as a high-elevation erosion surface. Similar, more extensive and possibly correlative erosion surfaces have been described in central Honduras south of the Nombre de Dios range and Aguan Valley (Helbig, 1959; Manton, 1987; Rogers et al., 2002) (morphologic zones 2 and 3 in Fig. 3A).

TABLE 1. NAMED FAULTS OF THE CORDILLERA NOMBRE DE DIOS, HONDURAS

Fault	Number on figure	Sense of motion	Observation	Original reference
La Ceiba	1	Left-lateral; normal	Forms front of Nombre de Dios range; western end terminates at edge of Sula Rift; aligns with Jocoton-Chamelcon faults to west; offsets Río Lis Lis valley	Muehlberger (1976)
Aguan	2	Left-lateral?	Boundary between metamorphic and sedimentary rocks; contiguous with lineation in alluvium of western Aguan valley	Elvir (1976)
Río Viejo	3	Left-lateral	Located along axis of range bounding high topography to north; offsets Río Papaloteca valley	Manton (1987)
Lepaca	4	Left-lateral	Offsets Río Lepaca and Río Pimienta valleys	This study
La Esperanza	5	Normal; left-lateral	Southern edge of Aguan Valley	Manton (1987)

Fluvial Response to Deformation on the North Coast

To assess the activity of the five major faults (Table 1), along with previously unrecognized lineaments identified on LANDSAT imagery, we use the longitudinal profiles of rivers and streams compiled from 1:50,000 scale topographic maps with 20 m contour intervals. The east-northeast faults and lineaments of the Nombre de Dios range intersect the north- and south-flowing rivers at high angles, providing an ideal geometry for examining the effects of recent tectonic activity on the rivers (Fig. 5B). In order to test whether these lineaments represent active faults or whether they are inactive faults or some other form of layering (e.g., foliations, bedding planes, sills and dikes in basement units, joints), we compiled twelve longitudinal and representative river and stream profiles across the range from the modern drainage divide to the mountain front (Fig. 6).

North-flowing streams have convex reaches in the longitudinal profiles indicative of tectonic uplift in the Nombre de Dios range that is outpacing the rate of fluvial downcutting (cf. Hovius, 2000) (Fig. 6). Also plotted on the profiles are the derived gradient for each segment of the river calculated by taking the first derivative of the profile curve:

$$s = -dH/dL, \qquad (1)$$

where s is slope, H is height, and L is distance (Hack, 1957). Parts of the stream with local areas of steep gradient are shown as the spikes or knickpoints. All knickpoints are plotted on the tectonic map of the Nombre de Dios range in Figure 5B in order to compare the locations of the knickpoints with major, mapped faults and with lineaments interpreted from the LANDSAT imagery.

The alignment on many of the profiles of lineaments and knickpoints suggests that these lineaments may be active faults that are uplifting the stream channel. Moreover, fault lineaments commonly occur downstream from convex reaches, indicating sections of the river undergoing active uplift (cf. Fig. 6, profiles 3, 5, 6, 7, 8, 9, 10, and 11). This suggests that a range-bounding fault in the knickpoint area is contributing to the upstream uplift of the upthrown block of the range. The coincidence of knickpoints, reaches of convex river profiles, lineaments, and known faults makes it unlikely that the knickpoints are solely the result of resistant lithologies in the river channels. Moreover, as annual rainfall in the Nombre de Dios range exceeds 3 m/yr it is likely that the erosive power of the rivers and streams exceeds rock resistance.

The parallel trend of the lineaments with known, active faults, like the range-bounding La Ceiba and Río Viejo fault zones, suggests that the Nombre de Dios range is being pervasively and internally sheared (Fig. 5B). The pervasiveness of the shearing may reflect the fact that the trend of active faults is roughly parallel to the foliations and strike of older basement structures (Manton and Manton, 1999).

We identify lateral offsets of river channels in order to constrain the horizontal slip in areas exhibiting vertical uplift. Four examples of left-lateral offsets ranging from 1.5 to 2.4 km are shown on Figure 7. Bedrock-confined channels of the Río Papaloteca and Río Lis Lis are deflected left-laterally where the rivers cross the Río Viejo fault (2.4 km, Fig. 7A) and the La Ceiba fault (1.5 km, Fig. 7B), respectively, along the steep, northern mountain front of the Nombre de Dios range. The bedrock-confined channels of the Río Lepaca and Río Pimineta are deflected left by 1.7 km and 2.1 km, respectively, where they cross the Lepaca fault along the southern mountain front of the range (Figs. 7C and 7D). These data suggest that pervasive left-lateral shearing is active and accompanies the active vertical uplift constrained by the river profiles shown in Figure 6.

Meandering alluvial rivers respond to tilting perpendicular to the channel by migrating in the direction of tilt (Leeder and Alexander, 1987). The Río Aguan has migrated south, toward the La Esperanza fault, in the eastern part of its valley, indicating that the fault is downthrown to the northeast and is active (Fig. 8). In the central part of the Aguan valley, the Río Aguan flows against the northern fault-bounded valley wall just south of the Lepaca uplift, indicating that the fault bounding the northern edge of the valley is downthrown to the south.

We constrain the active tilting of multiple fault blocks in the Nombre de Dios range by quantifying drainage basin asymmetry to determine the direction that a drainage basin has been tilted (Gardner et al., 1987). Drainage basin asymmetry is calculated as an asymmetry factor (*AF*) by the equation:

$$AF = 100(Ar/At), \qquad (2)$$

where Ar is the drainage area on the downstream right-side of the trunk stream and At is the total drainage area. Drainage basin

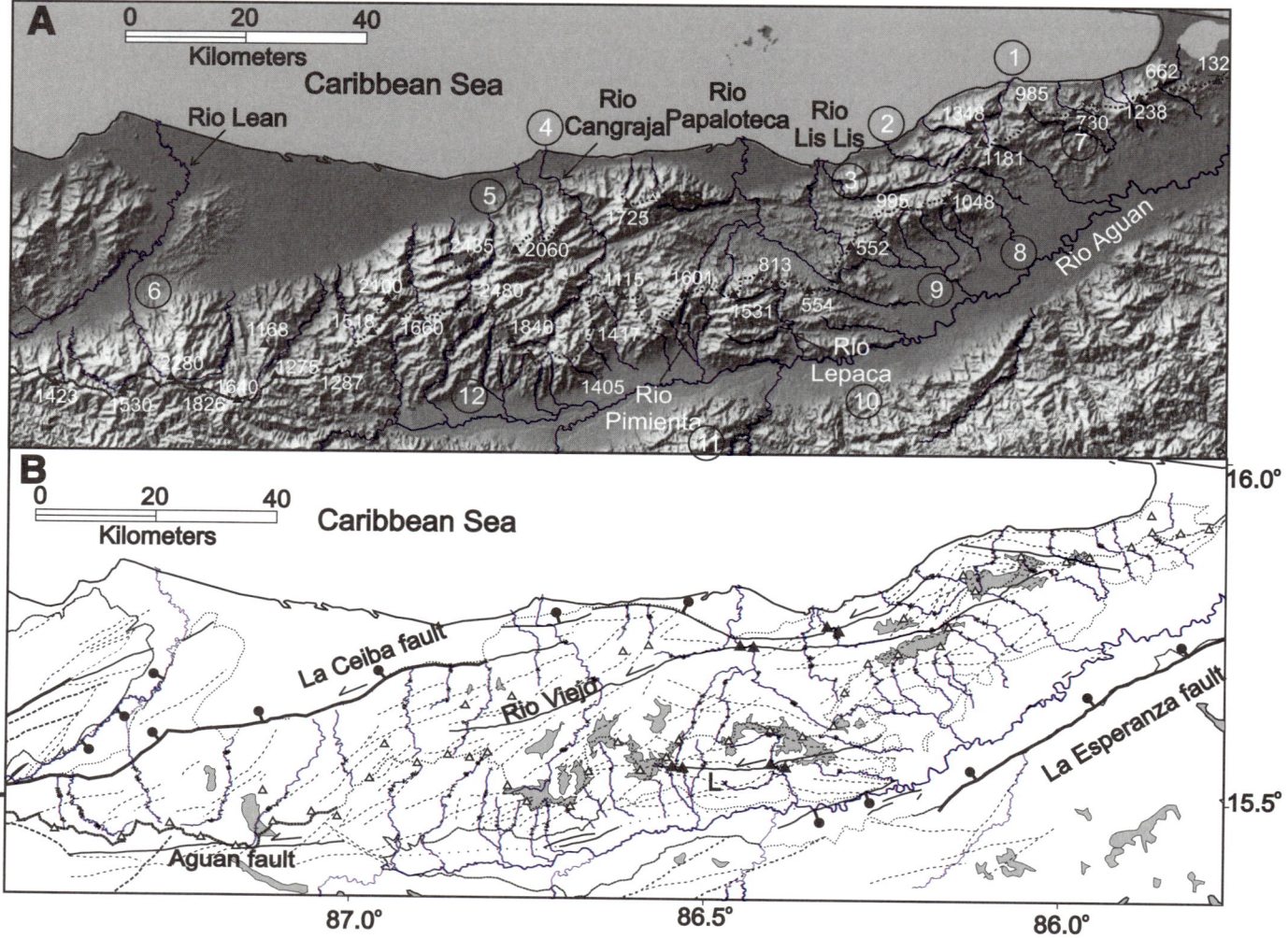

Figure 5. (A) Geomorphic features of east-northeast–trending Nombre de Dios range and Aguan Valley of northern Honduras (cf. Fig. 3 for location of map area). Drainage divides are shown by heavier dotted lines. Culminations along drainage divides are shown by triangles with numbers indicating their height in meters above sea level. Note that rivers draining northward into the Caribbean are significantly longer than rivers draining southward into the Aguan Valley. Circled numbers are keyed to 12 stream profiles show in Figure 6. (B) Major lineaments and mapped faults (heavy black line, ticks on downthrown side of fault) and all lineaments interpreted from LANDSAT imagery. Steep, topographic mountain fronts along Caribbean coast and in the Aguan Valley are shown as light, dotted lines. Gray areas show low-relief surfaces preserved along the drainage divide of the ranges that are interpreted high-level erosional surfaces. Gray areas north of the modern divide are inferred as erosional surfaces formed on the previous paleo-divide prior to the southward shift to the modern divide. Apparent left-lateral offsets of river channels seen on imagery are indicated by heavy black triangles on the Río Papaloteca, Río Balfante, Río Pimenta, and Río Lepaca. Knickpoints, or localized areas of steeper river gradients from Figure 6, are shown as small black dots along rivers. L—Lepaca fault.

asymmetry for 21 north-draining watersheds and 23 south-draining watersheds of the Sierra Nombre de Dios are displayed in Figure 8.

North-draining watersheds of the western Nombre de Dios range are tilted westward, while in the central section of the range, watersheds are tilted to eastward (Fig. 8). The watersheds of the eastern part of the range near Trujillo, draining to the Caribbean Sea, are tilted westward with the exception of the easternmost watersheds. Asymmetry of watersheds draining to the Aguan valley in the western part of the range is complicated by the southward migration of the drainage divide (Fig. 5A).

The central watersheds draining to the Aguan valley originate in or cross a topographic uplift centered on the Lepaca fault zone (Fig. 8). Drainage basin asymmetry reflects an eastward-plunging anticlinal uplift centered along the left-lateral Lepaca fault. Half of the eastern watersheds draining to the Aguan valley display a tilt to the west.

Topographic Development of Nombre de Dios Range

We propose several fault blocks coincident with and underlying the regions of topographic uplifts named on Figure 8:

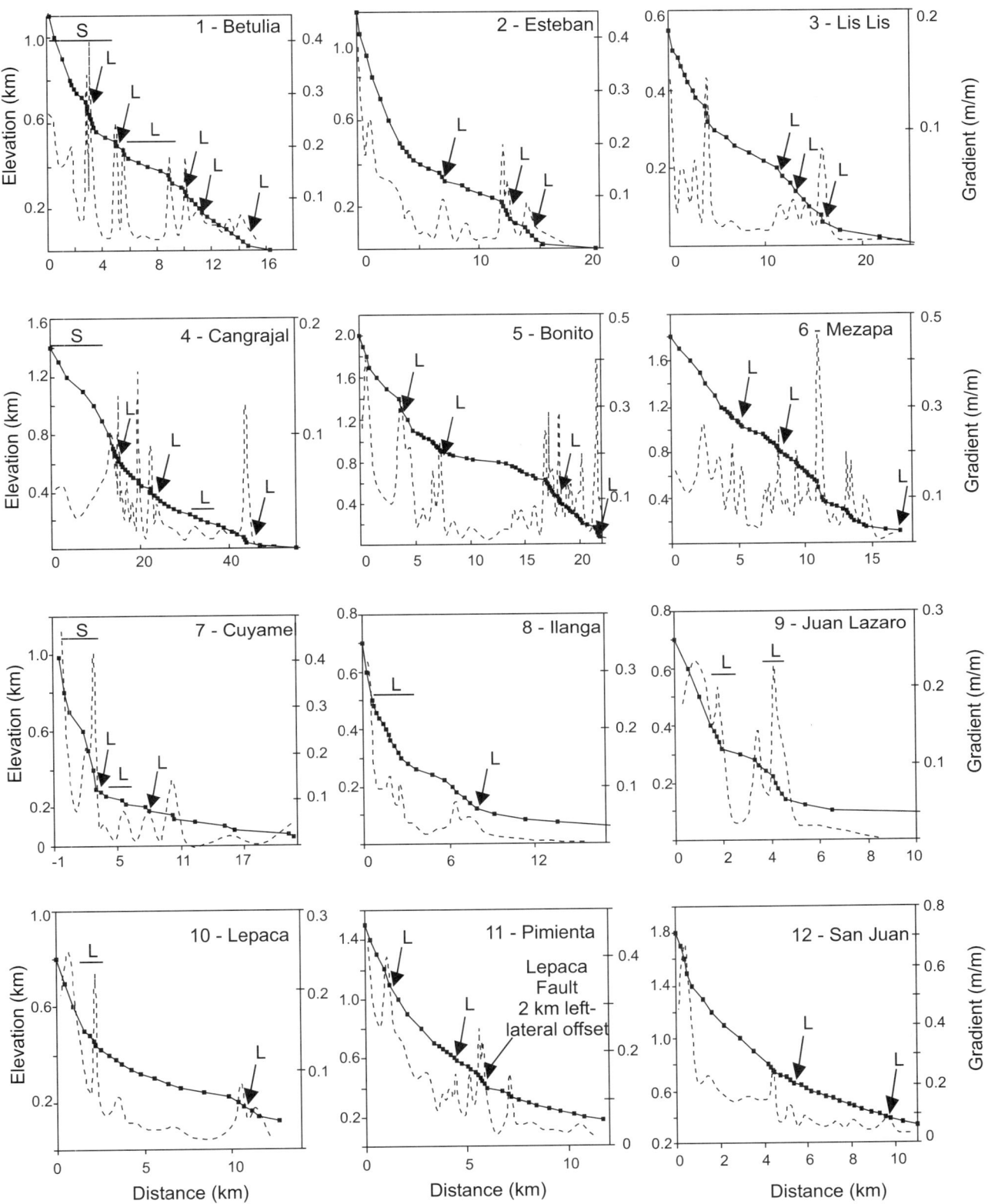

Figure 6. Stream profiles from the drainage divide to mountain fronts compiled from 1:50,000 scale topographic maps of Sierra Nombre de Dios (locations of streams, topographic divides, and topographic mountain fronts shown in Fig. 5). Solid lines show elevation of stream channels with crossings of 20 m contour interval indicated by small squares. Gradient peaks (thin line) are knickpoints. Knickpoints coinciding with mapped lineaments are labeled with "L" for "lineament" (lineaments displayed on Fig. 5B). "S" shows locations where streams cross the topographic surfaces shown in gray in Figure 5.

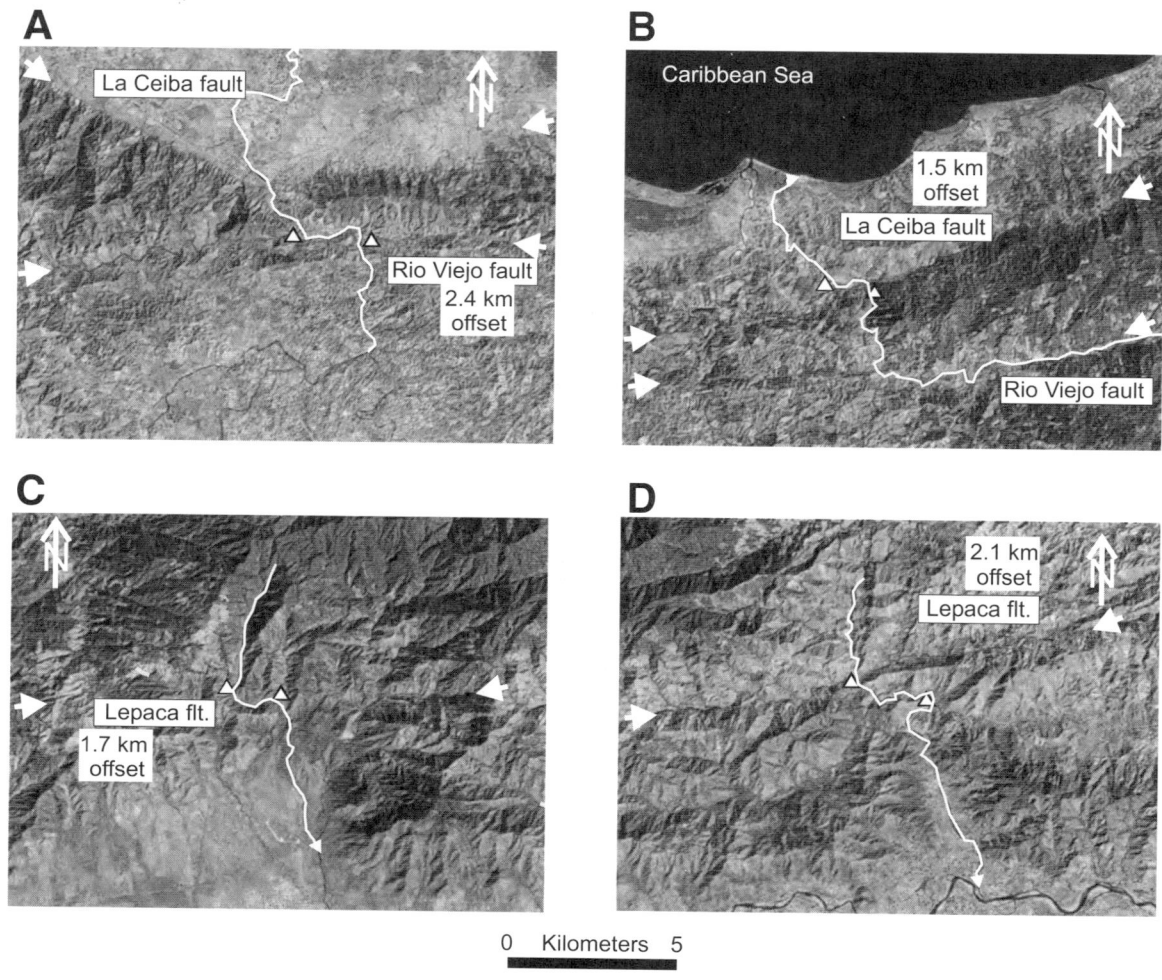

Figure 7. Four examples from LANDSAT imagery showing apparent left-lateral offsets of rivers in the Sierra Nombre de Dios (white arrows indicate fault traces shown in Fig. 5B and white triangles show magnitude of apparent left-lateral offset; locations of rivers shown in Fig. 5). All rivers shown occupy deeply incised, bedrock canyons. (A) Apparent left-lateral offset of 2.4 km of Río Papaloteca by the Río Viejo fault. (B) Apparent left-lateral offset of 1.5 km of Río Lis Lis by the La Ceiba fault. Note that the upper reaches of this river follow the trace of the Río Viejo fault zone. (C) Apparent left-lateral offset of 1.7 km of Río Lépaca by the Lepaca fault zone. (D) Apparent left-lateral offset of 2.1 km of southward-flowing Río Pimienta by the Lepaca fault. flt.—fault.

1. The Pico Bonito restraining bend topographic uplift is a restraining bend formed at a right-step between the La Ceiba and Río Viejo left-lateral fault zones. Active uplift is supported by the reaches of convex river profiles shown in Figure 6, and left-lateral displacement is inferred from left-lateral river channel offsets compiled in Figures 7A–7B. Asymmetric watersheds shown in Figure 8 indicate that the bend area plunges westward.
2. La Ceiba topographic uplift is a peripheral effect related to the smaller, fault-controlled Pico Bonito restraining bend (Fig. 8). Tilt directions are variable but generally trend about a north axis.
3. Lepaca topographic uplift relates to left-lateral motion along the Lepaca fault and the formation of a large anticline parallel to the fault trace. The fault may accommodate motion along the largely buried fault along the northern edge of the Aguan Valley. Tilt directions vary across the structure.
4. Trujillo topographic uplift is related to left-lateral oblique motion along the east-northeast segment of the Río Viejo fault zone.

Transition between the Western Rifts and the North Coast Ranges

Normal faults defining the eastern edge of the Sula rift of the Honduras depression abruptly truncate the east-northeast–striking La Ceiba fault zone (Fig. 4), and faults of the eastern boundary of the Sula rift exhibit more northeast trends. The El Negrito and

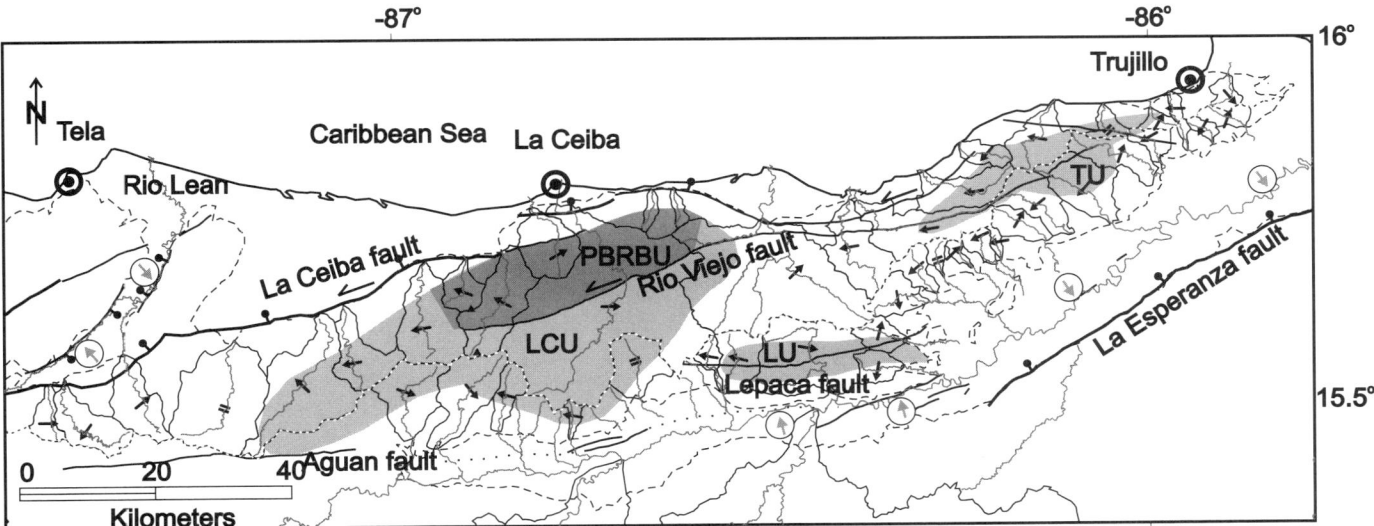

Figure 8. Comparison of topographically elevated areas associated with mapped faults of the Nombre de Dios range. The Pico Bonito restraining bend uplift (PBRBU, shown by dark shading) occurs at a right-step between the right-lateral Río Viejo and La Ceiba fault zones. The La Ceiba topographic uplift (LCU, shown by lighter shading) borders the south flank of the PBRBU. The Lepaca topographic uplift (LU, lighter shading) is associated with the Lepaca fault zone, and the Trujillo topographic uplift (TU, lighter shading) is associated with the eastern extension of the Río Viejo fault zone. Tectonic tilt directions are inferred from asymmetric watersheds (black arrows show direction of inferred tilting of the watershed; i.e., the larger half of the watershed is tilted toward the narrower half of the watershed). An equals sign indicates watersheds of the Sierra Nombre de Dios whose centerlines (trunk streams) are symmetrical. Arrows in open circles show tilt directions within the meandering alluvial rivers of the Aguan Valley and Lean Valley.

Morazon half-grabens form outliers to the east of the Sula rift but exhibit orientations more to the northeast. The RADARSAT image shown in Figure 9 documents the transition from the more northerly trends of normal faults that disrupt the Quaternary (0.5 Ma) volcanic field in the Sula rift to the more northeasterly trends of the El Negrito and Morazon rifts (average trends of rift-bounding normal faults indicated by white arrows in Fig. 9). The Lean rift is a full-graben northeast of the El Negrito and Morazon rifts and north of the La Ceiba fault (Fig. 4B). Avé Lallemant and Gordon (1999) have previously related the origin of the Lean graben to left-slip motion on the La Ceiba fault. We propose that the northeast trend of the northern Sula, El Negrito, Morazon, and Lean rifts forms a 35-km-wide rift province of intermediate orientation between the north-trending rifts of western Honduras and the east-northeast–trending rifts in the Nombre de Dios range and Aguan Valley (Fig. 4).

OFFSHORE STRUCTURE AND STRATIGRAPHY OF THE HONDURAN BORDERLANDS

The structure and late Neogene sedimentation of the Honduran borderlands is described in papers by Kornicker and Bryant (1969) and Pinet (1971, 1972, 1975) using single-channel seismic profiling and shallow coring. We utilize deeper penetration, multi-channel seismic reflection profiles, sidescan images of the seafloor, and offshore exploration wells to better define the distribution and structure of the elongate basins and ridges underlying the Honduran borderlands. We mapped a total of 21 different basins using seismic and sidescan sonar (Fig. 2B), although more densely spaced seismic profiling may reveal that some of the basins are contiguous. The basins form a belt that narrows toward the Mid-Cayman spreading center then widens toward the west. There are two parallel basin axes, the Bonacca and Patuca, separated by an elongate basement high.

Geologic and Bathymetric Setting of the Honduran Borderlands

GEOSAT (geodetic satellite) marine gravity data compiled in Figure 10A shows the large-scale basement structure of the Honduran borderlands. To the north of the borderlands, the Cayman trough shows a large gravity low associated with thin oceanic crust produced at the Mid-Cayman spreading center between 49.3 and 0 Ma (Leroy et al., 2000). A gravity high is centered on the active spreading center of the short, 100-km-long spreading center. The Nicaragua carbonate platform, located south of the Honduran borderlands, is characterized by a smooth gravity high. The Honduran borderlands exhibits a disrupted gravity field produced by the existence of elongate, fault-bounded, margin-parallel basins and ridges characteristic of continental borderland provinces (Gorsline and Teng, 1989). Twin gravity highs or basement ridges extend along the southern, faulted margin of the Cayman trough while a gravity low basin extends along the edge of the Nicaragua Rise. A gravity or basement high, the Patuca

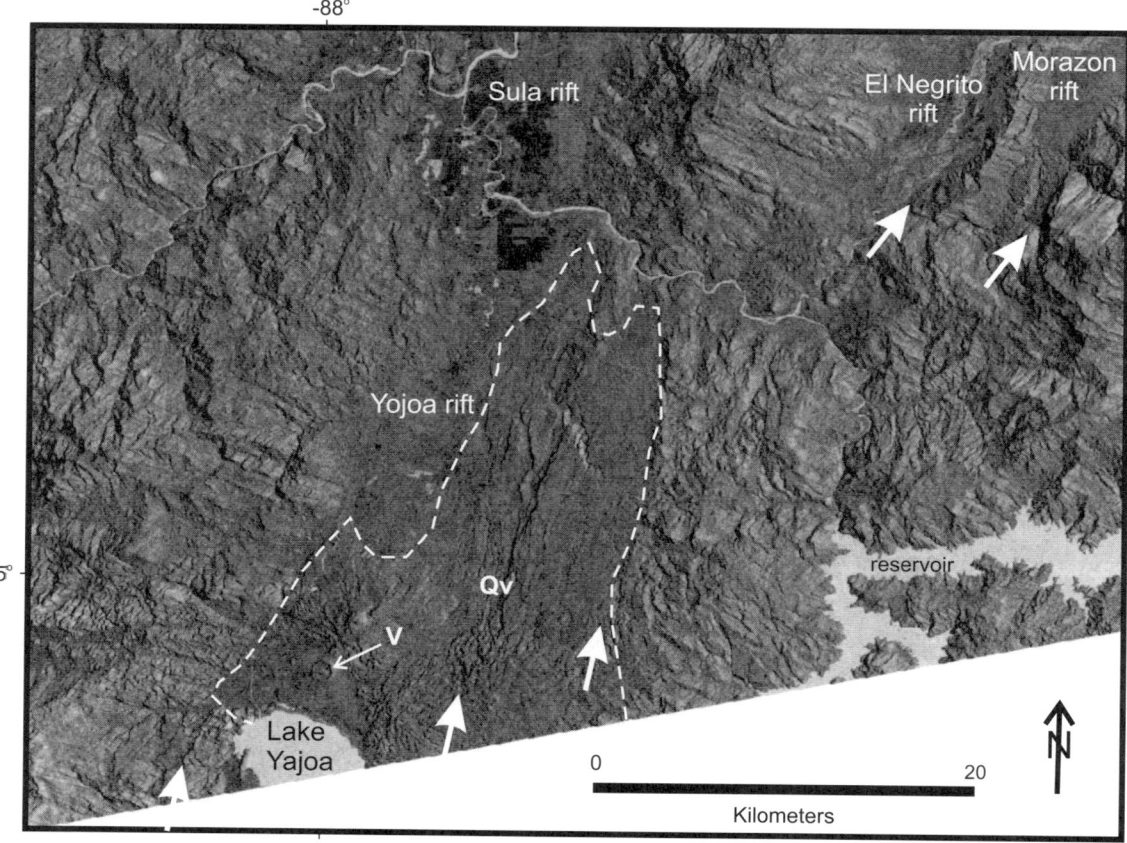

Figure 9. RADARSAT satellite image showing north-northeast–trending, late Quaternary fault scarps (white arrows) in Yojoa Quaternary basalt field (outlined by white dashed line) at the north end of Lake Yojoa within the Yojoa rift (see Fig. 4 for location of image). These scarps are mainly downthrown to the west. K-Ar age dating by ENEE (1987) shows age of field is 0.5 Ma (middle Quaternary). V—volcano; Qv—Quaternary volcanic flows.

Ridge, is present within this basinal structure and projects into the Nombre de Dios range of northern Honduras. The southern part of the Patuca Basin projects into the Aguan Valley.

The predicted bathymetry (Smith and Sandwell, 1997) of the region shown in Figure 10B is compiled along with the onshore topography. Kornicker and Bryant (1969) considered the basement and bathymetric high along the southern edge of the Cayman trough as a single ridge (Bonacca). Because regional gravity and bathymetry data show two parallel ridges, we name the features the North and South Bonacca Ridges. Kornicker and Bryant (1969) also considered the Tela Basin north of the Nombre de Dios range of northern Honduras a continuous feature extending to the northeast of Honduras, which we renamed the Patuca Basin to more accurately reflect the complex borderlands basins seen on Figure 10. The Bonacca Basin separates the North and South Bonacca Ridges and the Patuca Basin occurs between the South Bonacca Ridge and the Nicaragua Rise. The Patuca and Tela Basins are roughly collinear but are separated by a bathymetric and structural saddle northeast of Honduras. Figure 11 shows three bathymetric profiles across the Honduran borderlands correlated with the newly named basement and bathymetric features.

The twin North and South Bonacca Ridges can be identified on profiles A and B but merge on profile C (Fig. 11) to form the single, emergent Bay Islands of northern Honduras. The Swan Islands restraining bend, an active, localized restraining bend structure on the Swan Islands fault zone (Mann et al., 1991) (Fig. 2A), appears on profiles A and B but disappears along-strike on profile C (Fig. 11). The Patuca Basin forms a deep basin on profile A, which is interrupted by the appearance of the Patuca basement high on profile B. The northern half of the Patuca Basin continues westward into the Tela Basin, while the Patuca high projects into the Nombre de Dios range and the southern half of the Patuca Basin projects into the Aguan Valley. The overall effect of the parallel ridges and valleys of the Honduran borderlands is widening from east to west (Fig. 11).

Evidence of Active Tectonics and Structure

We use previously unpublished multichannel seismic (MCS) profiles across the Honduran borderlands collected during the University of Texas Institute for Geophysics (UTIG) Caribbean Tectonics (CT1) survey in 1979 combined with SeaMARC II

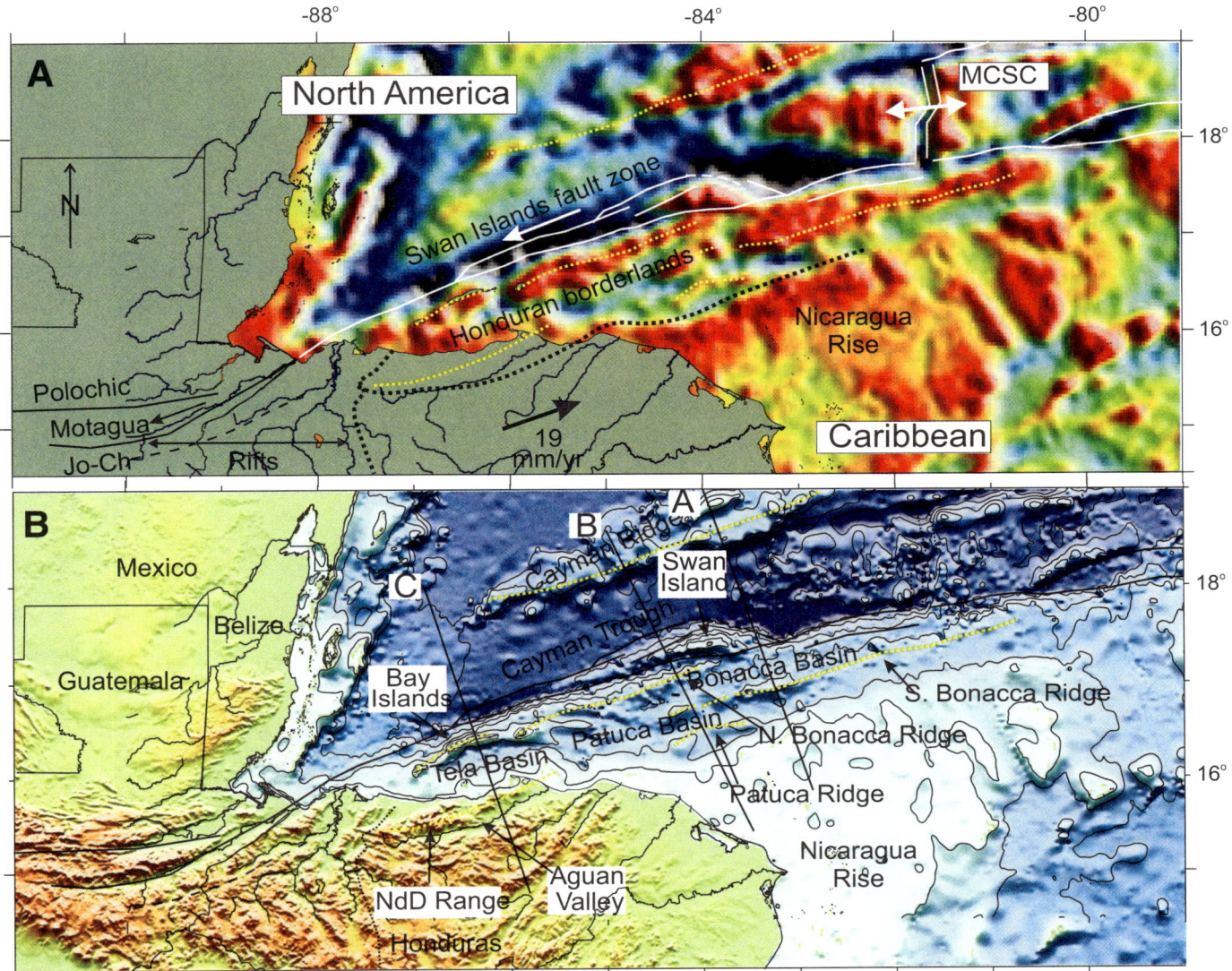

Figure 10. (A) Free-air marine gravity map of offshore areas of Honduras, Guatemala, and Belize in the Cayman trough, Mid-Cayman spreading center (MCSC), Honduran borderlands, and Nicaragua Rise (gravity data from Sandwell and Smith, 1997). Hot colors represent relatively higher gravity values above basement ridges; cooler colors show sediment-filled basins between basement ridges. Fault locations taken from fault compilation shown in Figure 2. Arrow shows Caribbean plate motion vector in Honduras relative to a fixed North America plate. Dotted lines correspond to crests of major basement ridges identified by name in Figure 10B. The northern edge of the Honduran borderlands is defined by the Swan Islands fault zone; the southern edge of the borderlands is defined by the Nicaragua Rise offshore and the Aguan Valley onshore. Jo-Ch—Jocotan-Chamelecon faults. (B) GTOPO30 digital elevation model of Central America merged with gravity-derived GTOPO5 bathymetry of offshore areas of Honduras from Smith and Sandwell (1997) (bathymetric contours are 1000 m, except for 100 m contour near shore). Yellow lines represent crests of bathymetric ridges and correspond to the yellow lines shown in the gravity map. NdD—Nombre de Dios.

sidescan sonar images and single-channel seismic lines of the seafloor and shallow subsurface collected in 1989 (Rosencrantz and Mann, 1991; Mann et al., 1991) to document and map active faults and their relation to observed bathymetry and structure. We consider faults breaking young seafloor sediments to be "active." The sidescan imagery was used in conjunction with the MCS profiles to verify that the faults extend to the seafloor and to map curving and anastomosing faults between MSC lines. The locations of the MCS and single-channel seismic (SCS) seismic lines used in the study are shown on Figure 2B; the footprint of the sidescan image is shown on Figure 2A.

MCS profiles CT1-8a and CT1-8b trend north-northwest and provide a cross section of ~4 s two-way travel time (TWTT) of the eastern part of the Honduran borderlands southeast of the

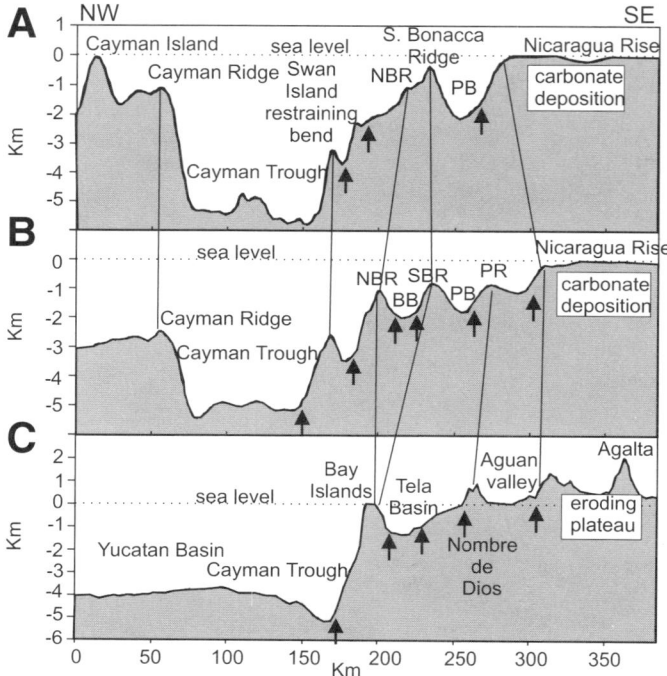

Figure 11. Bathymetric profiles A, B, and C across the Honduran borderlands and northern Honduras showing along-strike correlation of the main east-northeast–trending bathymetric ridges and flanking basins (cf. Figure 10B for locations of profiles). Arrows show positions of major active faults from Figure 2. Lines from profile to profile display features that correlate. Note the overall east to west widening of the basins consistent with the predictions for east to west increase in extension shown in Figure 18A. BB—Bonacca Basin; NBR—North Bonacca Ridge; PB—Patuca Basin; PR—Patuca Ridge; SBR—South Bonacca Ridge.

Swan Islands restraining bend (Fig. 2B). Seven active faults are recognized: two are parallel strands of the Swan Islands fault zone (see Rosencrantz and Mann, 1991), and five bound the Bonacca and Patuca Basins and separate them from the intervening basement highs, the North and South Bonacca Ridges (Fig. 12A). Correlation of line CT1-8 with the sidescan image in Figure 13B shows that a linear fault along the crest of the South Bonacca ridge forms a major scarp on the seafloor, indicating an active feature. This fault forms one edge of a large basement ridge with normal faults dipping away from its center (Fig. 13C).

MCS profile CT1-6a and CT1-6b trends north to a depth of ~5 s TWTT in the central part of the Honduran borderlands southwest of the Swan Islands restraining bend (Fig. 2B). Five active faults are recognized: three are parallel strands of the Swan Islands fault zone (highlighted by white arrows in Fig. 14B) and two bound the South Bonacca Ridge (Fig. 12A). Correlation of line CT1-6 with the sidescan image in Figure 14B shows linear fault scarps along the en echelon crest of the North Bonacca ridge. A southeast-dipping normal fault forms a half-graben (Bonacca Basin) filled with asymmetric wedge-shaped seismic unit 2 (Fig. 14C). Normal faults bounding the symmetrical Patuca graben at the south end of the seismic line also penetrate the seafloor (Fig. 14D).

MCS profile CT1-3b is the westernmost MCS line and trends northward across the borderlands (Fig. 2A). This line displays the steep scarp of the Swan Islands fault zone (indicated by the white arrows in Fig. 15B) at the edge of the Cayman trough, a series of six half-grabens formed along southward dipping normal faults, and tilting and faulting of sequence 3, suggesting that the rift process is ongoing (Fig. 12C).

Both the Bonacca and Patuca rifts are much wider on line CT1-6ab than on line CT1-8ab to the west, supporting the basic observation of the north-south widening of the Honduran borderlands from east to west seen on the regional bathymetric profiles in Figure 11.

SCS line 71, collected in 1989 along with the sidescan data, trends east to west across the Tela Basin (Fig. 16A). Line 71 images two active fault scarps seen as faint lineaments on the sidescan image (Fig. 16B). In the subsurface, these faults control small east-northeast–trending rifts within the Tela Basin (Fig. 16C).

Stratigraphic Development of the Honduran Borderlands

Three distinct seismic sequences are identified on the seismic profiles and are numbered 1, 2, and 3 on Figure 12. The lowest sequence, sequence 1, consists of tabular, tilted reflectors up to 1.0 seconds TWTT in thickness (Figs. 13C and 14C). The base of this sequence merges with the acoustic basement due to loss of energy with depth and therefore the nature of the contact between sequence 1 and basement is not clear. Seismic facies associated with sequence 1 include a semi-continuous high-amplitude facies consisting of subparallel, semi-continuous high-amplitude reflections with high-amplitude reflections dominant suggesting sand-rich channel-lobe complexes. Based on its geometry, we interpret tabular sequence 1 as a pre-rift unit deposited prior to the formation of the full and half-grabens seen on the profiles. The parallel, horizontal reflectors indicate that this unit was deposited as a flat layer above a continental or island arc crust at the northern edge of the Nicaragua Rise.

Sequence 2 consists of a wedge-shaped, isolated, fault-bounded package of reflectors indicative of long-term vertical motion on the adjacent fault scarps (Figs. 13C, 13D, 14C, 14D, and 15C). Seismic facies associated with sequence 2 include a semi-continuous high amplitude alternating with low-amplitude facies containing concave-up high-amplitude reflections and wedge-shaped external configuration, suggesting partially channelized turbidite lobe complexes. For example, on Figure 13C, the wedge-shape of sequence 2 is imaged adjacent to the presently active basement block (South Bonacca Ridge) and linear fault shown in Figure 13B. On Figure 13D crossing the Patuca Basin, sequence 2 forms two similar wedges against normal faults dipping to the southeast. We interpret wedge-shaped sequence 2

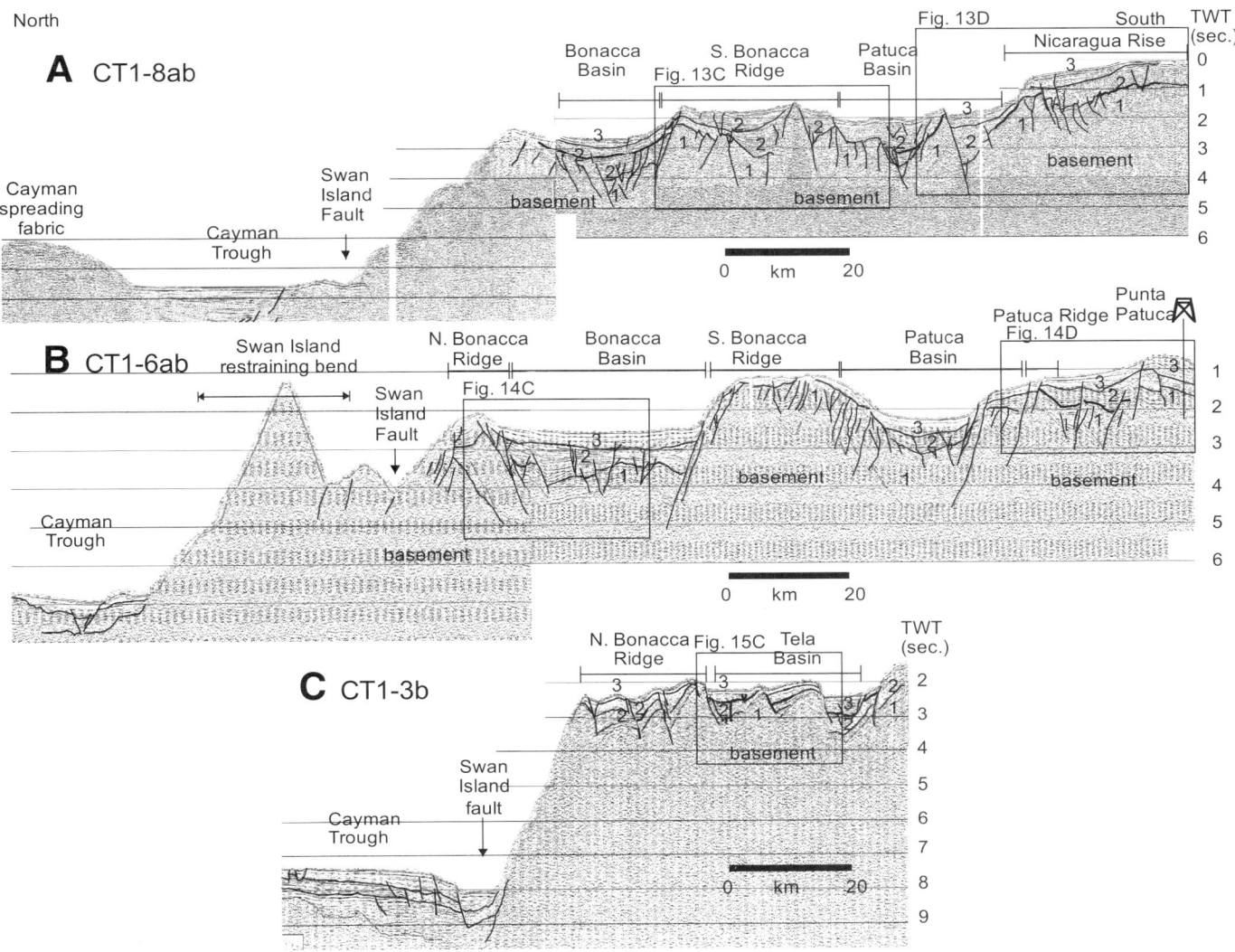

Figure 12. (A) Interpreted University of Texas Institute for Geophysics multi-channel seismic (MCS) line CT1-8ab across half-grabens and intervening basement horsts of the Honduran borderlands and the southern part of the Cayman trough. Depth is in two-way time (TWT). Three main unconformity-bound sequences are Sequence 1—a lower, pre-rift sequence; Sequence 2—a syn-rift sequence; and Sequence 3—a late syn-rift sequence. Boxes show locations of blowups shown in Figures 13C and 13D. (B) Interpreted UTIG MCS line CT1-6ab. Boxes show locations of blowups shown in Figures 14C and 14D. (C) Interpreted UTIG MCS line CT1-3b. Box shows location of blowup shown in Figure 15C. Note the top (east) to bottom (west) increase in the width of the Bonacca Basin and deepening of the Patuca Basin.

as a syn-rift unit that accompanied the main phase of rifting and fault scarp formation.

Sequence 3 is the uppermost unit and provides a tabular fill of the basinal areas of the Bonacca and Patuca Basins (Figs. 13C, 13D, 14C, and 14D). A progradational, shingled facies displays a basinward progradation geometry of reflectors that changes downdip into semi-continuous high amplitude facies. Sequence 3 is interpreted as a post-rift unit. Fault control on tabular sequence 3 is less strong than on the wedge-shaped units of sequence 2. Nevertheless, some fault scarps penetrate unit 3 to the seafloor and indicate that faulting continues today although generally not forming rifts with clastic wedge units (cf. Figs. 13C and 13D). The source of sediment for sequence 3 is either peri-platform carbonate ooze derived from carbonate production on the Nicaragua Rise (cf. Fig. 13D) or terrigenous clastic material derived from major river systems of northern and eastern Honduras (Fig. 10B).

This proposed rift interpretation is similar to that proposed by Pinet (1975) based on his study of single-channel seismic reflection study in the area of the Tela basin (Fig. 2A). Pinet (1975) documented two stratigraphic sequences above an angular unconformity disrupted by normal faults that form horsts and grabens

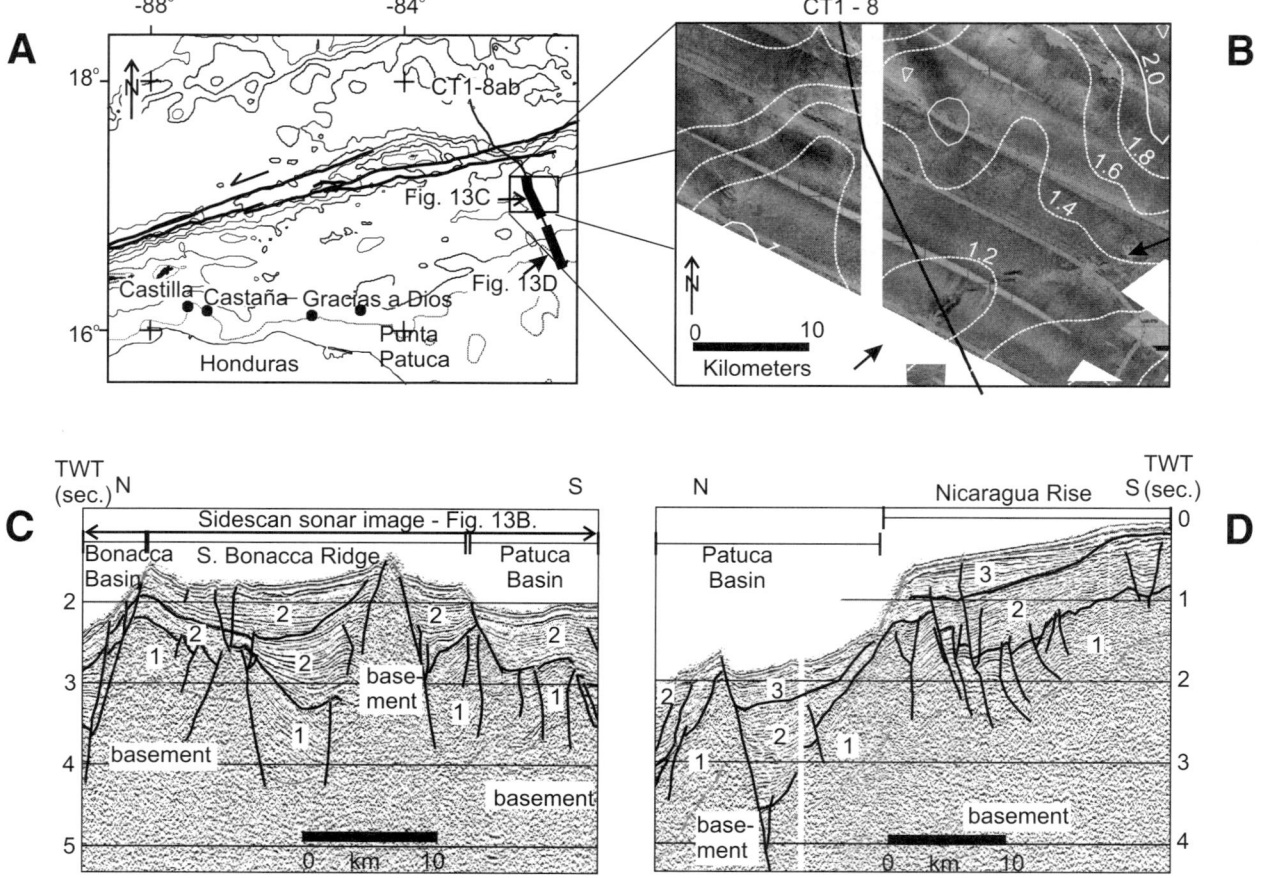

Figure 13. (A) Bathymetric map showing track of University of Texas Institute for Geophysics multi-channel seismic (MCS) line CT1-8ab relative to the Swan Islands fault zone, the Honduran coastline, and four offshore exploration wells. Bold segments of track indicate locations of seismic lines shown in C and D. Boxed area shows location of the SeaMARC II sidescan sonar image of seafloor expression of South Bonacca Ridge shown in B. (B) Sidescan image of seafloor overlying area imaged by line CT1-8ab showing crest of the South Bonacca Ridge (indicated by black arrows). Bathymetric contours are in kilometers (200 m contour). (C) Segment of MCS line CT1-8ab showing South Bonacca Ridge and flanking Bonacca Basin to the north and Patuca Basin to the south. Note fanning of syn-rift seismic sequence 2 indicative of vertical throw on the South Bonacca horst. (D) Segment of MCS line CT1-8ab showing fault scarp cutting youngest sequence 3 in the Patuca Basin. Note contrast in asymmetry between sequences 2 and 3 indicating that the recent phase of extension may be weaker than the phase responsible for sequence 2. TWT—Two-way time.

filled by turbidites (Fig. 2A). Noting that these sequences are uniform in thickness, Pinet (1975) inferred that the block faulting postdates the deposition and proposed a late Miocene–Pliocene age of deposition and a Pliocene normal faulting event based on correlation to onshore geologic history.

Four oil exploration wells (Fig. 17) drilled at shelfal to slope water depths (66–337 m below sea level [mbsl]) along the southern edge of the Honduran borderlands adjacent to the Patuca rift basin (Fig. 2B) allow correlation between the dated units in the wells and seismic units 1–3. Sonic log data from Punta Patuca-1 was converted to two-way travel time and this time section was then correlated to the south end of MSC profile CT1-6ab crossing the Patuca basin (Fig. 14D). The Punta Patuca-1 well was drilled in an area outside of the main zone of rifting in the deeper water Patuca basin. Despite this, all four wells show a similar history, suggesting that the general timing and lithologies of the wells provide a reliable insight into the rift history of the Honduran borderlands. The four wells, drilled to depths of 2397–3790 m, penetrated a basement of variable lithologies (Fig. 17). Above a basal unconformity, the overlying rock types range from Early Cretaceous through late Eocene (Fig. 17). Clastic rocks of middle Miocene age overlie the unconformity and range in water depth from subaerial (?) redbeds to coastal or coastal shelf. We interpret this contact as middle to early late Miocene transition from the acoustic basement to the overlying tabular pre-rift seismic unit 1 (Fig. 12). Syn-rift seismic unit 2

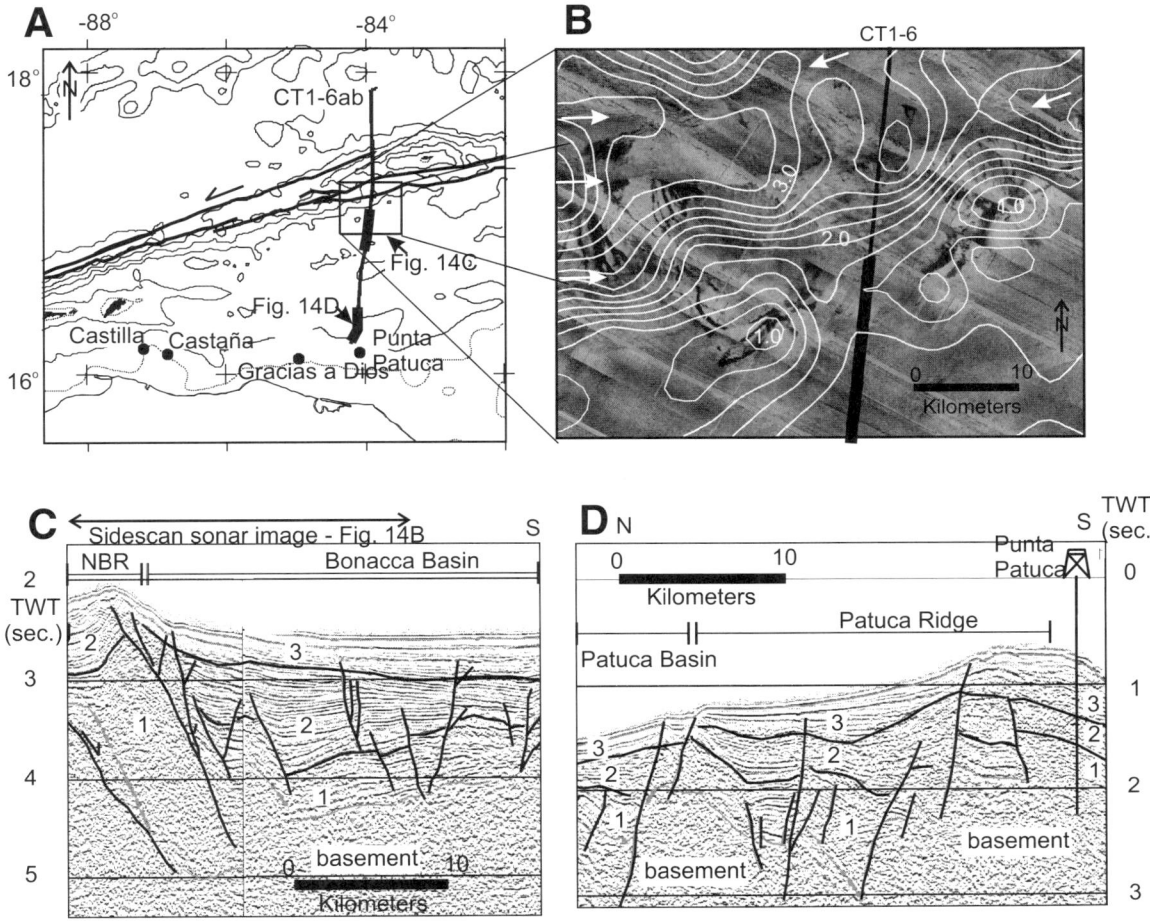

Figure 14. (A) Bathymetric map showing track of University of Texas Institute for Geophysics multi-channel seismic (MCS) line CT1-6ab. Bold segments of track indicate locations of seismic lines shown in C and D. Boxed area shows location of the SeaMARC II sidescan sonar image of seafloor expression of North Bonacca Ridge shown in B. (B) Sidescan image of seafloor overlying area imaged by line CT1-6ab showing fault strands of the active, left-lateral Swan Islands fault zone indicated by white arrows and crest of the North Bonacca Ridge. (C) Segment of MCS line CT1-6ab showing North Bonacca Ridge (NBR) and flanking Bonacca Basin to the south. Note fanning of syn-rift seismic sequence 2. Emergent horst block and faults with seafloor expression indicates active faults. (D) Segment of MCS line CT1-6ab showing sequences 1–3 and position of Patuca Ridge on which the Punta Patuca well was drilled in 1978 (cf. Fig. 17; well location was projected 9 km onto this line). Note faults extending to seafloor. TWT—Two-way time.

correlates to a late Miocene section of sandstone and shale with minor shallow marine limestone in the Punta Patuca well. Post-rift seismic sequence 3 correlates to a Plio-Pleistocene section of mudstone deposited at bathyal depths.

The Castaña-1 drilled near the mouth of the Río Aguan and the bathymetric high that separates the Tela and Bonacca Basins penetrated 400 m of limestone, which is suggestive of localized clastic bypass and carbonate deposition on the bathymetric high. Numerous bentonite horizons along with volcanic fragments found in the middle Miocene clastic shelf strata of Punta Patuca-1 and Gracias A Dios-1 represent airfall and fluvial linkage to the ignimbrite province of the Miocene arc of Central America (Sigurdsson et al., 2000; Jordan et al., this volume).

ACTIVE DEFORMATION RELATED TO PLATE KINEMATICS—TWO STYLES OF TRANSTENSIONAL DEFORMATION

We have presented geologic, earthquake, marine geophysical, and remote sensing data to show that Neogene to Recent transtensional deformation produced at the North America–Caribbean plate boundary in northern Central America exhibits two distinct structural styles. East-west extension along faults normal to the plate boundary occurs south of the western 375-km-long plate boundary segment, producing basin and range morphology in western Honduras and southern Guatemala. To the east, NNW-SSE extension along faults subparallel to the

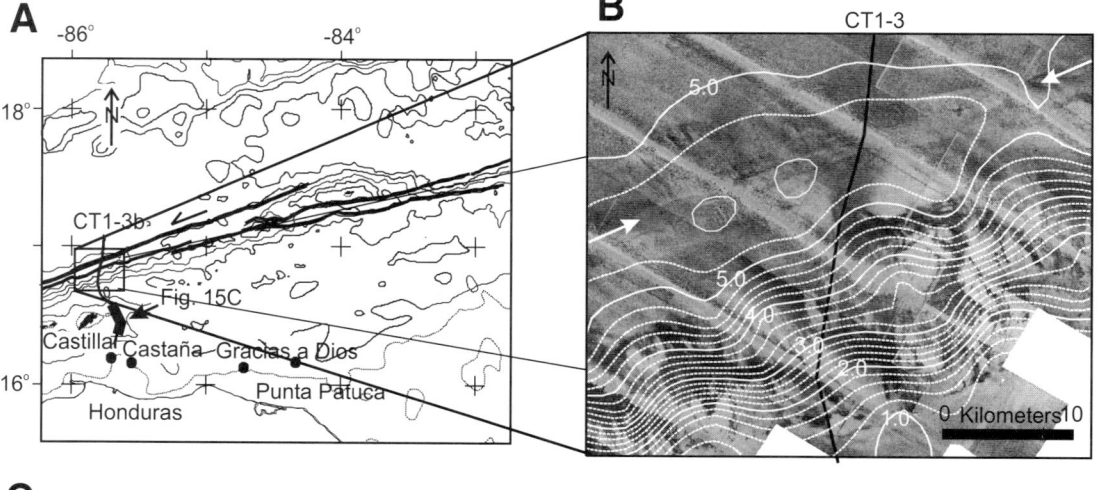

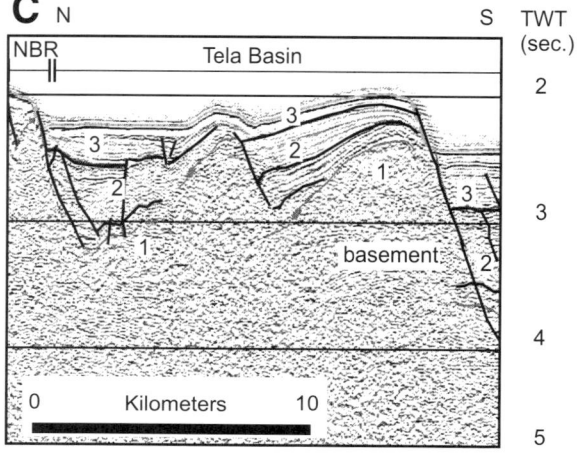

Figure 15. (A) Bathymetric map showing track of University of Texas Institute for Geophysics multi-channel seismic (MCS) line CT1-3b. Bold segment of track indicates location of seismic line shown in C. Boxed area shows location of the SeaMARC II sidescan sonar image of seafloor expression of North Bonacca Ridge shown in B. (B) Sidescan image of seafloor overlying area imaged by line CT1-3b showing main trace of the active, left-lateral Swan Islands fault zone indicated by white arrows and steep submarine escarpment leading up to Bay Islands of Honduras to the south. (C) Segment of MCS line CT1-3b showing North Bonacca Ridge (NBR) and flanking Tela Basin to the north. TWT—Two-way time.

plate boundary occurs south of a 600-km-long plate boundary segment, producing margin-parallel ridges and basins onshore in northern Honduras and in the offshore Honduran borderlands region (Fig. 18A).

In the western area, the north trending normal faults have been interpreted as (1) intraplate deformation and block rotations about a highly arcuate, convex-southward left-lateral strike-slip fault system (Plafker, 1976; Burkart and Self, 1985); and (2) fault termination structures related to the termination of left-lateral slip of the Motagua fault zone (Langer and Bollinger, 1979; Guzman-Speziale, 2001). While transtensional deformation has not been proposed previously for the Honduran borderlands and Nombre de Dios range and Aguan Valley of northern Honduras, we show that east-northeast–trending faults are oblique-slip and extend in a zone ~150 km south of the Swan Islands fault.

We compare the orientation of active faults with the trend of GPS-derived Caribbean plate motion vectors (Fig. 18). The three plate vectors shown are for points along the Motagua–Swan Islands fault system at longitudes 89°W, 85°W, and 81°W and are decomposed to show the extensional and strike-slip component of the plate vector. The extensional component of motion is controlled by the angular divergence (Fig. 18B) between the plate vector and the trend of the plate boundary fault. The extensional component of the plate vector increases from 0.2 mm/yr at longitude 81°W near the Mid-Cayman spreading center to 4.8 mm/yr at longitude 89°W in the Motagua Valley of Guatemala and is consistent with the widening of plate margin deformation from east to west.

The angular difference (Fig. 18B) remains relatively constant and <5° in the eastern area of the Mid-Cayman spreading center that is characterized by strike-slip faulting on well-defined faults (Rosencrantz and Mann, 1991). The value increases to amounts >5° in the central area adjacent to the offshore Honduran borderlands and increases up to 20° in the area of the Motagua Valley. East-west extension on faults normal to the plate margin occurs when the angle of divergence between the main plate boundary fault (Motagua fault zone) and the GPS-derived Caribbean plate vector is ≥10° (Fig. 18). NNW-SSE transtension on faults subparallel to the plate boundary occurs when the angle of divergence between the plate boundary fault and the Caribbean motion vector is between 5 and 10°. A narrow, 35-km-wide

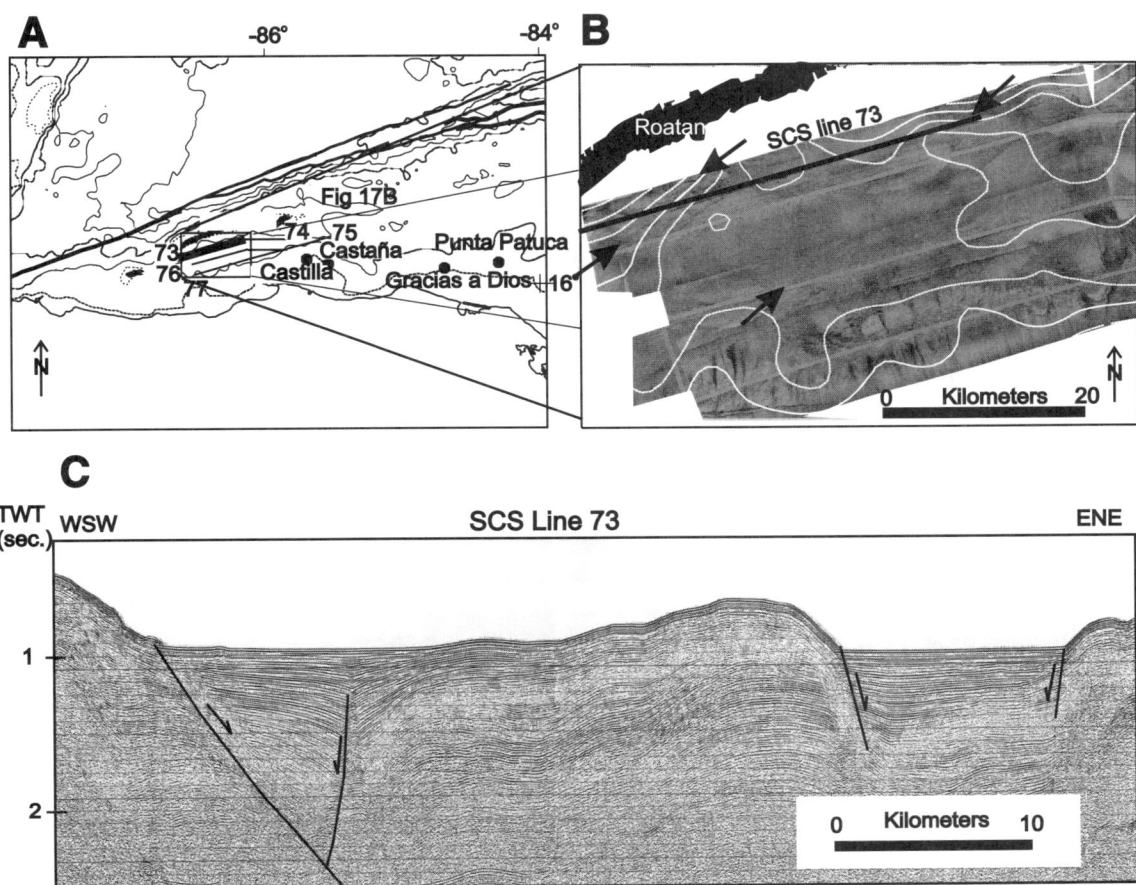

Figure 16. (A) Bathymetric map showing track of University of Texas Institute for Geophysics single-channel seismic (SCS) line 73. Bold segment of track indicates location of seismic line shown in C. Boxed area shows location of the SeaMARC II sidescan sonar image of seafloor expression of the Tela Basin adjacent to Roatan, Bay Islands, Honduras, shown in B. (B) Sidescan image of seafloor overlying area imaged by line 73 showing traces of two faults breaking the seafloor indicated by white arrows that trend east-northeastward. (C) Blowup of segment of SCS line 73 showing rift structures affecting the youngest units of the basin (Plio-Pleistocene turbidite fill penetrated in the wells on the east flank of the deep basin). Lack of penetration on SCS line 73 does not allow recognition of sequences 1, 2, and 3 as observed on UTIG multi-channel seismic lines.

tectonic transition area in north-central Honduras separates the north-trending rifts of western Honduras and from the boundary-parallel rifts in northeastern Honduras and the offshore Honduran borderlands (Fig. 18).

Estimated partitioning of the Caribbean–North America velocity components into parallel (strike-slip faulting) and perpendicular (normal faulting) to the trend of the plate boundary faults is derived by rotating the predicted plate velocities using the DeMets et al. (2000) pole information onto local fault trends at the locations separated by ~50 km (Fig. 18C). Values of strike-slip remain relatively constant around 18 mm/yr except for a decrease to 17 mm/yr in the area of maximum predicted extension near the eastern end of the Motagua Valley (Fig. 18C). Extension varies but clusters around the median 0° line in pure strike-slip areas of the Oriente fault and Mid-Cayman spreading center. To the west, extension increases to a maximum of 6 mm/yr in the Motagua Valley near longitude 88.5°W (except near the Swan Islands restraining bend near longitude 84°W).

NNW-SSE extension along faults subparallel to the plate boundary in the Honduran borderlands and northern coast of Honduras coincides with extensional partitioning rates <5 mm/yr. E-W extension on faults normal to the plate boundary in western Honduras and Guatemala coincides with extensional partitioning rates >5 mm/yr.

GEOLOGIC EVOLUTION OF THE BORDERLAND REGION

We address the Neogene evolution of transtension south of the North America–Caribbean plate boundary in two ways. First, we quantify the amount of extension that has occurred in both the offshore borderlands region and also in the western

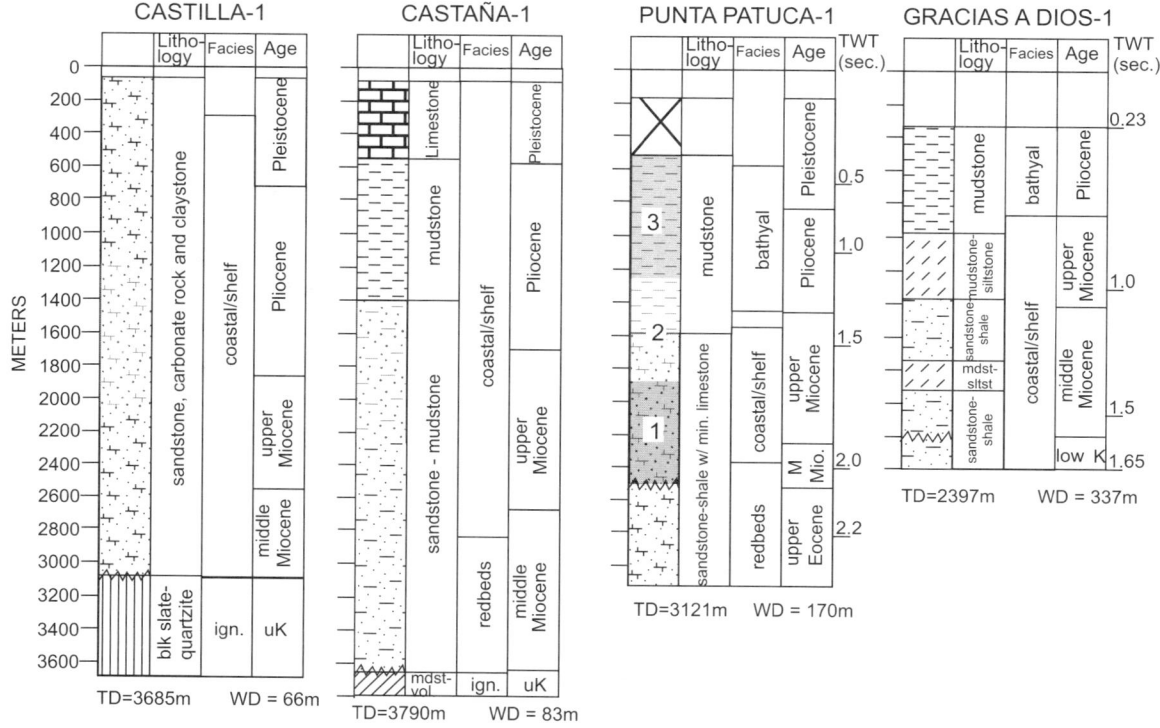

Figure 17. Well logs from four offshore exploration wells from the Honduran borderlands (well logs were provided by and used with written permission from the Dirección de Energía of Honduras). Sonic log data available for Punta Patuca-1 and Gracias a Dios-1 wells was used to convert well logs to two-way time sections that were then correlated to multi-channel seismic time sections. This correlation showed that pre-rift sequence 1 is middle to early late Miocene in age, syn-rift sequence 2 is late Miocene, and late syn-rift sequence 3 is Pliocene and younger. Note the large unconformity at the base of the middle Miocene.

rifts. Next, we utilize these results along with regional geologic and plate kinematic constraints to develop a quantified plate reconstruction of this transtensional margin as it developed during the Neogene.

Regional Extension

Individual rifts of the Honduran borderlands widen from east to west (Fig. 11). This trend can be explained by more extension in the west during the development of the rifts, as suggested by the distribution of instantaneous strain displayed in Figure 18. To estimated the finite strain during evolution of the rifts, we restore the three seismic lines (CT1-8ab, CT1-6ab, and CT1-3 from Fig. 12) using the vertical shear fault restoration method (Rowan and Kligfield, 1989). We take the top of the middle Miocene–age seismic sequence 1 (Fig. 12) as the pre-rift datum for the reconstruction for the three MCS lines. Figure 19 displays the restored seismic lines along with their finite elongation values of 11.8% for CT1-8ab (eastern line, Fig. 19A), 16.7% for CT1-6ab (central line, Fig. 19B) and 21.3% for CT1-3 (western line, Fig. 19C). These values support the prediction of progressive east to west increase in extension indicated by the instantaneous GPS-based strain calculations (Fig. 18).

We make a second estimate of extension across the Honduras borderlands by extrapolating the instantaneous GPS-based Caribbean plate vector (Fig. 18) to 12 Ma (the late Miocene age of initiation of rift based on well data). Assuming full strain partitioning, 24.7% extension occurred across the 150-km-wide Honduran borderlands at longitude 85°W (roughly the location of line CT1-3ab and CT1-6ab) (Fig. 19D). While these numbers are very close, the smaller extension values (11.8%–21.3%) obtained by the restoration of seismic lines compared to the GPS-based estimate (24.7%) may be the result of (1) incomplete partitioning (i.e., more motion on strike-slip faulting of the Swan Islands fault zone or on strike-slip faults within the rifts and basement ridges); (2) extension south of the region covered by the seismic lines; and (3) that the trend of the seismic lines are somewhat oblique rather than perpendicular to the trend of faults (Fig. 2A). Our analyses are consistent with Miocene age north-south extension determined from a paleostress analysis on striated fault planes from outcrops on Roatan Island, the largest of the Bay Islands, by Avé Lallemant and Gordon (1999).

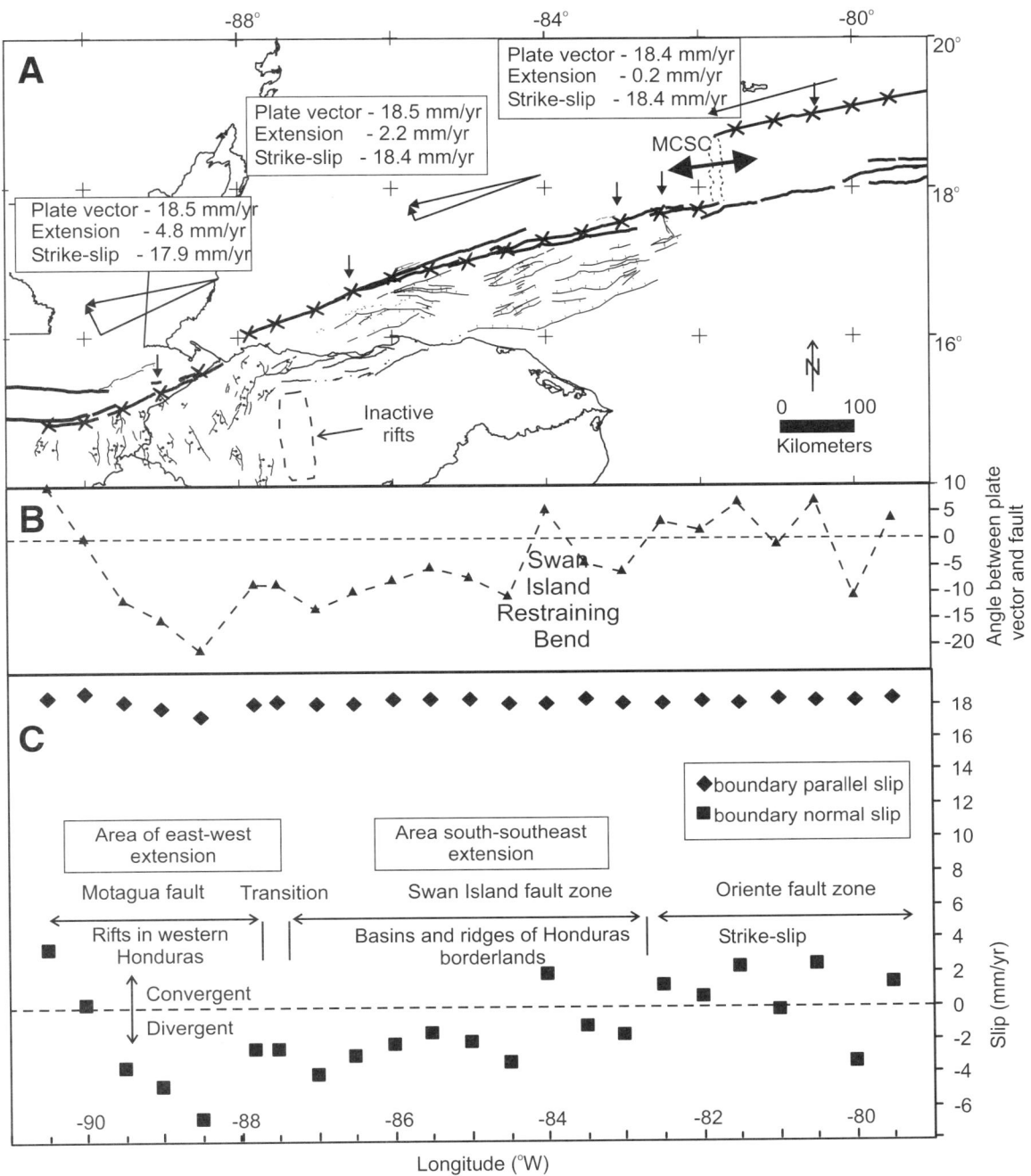

Figure 18. (A) Distribution of active faults along and south of the North America–Caribbean plate margin. GPS-derived Caribbean plate velocity (DeMets et al., 2000) was calculated at 30 min increments (Xs on map) along the main North America–Caribbean plate boundary faults (Motagua–Swan Islands–Mid-Cayman spreading center–Oriente system). The three plate vectors shown are for points along the fault system at longitudes 89°W, 85°W, and 81°W and are decomposed to show the extensional and strike-slip component of the plate vector. The extensional component of motion is controlled by the angular divergence between the plate vector and the trend of the plate boundary fault. Vertical arrows show location North America–Caribbean velocity predictions of DeMets et al. (2000). (B) Plot of the angular difference between the North America–Caribbean GPS-derived plate vector from DeMets et al. (2000) and the trend of the Motagua–Swan Islands–Mid-Cayman spreading center–Oriente fault system taken at the points along the fault system marked by "X" in Figure 18A. An angle of 0° would indicate predicted pure strike-slip; angles above the dashed line indicate convergence as observed at the Swan Islands restraining bend, and angles below the dashed line indicate divergence. (C) Plot of the Caribbean velocity components decomposed into rates of plate-boundary parallel slip (diamonds) and boundary-perpendicular slip (squares). MCSC—Mid-Cayman spreading center.

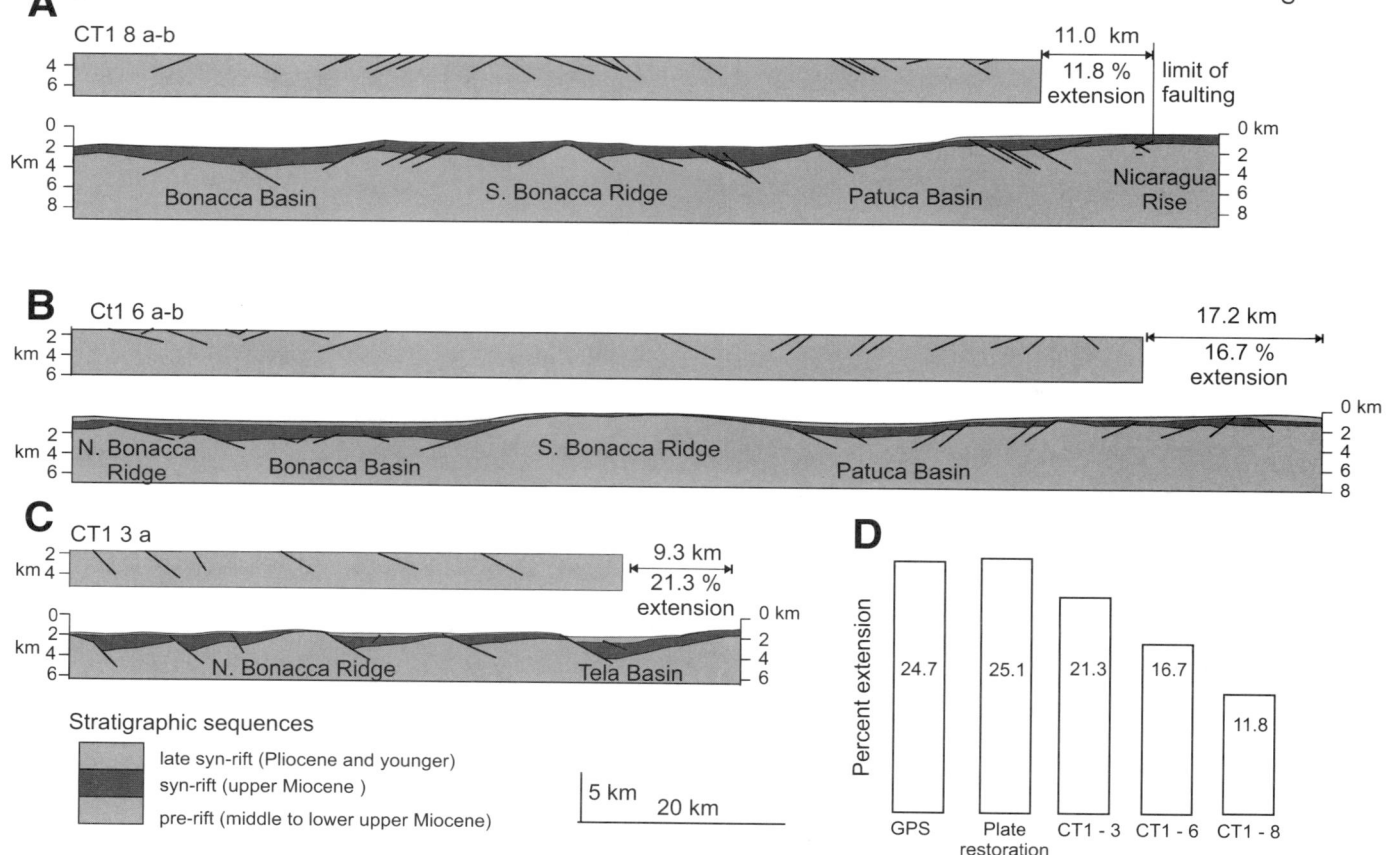

Figure 19. Interpretation of depth converted and restored University of Texas Institute for Geophysics multi-channel seismic (MCS) lines across the Honduran borderlands (cf. Fig. 2B for locations of three seismic lines). Interval velocities derived from well velocities in Honduran offshore exploration wells whose logs are shown in Figure 17 were used to convert time sections to depth sections (average velocity of pre-rift sequence 1:3.3 km/sec; syn-rift sequence 2:3.0 km/sec; late syn-rift sequence 3:2.2 km/sec). (A) Restoration of CT1-8ab MCS line restored to top of pre-rift sequence 1 yields 17.0 km or 18.3% extension compared to unrestored depth converted line. (B) Restoration of CT1-6ab MCS line restored to top of pre-rift sequence 1 yields 17.2 km or 16.7% extension compared to unrestored depth converted line. (C) Restoration of CT1-3ab MCS line restored to top of pre-rift sequence 1 yields 9.3 km or 21.3% extension compared to unrestored depth converted line. (D) Comparison of percent extension predicted from boundary normal slip (24.7%) known from present-day GPS measurements of the Caribbean plate (DeMets et al., 2000) with percent extension predicted from extrapolating the present-day GPS rate for the period of known rifting (25.1%) and with percent extension measured on the three seismic lines.

A third estimate of extension across the borderland is a byproduct of plate reconstruction that utilizes variable rates of Caribbean plate motion during the past 12 m.y. rather than extrapolation of the instantaneous GPS-derived plate motion vector. The extension calculated from the plate reconstruction process shows 25.1% extension across the Honduran borderlands (Fig. 19D), which is consistent with the other estimates of extension.

We make two estimates of extension across the 340-km-wide zone of north-trending rifts of Honduras and Guatemala (centered on longitude 89°W) using (1) the instantaneous GPS-derived plate motion vector and (2) the cumulative plate restoration process. The GPS-based method resulted in 45.0 km or 15.3% extension for the past 12 m.y. This is in close agreement with the extension estimate obtained through the plate restoration process, which yields an extensional amount of 47.8 km, or 16.3% extension.

Plate Reconstructions of Transtensional Environments in the Northwestern Caribbean

In order to address the question of how the features of the northwest corner of the Caribbean plate have evolved, we utilize quantitative plate restorations. In Figure 20, we present six quantitative plate reconstructions made to illustrate the complex evolution of this transtensional plate margin segment for the past 20 m.y. (early Miocene). The position of the North America plate is

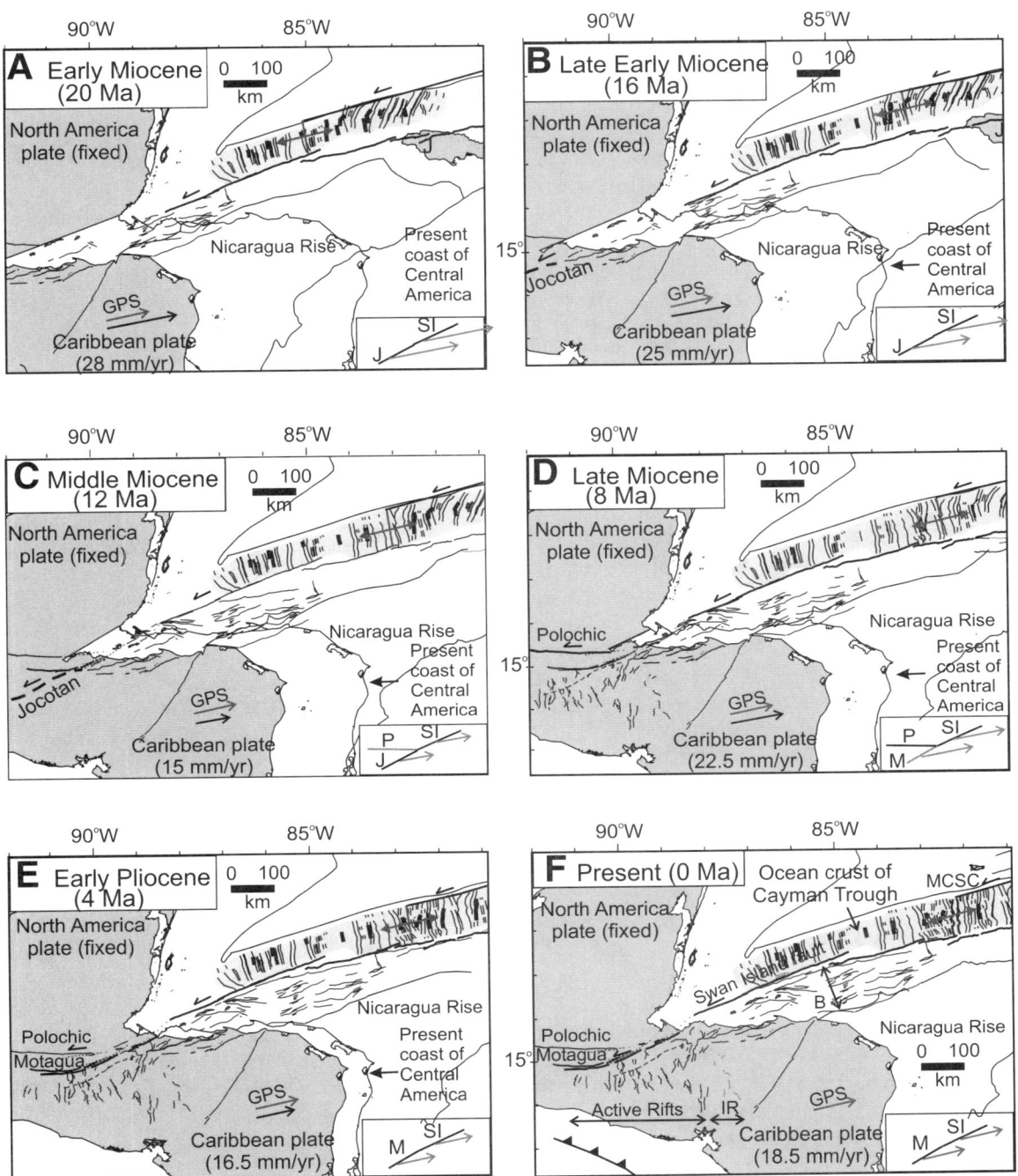

Figure 20. Plate tectonic reconstructions of northern Central America, the Cayman trough, and the Honduran borderlands with the North America plate held fixed. Plate boundary faults (P—Polochic fault; M—Motagua fault; SI—Swan Islands fault) are held fixed to the North America plate with the exception of the Jocatan (J) fault. Rotations are based on interpretations of ages of Cayman trough magnetic anomalies formed at Mid-Cayman spreading center (MCSC) by Rosencrantz (1994). Arrows show the direction and rate of the GPS present-day motion of the Caribbean plate from DeMets et al. (2000) compared to the predicted rate of past Caribbean plate motion from the Rosencrantz (1994) anomaly interpretation. Inset box summarizes the direction of the Rosencrantz (1994) direction of Caribbean plate motion relative to the plate boundary faults system inferred to have been active during that time period based on previous geologic studies summarized by Burkart (1994) and Donnelly et al. (1990). (A) Early Miocene (20 Ma): The Jocotan–Swan Islands fault system is active. (B) Late early Miocene (16 Ma): The Jocotan–Swan Islands fault system is active. (C) Middle Miocene (12 Ma): The Jocotan–Swan Islands fault system is active. (D) Late Miocene (8 Ma): The Polochic–Swan Islands fault system is active. (E) Early Pliocene (4 Ma): The Polochic–Swan Islands fault system is active. (F) Present-day (0 Ma): The Motagua–Swan Islands fault system is active. Note that the present-day coast of Central America is shown to provide an indication of cumulative lateral offset between the two plates. IR—inactive rifts.

fixed and the present-day coastline of Central America is shown in all the reconstructions as a frame of reference. Two vectors are shown on each reconstruction: (1) the current GPS-derived Caribbean plate motion vector (DeMets et al., 2000), and (2) the plate vector derived from magnetic anomalies in the surrounding oceanic basins, including the Cayman trough (Rosencrantz, 1994). Reconstruction results in a 20 m.y. average Caribbean plate vector of 21.4 mm/yr at azimuth N78E relative to North America. That is slightly greater than the 18.5 mm/yr at azimuth N77E GPS measured rate at 15°N, 85°W. A small inset map at the bottom of each reconstruction summarizes the major faults that are known from geologic studies to be active during that time interval along with the angle between these faults and the magnetic anomaly–derived plate vector inferred for that time period.

In the late early Miocene (16 Ma) and early Miocene (20 Ma) the Jocotan fault was the main plate boundary fault (Ritchie, 1976) (Figs. 20A and 20B). Our reconstruction suggests that the Jocotan fault formed a much straighter continuation of the Swan Islands fault than either the Polochic or Motagua would at a later time. This alignment places the Chamelecon and the La Ceiba fault in a position to be part of the plate boundary fault. No evidence for the north-trending rifts of the western area is known from this time. However, it is possible that borderland-style deformation "rifts parallel to margin" may have extended south of the plate margin because the geometry between the bounding fault and the plate motion vector is similar to that of today.

In the middle Miocene (12 Ma), the Jocotan fault was still the main plate boundary fault (Ritchie, 1976) (Fig. 20C). A major middle Miocene unconformity occurs in the Honduran borderland between deformed basement to Eocene rocks and overlying middle Miocene shallow-water clastic sedimentary rocks (Fig. 18). Subsidence in the Honduran borderland region is evidenced by deposition of the pre-rift sequence and may indicate the onset of transtension in this region.

By the late Miocene (8 Ma), the plate boundary shifted to the north and the Polochic fault was active (Burkart, 1983) (Fig. 20D). The north-south–trending rifting in western Honduras commenced shortly before this prior time (after 10.5 Ma according to Gordon and Muehlberger, 1994). We suggest that north-south rifts of the western area were initiated by the plate boundary moving from the Jocotan to the Polochic fault, thereby increasing the divergence between the Caribbean plate motion vector and the plate margin fault. According to the offshore wells, the late Miocene is the when the rift basins of the Honduran borderlands were filling with syn-rift sediments (Fig. 12).

During the early Pliocene (4 Ma), the plate margin fault geometry (Fig. 20E) remained the same as in the late Miocene. In the Honduran borderlands, well and seismic data indicates that the rift phase that began in the late Miocene is ongoing. According to Burkart (1983), the plate boundary shifted from the Polochic fault to the Motagua fault shortly after this time. This switch to the arcuate Motagua fault increased the angle of divergence and amount of strain partitioning in the area of north-south rifting.

At present, the main left-lateral faults of the plate boundary zone are the Motagua (Plafker, 1976) and Swan Islands fault zone (Rosencrantz and Mann, 1991) (Fig. 20F). The Polochic fault in Guatemala, although active, is no longer the plate boundary. Rifting of the Honduran borderlands is decreasing (seismic sequence 3 in Fig. 12). This decrease in rifting may be related to the eastward movement of this area into an area of lesser plate divergence. A zone of inactive, north-trending rifts exists in east-central Honduras (IR in Fig. 20F). These rifts were generated west of their present position and rafted eastward with the Caribbean plate.

CONCLUSIONS

The main conclusions of this study are as follows:
1. Geologic, earthquake, marine geophysical, and remote sensing data from on- and offshore areas show that Neogene to Recent transtensional deformation along the North America–Caribbean plate boundary in northern Central America exhibits two distinct structural styles. In the western area, east-west extension affects a 375-km-long plate boundary segment with basin and range morphology in western Honduras and northern Guatemala. In the eastern study area, NNW-SSE transtension occurs along a 600-km-long plate boundary segment of margin-parallel ridges and basins in the Nombre de Dios range and Aguan Valley of northern Honduras and offshore Honduran borderlands region (Fig. 2).
2. Comparison of fault orientations and the trend of the GPS-derived Caribbean plate motion vector show that east-west extension along faults normal to the plate boundary occurs in the western area when the angle of divergence between the Motagua plate boundary fault and the plate motion vector is ≥10° (Fig. 18). This oblique opening model differs from previous interpretations of Burkart and Self (1985) that invoke block rotations about the arcuate Motagua fault zone.
3. In northern Honduras and its borderland, NNW-SSE extension along faults subparallel to the plate margin coincide with an angle between the Swan Islands plate bounding fault and the plate motion vector of between 5 and 10°.
4. A narrow, 35-km-wide tectonic transition area in north-central Honduras separates the north-trending rifts of western Honduras from the plate boundary–parallel rifts in northeastern Honduras and the offshore Honduran borderlands (Fig. 2).
5. Faults of the offshore Honduran borderlands extend onshore into the Nombre de Dios range and Aguan Valley of northern Honduras, where subaerial transtensional deformation is similar to deformation of the submarine Honduran borderlands (Fig. 5). Tectonic geomorphology studies show pervasive oblique-slip faulting with evidence for late Quaternary left-lateral offsets (Fig. 7) and active uplift of stream networks (Fig. 6).

6. Seismic data tied to wells in the Honduran borderlands (Fig. 18) show that plate boundary–related submarine faults in this region are active, transtensional features that initiated in the middle Miocene with filling of asymmetric half-grabens and continued through the Pliocene-Pleistocene.

7. Quantitative plate reconstructions suggest that the north-trending rifts of the western region developed in response to increased interplate divergence as the plate boundary shifted from the Jocotan fault to the Polochic fault prior to 8 Ma.

ACKNOWLEDGMENTS

Funding for R. Rogers and P. Mann was provided by the Petroleum Research Fund of the American Chemical Society (grant 33935-AC2 to P. Mann). We thank Dirección de Energia of Honduras and Japanese Petroleum Exploration Co. Geoscience Institute for releasing data for this study. P. Emmet, M. Gordon, and C. DeMets provided comments on an earlier version of the manuscript. We thank J. Marshall, T. Wawrzyniec, and M. Guzman-Speziale for their reviews. Lisa Gahagan at Institute for Geophysics provided plate reconstructions. This is University of Texas Institute for Geophysics contribution no. 1869.

REFERENCES CITED

Avé Lallemant, H., and Gordon, M., 1999, Deformation history of Roatan Island: Implications for the origin of the Tela basin (Honduras), in Mann, P., ed., Caribbean basins, Sedimentary basins of the world series: Amsterdam, Elsevier, v. 4, p. 197–218.

Ben-Avraham, Z., 1992, Development of asymmetric basins along continental transform margins: Tectonophysics, v. 215, p. 209–220, doi: 10.1016/0040-1951(92)90082-H.

Ben-Avraham, Z., and Zoback, M., 1992, Transform-normal extension and asymmetric basins: An alternative to pull-apart models: Geology, v. 20, p. 423–426, doi: 10.1130/0091-7613(1992)020<0423:TNEAAB>2.3.CO;2.

Burkart, B., 1983, Neogene North America–Caribbean plate boundary across northern Central America: Offset along the Polochic fault: Tectonophysics, v. 99, p. 251–270, doi: 10.1016/0040-1951(83)90107-5.

Burkart, B., 1994, Northern Central America, in Donovan, S., and Jackson, T., eds., Caribbean geology: An introduction: Jamaica, University of the West Indies Publisher's Association, p. 265–284.

Burkart, B., and Self, S., 1985, Extension and rotation of crustal blocks in northern Central America and effect on the volcanic arc: Geology, v. 13, p. 22–26, doi: 10.1130/0091-7613(1985)13<22:EAROCB>2.0.CO;2.

Cáceres, D., Monterroso, D., and Tavakoli, B., 2005, Crustal deformation in northern Central America: Tectonophysics, v. 404, p. 119–131, doi: 10.1016/j.tecto.2005.05.008.

Calais, E., and Mercier de Lépinay, B., 1993, Semiquantitative modeling of strain and kinematics along the Caribbean–North America strike-slip plate boundary zone: Journal of Geophysical Research, v. 98, p. 8293–8308.

Calais, E., Mazabraud, Y., Mercier de Lépinay, B., Mann, P., Jansma, P., and Mattioli, G., 2002, Oblique collision and strain partitioning from GPS measurements in the northeastern Caribbean: Geophysical Research Letters, v. 29, no. 18, p. 3-1–3-4.

Case, J., and Holcombe, T., 1980, Geologic-tectonic map of the Caribbean region: U.S. Geological Survey Miscellaneous Investigations Series I-1100, scale 1:2,500,000.

Claypool, A., Klepeis, K., Dockrill, B., Clarke, G., Zwingmann, H., and Tulloch, A., 2002, Structure and kinematics of oblique continental convergence in northern Fiordland, New Zealand: Tectonophysics, v. 359, p. 329–358, doi: 10.1016/S0040-1951(02)00532-2.

Deng, J., and Sykes, L., 1995, Determination of Euler pole for contemporary relative motion of Caribbean and North American plates using slip vectors of interplate earthquakes: Tectonics, v. 14, p. 39–53, doi: 10.1029/94TC02547.

DeMets, C., Jansma, P., Mattioli, G., Dixon, T., Farina, F., Bilham, R., Calais, E., and Mann, P., 2000, GPS geodetic constraints on Caribbean-North American plate motion: Geophysical Research Letters, v. 27, p. 437–440, doi: 10.1029/1999GL005436.

DeMets, C., Mattioli, G., Jansma, P., Rogers, R., Tenorio, C., and Turner, H.L., 2007, this volume, Present motion and deformation of the Caribbean plate: Constraints from new GPS geodetic measurements from Honduras and Nicaragua, in Mann, P., ed., Geologic and Tectonic Development of the Caribbean Plate Boundary in Northern Central America: Geological Society of America Special Paper 428, doi: 10.1130/2007.2428(02).

Donnelly, T., Horne, G., Finch, R., and López-Ramos, E., 1990, Northern Central America: The Maya and Chortis blocks, in Dengo, G., and Case, J., eds., The Caribbean region: Geological Society of America, Geology of North America, v. H, p. 37–76.

Dupré, W.R., 1970, Geology of the Zambrano Quadrangle, Honduras, Central America [M.A. thesis]: Austin, Texas, University of Texas at Austin, 128 p.

ENEE (Empressa Nacional de Energia Electrica), 1987, Geologic investigations of proposed hydrothermal sites, Honduras: Tegucigalpa, Honduras, Honduras Electrical Company unpublished report, 24 p.

Everett, J., 1970, Geology of the Comayagua Quadrangle, Honduras, Central America [Ph.D. dissertation]: Austin, Texas, University of Texas at Austin, 152 p.

Gardner, T., Back, W., Bullard, T., Hare, P., Kesel, R., Lowe, D., Menges, C., Mora, S., Pazzaglia, F., Sasowsky, I., Troester, J., and Wells, S., 1987, Central America and the Caribbean, in Graf, W., ed., Geomorphic systems of North America: Boulder, Colorado, Geological Society of America Centennial Special Volume, v. 2, p. 343–402.

Gordon, M., 1994, Evolution of Neogene microtectonic phases on the Chortís block (Northern Central America): Geological Society of America Abstracts with Programs, v. 26, no. 7, p. A-209.

Gordon, M., and Muehlberger, W., 1994, Rotation of the Chortís block causes dextral slip on the Guayape fault: Tectonics, v. 13, p. 858–872, doi: 10.1029/94TC00923.

Gorsline, D.S., and Teng, L.S.-Y., 1989, The California continental borderland, in Winterer, E.L., Hussong, D.M., and Decker, R.W., eds., The Eastern Pacific Ocean and Hawaii: Boulder, Colorado, Geological Society of America, Geology of North America, v. N, p. 471–487.

Guzman-Speziale, M., 2001, Active seismic deformation in the grabens of northern Central America and its relationship to the relative motion of the North America–Caribbean plate boundary: Tectonophysics, v. 337, p. 39–51, doi: 10.1016/S0040-1951(01)00110-X.

Hack, J., 1957, Studies of longitudinal stream profiles in Virginia and Maryland: U.S. Geological Survey Professional Paper 294B, p. 45–97.

Helbig, K., 1959, Die landschaften von Nordost—Honduras: Petermanns Geographischen Mitteilungen, Veb Hermann Haack, Geographisch-Kartographische Anstalt, Gotha, Ergänzungsheft Nr. 268, 270 p.

Hovius, N., 2000, Macroscale process systems of mountain belt erosion, in Summerfield, M., ed., Geomorphology and global tectonics: Chichester, UK, John Wiley & Sons, p. 77–105.

Jansma, P., Lopez, A., Mattioli, G., DeMets, C., Dixon, T., Mann, P., and Calais, E., 2000, Neotectonics of Puerto Rico and the Virgin Islands, northeastern Caribbean, from GPS geodesy: Tectonics, v. 19, p. 1021–1037, doi: 10.1029/1999TC001170.

Jones, R., and Tanner, P., 1995, Strain partitioning in transpression zones: Journal of Structural Geology, v. 17, p. 793–802, doi: 10.1016/0191-8141(94)00102-6.

Jordan, B.R., Sigurdsson, H., Carey, S., Lundin, S., Rogers, R., and Barquero-Molina, M., 2007, this volume, Petrogenesis of Central American Tertiary ignimbrites and associated Caribbean Sea tephra, in Mann, P., ed., Geologic and tectonic development of the Caribbean margin boundary in northern Central America: Geological Society of America Special Paper 428, doi: 10.1130/2007.2428(07).

Kanamori, H., and Stewart, G., 1978, Seismological aspects of the Guatemala earthquake of February 4, 1976: Journal of Geophysical Research, v. 83, p. 3427–3434.

King, A., 1972, Mapa Geológica de Honduras, Hoja de Talanga (Geologic Map of Honduras, Talanga sheet): Tegucigalpa, Honduras, Instituto Geográfico Nacional, 1 sheet, scale 1:50,000.

King, A., 1973, Mapa Geológica de Honduras, Hoja de Cedros (Geologic Map of Honduras, Cedros sheet): Tegucigalpa, Honduras, Instituto Geográfico Nacional, scale, 1:50,000.

Kornicker, L., and Bryant, W., 1969, Sedimentation on continental shelf of Guatemala and Honduras, Tectonic relations of northern Central America and the western Caribbean—The Bonacca Expedition: American Association of Petroleum Geologists Memoir 11, p. 244–257.

Kozuch, M., 1991, Mapa Geológica de Honduras: Tegucigalpa, Honduras, Instituto Geográfico Nacional, 3 sheets, scale 1:500,000.

Langer, C., and Bollinger, G., 1979, Secondary faulting near the terminus of a seismogenic strike-slip fault; aftershocks of the 1976 Guatemala earthquake: Bulletin of the Seismological Society of America, v. 69, p. 427–444.

Leeder, M., and Alexander, J., 1987, The origin and tectonic significance of asymmetrical meander-belts: Sedimentology, v. 34, p. 217–226, doi: 10.1111/j.1365-3091.1987.tb00772.x.

Leroy, S., Mauffret, A., Patriat, P., and Mercier de Lepinay, B., 2000, An alternative interpretation of the Cayman Trough evolution from a reidentification of magnetic anomalies: Geophysical Journal International, v. 141, p. 539–557, doi: 10.1046/j.1365-246x.2000.00059.x.

Mann, P., and Burke, K., 1984, Neotectonics of the Caribbean: Reviews of Geophysics and Space Physics, v. 22, p. 309–362.

Mann, P., Tyburski, S., and Rosencrantz, E., 1991, Neogene development of the Swan Islands restraining bend complex, Caribbean Sea: Geology, v. 19, p. 823–826, doi: 10.1130/0091-7613(1991)019<0823:NDOTSI>2.3.CO;2.

Mann, P., Calais, E., Ruegg, J., DeMets, C., Dixon, T., Jansma, P., and Mattioli, G., 2002, Oblique collision in the northeastern Caribbean from GPS measurements and geological observations: Tectonics, v. 21, 1057, doi: 10.1029/2001TC001304.

Manton, W., 1987, Tectonic interpretation of the morphology of Honduras: Tectonics, v. 6, p. 633–651.

Manton, W., and Manton, R., 1999, The southern flank of the Tela basin, Republic of Honduras, in Mann, P., ed., Caribbean basins: Sedimentary basins of the world series: Amsterdam, Elsevier, v. 4, p. 219–236.

Markey, R., 1995, Mapa Geológica de Honduras, Hoja de Morocelli (Geologic Map of Honduras, Morocelli sheet): Tegucigalpa, Honduras, Instituto Geográfico Nacional, 1 sheet, scale 1:50,000.

Marshall, J., Fisher, D., and Gardner, T., 2000, Central Costa Rica deformed belt; kinematics of diffuse faulting across the western Panama Block: Tectonics, v. 19, p. 468–492, doi: 10.1029/1999TC001136.

Meckel, T., Coffin, M., Mosher, S., Symonds, P., Bernard, G., and Mann, P., 2003, Underthrusting at the Hjort Trench, Australian-Pacific plate boundary: Incipient subduction? Geochemistry, Geophysics, Geosystems, v. 4, no. 12, 1099, doi: 10.1029/2002GC000498.

Molnar, P., and Sykes, L., 1969, Tectonics of the Caribbean and Middle America regions from focal mechanisms and seismicity: Geological Society of America Bulletin, v. 80, p. 1639–1684.

Muehlberger, W., 1976, The Honduras Depression: Publicaciones Geológicas del ICAITI (Guatemala): Issue 5, p. 43–51.

Pinet, P., 1971, Structural configuration of the northwestern Caribbean plate boundary: Geological Society of America Bulletin, v. 82, p. 2027–2032.

Pinet, P., 1972, Diapirlike features offshore Honduras: Implications regarding tectonic evolution of Cayman Trough and Central America: Geological Society of America Bulletin, v. 83, p. 1911–1922.

Pinet, P., 1975, Structural evolution of the Honduras continental margin and the sea floor south of the western Cayman Trough: Geological Society of America Bulletin, v. 86, p. 830–838, doi: 10.1130/0016-7606(1975)86<830:SEOTHC>2.0.CO;2.

Pinet, P., 1976, Morphology off northern Honduras, northwest Caribbean Sea: Deep Sea Research, v. 23, p. 839–847.

Plafker, G., 1976, Tectonic aspects of the Guatemala earthquake of 4 February 1976: Science, v. 193, p. 1201–1208, doi: 10.1126/science.193.4259.1201.

Ritchie, A., 1976, Jocotan Fault; possible western extension: Publicaciones Geológicas del ICAITI (Guatemala): Issue 5, p. 52–55.

Rogers, R., 1995, Mapa Geológica de Honduras, Hoja de Valle de Jamastran (Geologic Map of Honduras, Valle de Jamastran sheet): Tegucigalpa, Honduras, Instituto Geográfico Nacional, 1 sheet, scale 1:50,000.

Rogers, R., and O'Conner, E., 1993, Mapa Geológica de Honduras: Hoja de Tegucigalpa (segunda edición): Tegucigalpa, Honduras, Instituto Geográfico Nacional, 1 sheet, scale 1:50,000.

Rogers, R., Kárason, H., and van der Hilst, R., 2002, Epeirogenic uplift above a detached slab in northern Central America: Geology, v. 30, p. 1031–1034, doi: 10.1130/0091-7613(2002)030<1031:EUAADS>2.0.CO;2.

Rogers, R.D., Mann, P., Emmet, P.A., and Venable, M.A., 2007, this volume, Colon fold belt of Honduras: Evidence for late Cretaceous collision between the continental Chortis block and intraoceanic Caribbean arc, in Mann, P., ed., Geologic and tectonic development of the Caribbean margin boundary in northern Central America: Geological Society of America Special Paper 428, doi: 10.1130/2007.2428(06).

Rogers, R.D., Mann, P., Scott, R., and Patino, L., 2007, this volume, Cretaceous intra-arc rifting, sedimentation and basin inversion in east-central Honduras, in Mann, P., ed., Geologic and tectonic development of the Caribbean margin boundary in northern Central America: Geological Society of America Special Paper 428, doi: 10.1130/2007.2428(05).

Rosencrantz, E., 1994, Opening of the Cayman Trough and the evolution of the northern Caribbean Plate boundary: Geological Society of America Abstracts with Programs, v. 27, no. 7, p. A-59.

Rosencrantz, E., and Mann, P., 1991, SeaMarc II mapping of transform faults in the Cayman Trough, Caribbean Sea: Geology, v. 19, p. 690–693, doi: 10.1130/0091-7613(1991)019<0690:SIMOTF>2.3.CO;2.

Rowan, M., and Kligfield, R., 1989, Cross-section restoration and balancing as aid to seismic interpretation in extensional terranes: AAPG Bulletin, v. 73, p. 955–966.

Sandwell, D., and Smith, W., 1997, Marine gravity anomaly from Geosat and ERS-1 satellite altimetry: Journal of Geophysical Research, v. 102, p. 10,039–10,054, doi: 10.1029/96JB03223.

Schwartz, D., Cluff, L., and Donnelly, T., 1979, Quaternary faulting along the Caribbean–North American plate boundary in Central America: Tectonophysics, v. 52, p. 431–445, doi: 10.1016/0040-1951(79)90258-0.

Sigurdsson, H., Kelley, S., Leckie, R., Carey, S., Bralower, T., and King, J., 2000, History of circum-Caribbean explosive volcanism: ^{40}Ar/^{39}Ar dating of tephra layers, in Leckie, R., and Sigurdsson, H., et al., eds., Proceedings of the Ocean Drilling Program, Scientific Results, Leg 165: College Station, Texas, Ocean Drilling Program, v. 165, p. 299–314.

Smith, W., and Sandwell, D., 1997, Global seafloor topography from satellite altimetry and ship depth soundings: Science, v. 277, p. 1957–1962.

Teyssier, C., Tykoff, B., and Markley, M., 1995, Oblique plate motion and continental tectonics: Geology, v. 23, p. 447–450, doi: 10.1130/0091-7613(1995)023<0447:OPMACT>2.3.CO;2.

Van Dusen, S., and Doser, D., 2000, Faulting processes of historic (1917–1962) M ≥ 6.0 earthquakes along the north-central Caribbean Margin: Pure and Applied Geophysics, v. 157, p. 719–736, doi: 10.1007/PL00001115.

Wadge, G., and Wooden, J.L., 1982, Late Cenozoic alkaline volcanism in the northwestern Caribbean: Tectonic setting and Sr isotopic characteristics: Earth and Planetary Science Letters, v. 57, p. 35–46, doi: 10.1016/0012-821X(82)90171-6.

Webb, D.S., and Perrigo, S.C., 1984, Late Cenozoic vertebrates from Honduras and El Salvador: Journal of Vertebrate Paleontology, v. 4, p. 237–254.

MANUSCRIPT ACCEPTED BY THE SOCIETY 22 DECEMBER 2006

Tectonic terranes of the Chortis block based on integration of regional aeromagnetic and geologic data

Robert D. Rogers*
Paul Mann
Institute for Geophysics, Jackson School of Geosciences, University of Texas at Austin, J.J. Pickle Research Campus, Bldg. 196 (ROC), 10100 Burnet Road (R2200), Austin, Texas 78758-4445, USA

Peter A. Emmet
Cy-Fair College, Fairbanks Center, 14955 Northwest Freeway, Houston, Texas 77904, USA

ABSTRACT

An aeromagnetic survey of Honduras and its northeastern Caribbean coastal area covering a continuous area of 137,400 km^2 was acquired by the Honduran government in 1985 and provided to the University of Texas at Austin for research purposes in 2002. We correlate regional and continuous aeromagnetic features with a compilation of geologic data to reveal the extent, structural grain, and inferred boundaries of tectonic terranes that compose the remote and understudied, Precambrian-Paleozoic continental Chortis block of Honduras. A regional geologic map and a compilation of isotopic age dates and lead isotope data are used in conjunction with and geo-referenced to the aeromagnetic map. These combined data provide a basis for subdividing the 531,370 km^2 Chortis block into three tectonic terranes with distinctive aeromagnetic expression, lithologies, structural styles, metamorphic grade, isotopically and paleontologically determined ages, and lead isotope values: (1) The Central Chortis terrane occupies an area of 110,600 km^2, exhibits a belt of roughly east-west–trending high magnetic values, and exposes small, discontinuous outcrops of Grenville to Paleozoic continental metamorphic rocks including greenschist to amphibolite grade phyllite, schist, gneiss, and orthogneiss that have been previously dated in the range of 1 Ga to 222 Ma; the northern 59,990 km^2 margin of the Central Chortis terrane along the northern Caribbean coast of Honduras exhibits an irregular pattern of east-west–trending magnetic highs and lows that correlates with an east-west–trending belt of early Paleozoic to Tertiary age metamorphic rocks intruded by Late Cretaceous and early Cenozoic plutons in the range of 93.3–28.9 Ma. (2) The Eastern Chortis terrane occupies an area of 185,560 km^2, exhibits belts of roughly northeast-trending high magnetic values, and correlates with outcrops of folded and thrusted Jurassic metasedimentary phyllites and schists forming a greenschist-grade basement; we propose that the Eastern and Central terranes are distinct terranes based on the strong differences in

---

*Present address: Department of Geology, California State University, Stanislaus, 801 W. Monte Vista Dr., Turlock, California 95832, USA; rrogers@geology.csustan.edu.

Rogers, R.D., Mann, P., and Emmet, P.A., 2007, Tectonic terranes of the Chortis block based on integration of regional aeromagnetic and geologic data, *in* Mann, P., ed., Geologic and tectonic development of the Caribbean plate boundary in northern Central America: Geological Society of America Special Paper 428, p. 65–88, doi: 10.1130/2007.2428(04). For permission to copy, contact editing@geosociety.org. ©2007 The Geological Society of America. All rights reserved.

their structural style and aeromagnetic grain, sedimentary thickness, metamorphic grade, and lead isotope values. (3) The Southern Chortis terrane occupies an area of 120,100 km², contains one known basement outcrop of metaigneous rock, exhibits a uniformly low magnetic intensity that contrasts with the rest of the Chortis block, and is associated with an extensive area of Miocene pyroclastic strata deposited adjacent to the late Cenozoic Central American volcanic arc. The outlines of the terranes as constrained by the aeromagnetic, lithologic, age, and lead isotope data are restored to their pre–early Eocene position along the southwestern coast of Mexico by a 40° clockwise rotation and 1100 km of documented post–early Eocene (ca. 43 Ma) left-lateral offset along the strike-slip faults of the northern Caribbean strike-slip plate boundary. The inner continental and outboard oceanic terranes of Chortis and the 120,100 km² Siuna terrane to the south trend roughly north-south and align with terranes of similar magnetic trend, lithology, age, and crustal character in southwestern Mexico. Additional progress in mapping and isotopic dating is needed for the proposed Chortis terranes in Honduras in order to constrain this proposed position against much better mapped and dated rocks in southwestern Mexico.

Keywords: tectonic terranes, Chortis block, Caribbean plate, tectonics, regional geology.

INTRODUCTION

Tectonic Significance

The Chortis block of northern Central America (Honduras, Nicaragua, El Salvador, southern Guatemala, and part of the Nicaragua Rise) forms the only emergent area of Precambrian to Paleozoic continental crust on the present-day Caribbean plate (DeMets et al., this volume) and has long been recognized as an important constraint on the tectonic origin and Cretaceous-Cenozoic displacement history of the Caribbean plate (Gose and Swartz, 1977; Pindell and Dewey, 1982; Case et al., 1984; Dengo, 1985; Case et al., 1990). (Fig. 1A). Largely due to the paucity of information from the Chortis block, three different tectonic models have been proposed to explain the tectonic origin of the Chortis block and its geologic relationship to Precambrian to Paleozoic rocks of similar lithologies and metamorphic grades from southwestern Mexico (Campa and Coney, 1983; Sedlock et al., 1993; Dickinson and Lawton, 2001; Solari et al., 2003; Keppie, 2004; Keppie and Morán-Zenteno, 2005; Talavera-Mendoza et al., 2005) and northern Guatemala (Ortega-Gutierrez et al., 2004).

The paucity of geologic and age isotopic information from Honduras that encompasses most of the Chortis block, is limited to a few studies, including Fakundiny (1970), Horne et al. (1976a), Simonson (1977), Sundblad et al. (1991), and Manton (1996), and can be attributed to the remoteness and inaccessibility of many parts of the country to geologic research (Fig. 1B). Because of its mountainous terrain, sparse population (62 people/km²), and poor road network, only ~15% of the geology of Honduras has been systematically mapped by field geologists at a scale of 1:50,000 (Fig. 2). The remaining 85% of the country remains either completely unmapped or subject to reconnaissance mapping and spot sampling usually confined to major roads (e.g., Mills et al., 1967; Finch, 1981; Manton and Manton, 1984, 1999).

Previous Tectonic Models for the Origin of the Chortis Block

Mexico-Derived Model for Chortis

The first model to explain the tectonic history of the Chortis block invokes large-scale strike-slip motion on the Cayman-Motagua-Polochic strike-slip fault system that presently forms the northern edge of the Chortis block (Figs. 1A and 1B). In this model, the Chortis block is detached from its pre–middle Eocene position along the southwestern coast of Mexico and moved eastward by ~1100 km of left-lateral strike-slip motion and ~30–40° of large-scale, counterclockwise rotation (Gose and Swartz, 1977; Karig et al., 1978; Pindell and Dewey, 1982; Gose, 1985; Rosencrantz et al., 1988; Pindell and Barrett, 1990; Sedlock et al., 1993; Dickinson and Lawton, 2001; Pindell et al., 2006) (Figs. 1A and 1B). Previous onland and offshore mapping and isotopic age dating studies have supported the presence of a linear, strike-slip–truncated margin along the southwestern coast of Mexico that became active in the early Cenozoic and terminated volcanic arc activity as the Cayman-Motagua-Polochic strike-slip fault lengthened eastward (Karig et al., 1978; Riller et al., 1992; Herrmann et al., 1994, Tardy et al., 1994, Schaaf et al., 1995). The 1100-km-wide Cayman trough pull-apart basin, bounded by strike-slip faults of the northern Caribbean plate boundary, opened as a result of large-scale, eastward strike-slip motion of the Caribbean plate (Rosencrantz et al., 1988; Leroy et al., 2000; Mann, 1999; Pindell et al., 2006) (Fig. 1). In this interpretation, the 400–600-km-wide Chortis block is an intermediate continental block separating the larger continental area of northern Central America (Maya block) from Pacific subduction zones (Karig et al., 1978; Gose, 1985; Schaaf et al., 1995) (Figs. 1A and 1B).

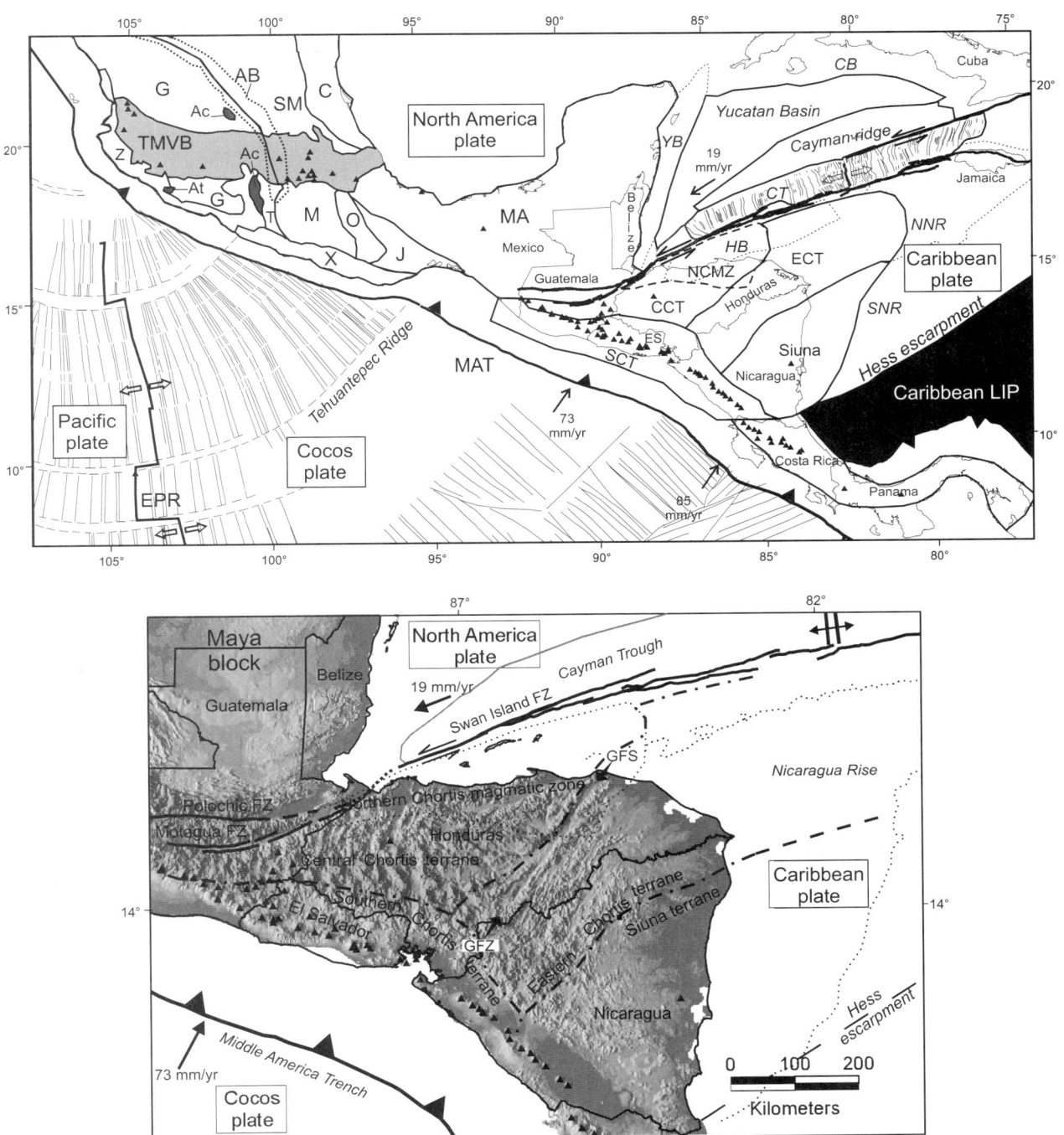

Figure 1. (A) Tectonic setting of northern Central America and southern Mexico showing the location and names of the Mexican terranes from previous workers and the Chortis terranes proposed in this paper. Chortis terrane abbreviations: CCT—Central Chortis terrane; ECT—Eastern Chortis terrane; SCT—Southern Chortis terrane; NCMZ—Northern Chortis metamorphic zone. Mexican terrane abbreviations: AB—Arperos basin, Mexico; Ac—Arcelia; At—Arteaga; C—Coahuila; G—Guerrero; J—Juarez; O—Oaxaca; M—Mixteca; MA—Maya block; SM—Sierra Madre; T—Teloloapan; X—Xolapa; and Z—Zihuatanejo. ES—El Salvador; TMVB—Trans-Mexican Volcanic Belt. Caribbean physiographic provinces shown as dashed lines and abbreviated as NNR—Northern Nicaraguan Rise; SNR—Southern Nicaraguan Rise; CB—Cuba; HB—Honduran borderlands; EPR—East Pacific Rise; MAT—Middle America trench; CT—Cayman trough; LIP—large igneous province; YB—Yucatan Basin. Triangles represent Quaternary volcanoes. Seafloor spreading anomalies shown for Pacific and Cocos plates and Cayman trough. Plate motions relative to a fixed Caribbean plate are from DeMets et al. (2000) and DeMets (2001). (B) Topographic map of northern Central America and the Chortis block showing physiographic expression of proposed terranes of the Chortis continental block. Note that structural grain and prominent faults like the Guayape fault system (GFS) are at right angles to the trend of the present-day Middle America trench and related volcanic arc (black triangles). Heavy dotted line offshore is 100 m bathymetric contour. FZ—fault zone.

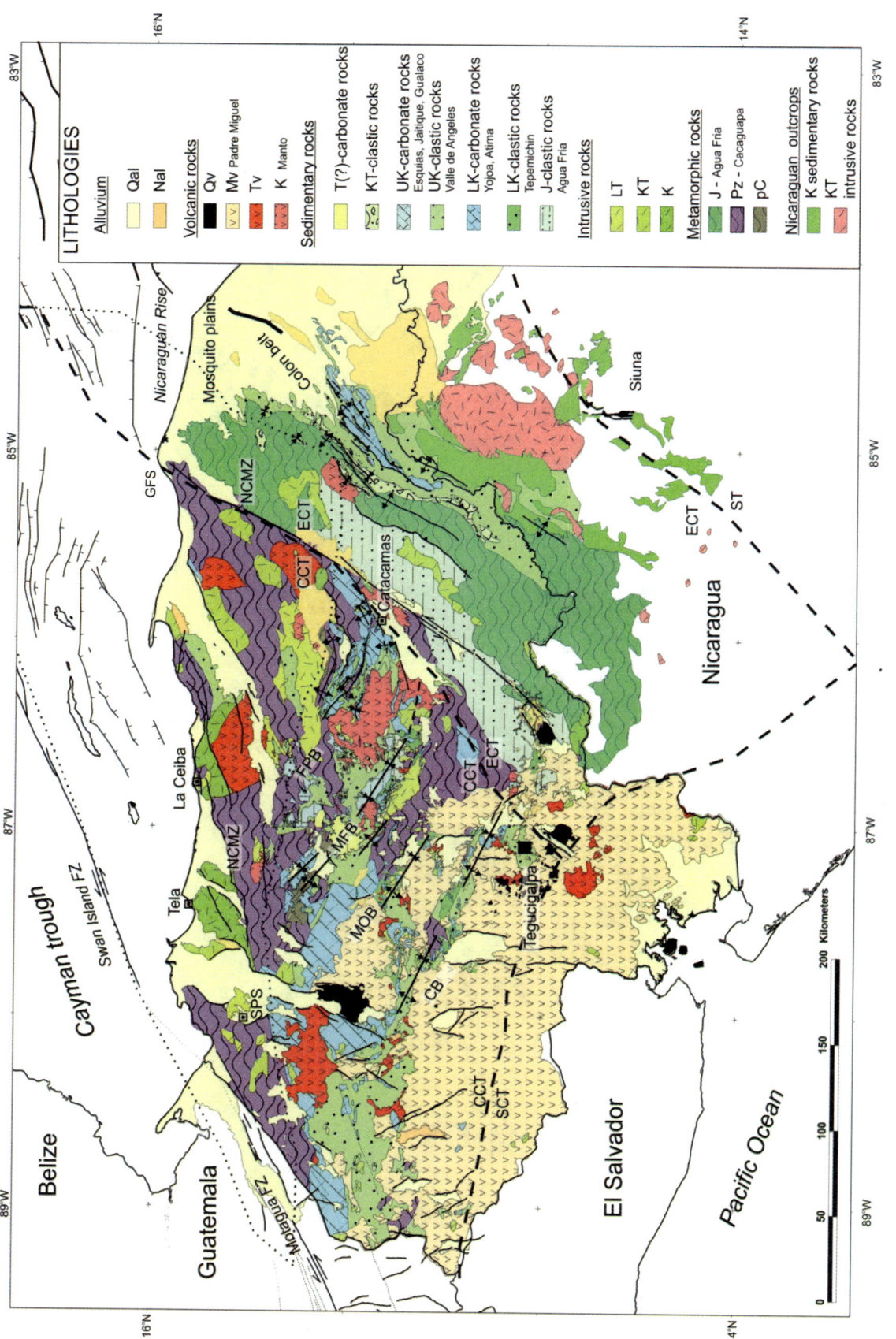

Figure 2. Geologic map of Honduras modified from national geologic compilation map by Kozuch (1991) combined with geologic map of northeastern Nicaragua modified from national geologic compilation map by the Instituto Nicaragüense de Estudios Territoriales (INETER, 1995). Dashed lines indicate terrane boundaries proposed in this paper using geologic and magnetic data. Paleozoic and Grenville-age basement occurs on the Central Chortis terrane. East of the Guayape fault system (GFS), mapping of the Eastern Chortis terrane reveals a basement composed of Jurassic Agua Fria strata and metasedimentary basement. Extensive Miocene ignimbrite deposits blanket much of western and southern Honduras and the boundary between the Central and Southern Chortis terranes. Lithologic abbreviations: Qal—Quaternary alluvium; Nal—Neogene alluvium; Qv—Quaternary volcanic rocks; Mv—Miocene volcanic rocks; Tv—Tertiary volcanic rocks; T—Tertiary; K—Cretaceous; KT—Cretaceous-Tertiary; UK—Upper Cretaceous; LK—Lower Cretaceous; J—Jurassic; lT—lower Tertiary; Pz—Paleozoic; and pC—Precambrian. CB—Comayagua belt; CCT—Central Chortis terrane; ECT—Eastern Chortis terrane; FPB—Frey Pedro belt; FZ—fault zone; MFB—Montaña La Flor belt; MOB—Minas de Oro belt; NCMZ—Northern Chortis magmatic zone; SCT—Southern Chortis terrane; SPS—San Pedro Sula; ST—Siuna terrane.

Pacific-Derived Model for Chortis

More recent mapping and isotopic dating based mainly in southern Mexico has led to a second, entirely different tectonic interpretation in which the Chortis block originates in the eastern Pacific Ocean and is ~700–800 km south of the truncated continental margin of southwestern Mexico during the Eocene (Keppie and Morán-Zenteno, 2005) (Fig. 1A). In this model, Chortis is far-traveled (1100 km) from 45 to 0 Ma and experiences ~40° of large-scale rotation in a clockwise sense. This sense of rotation is contrary to the overall, post-Cretaceous counterclockwise rotations measured by Gose and Swartz (1977) and Gose (1985) in rocks from Honduras.

In the Pacific-derived model, translation also occurs on the Cayman-Motagua-Polochic fault system, but the strike-slip fault is strongly concave southward about a distant pole of rotation in the southern hemisphere (cf. MacDonald, 1976, for a similar reconstruction). This fault geometry causes the Chortis block to travel in a northeastward arc from its original position far out in the eastern Pacific Ocean. In this model, a several hundred kilometer–wide zone of the southwestern margin of Mexico is removed by subduction erosion since the early Eocene. Continental rocks of southwestern Mexico and northern Central America are subject to the influence of subducting fracture zone ridges and other bathymetric highs that control the cessation and resumption of arc volcanism (Keppie and Morán-Zenteno, 2005).

Fixist Model for Chortis

A third type of model proposes that the Chortis block has occupied roughly its same position relative to southern Mexico and the rest of the Caribbean since Late Jurassic rifting between North and South America (Marton and Buffler, 1994; James, 2006) (Fig. 1A). This model is based on the interpretation of structural lineaments on the Chortis block and in surrounding areas that are inferred to be fossil rift and transform structures formed by the separation of North and South America during Late Jurassic time (James, 2006). The model also assumes that strike-slip displacement along the Motagua-Polochic-Cayman trough system (or between the Maya and Chortis blocks in Figs. 1A and 1B) is restricted to ~170 km of left-lateral offset.

OBJECTIVES OF THIS PAPER

In 2002, the Dirección General de Minas e Hidrocarburos of Honduras donated to the University of Texas at Austin the results of the national aeromagnetic survey of Honduras that was acquired by the Honduran government in 1985 as part of a countrywide program of exploration for minerals and petroleum (Dirección General de Minas e Hidrocarburos, 1985) (Fig. 3). These data provide an ideal tool for examining the distribution and lithologic-structural grain of the crystalline basement rocks of the sparsely mapped Chortis block because the source for most of the observed magnetic anomalies lies within the basement rocks, and the effects of the overlying sedimentary cover are essentially absent. The aeromagnetic data is particularly valuable for inferring the regional trends and continuity of understudied basement rocks of the heavily vegetated and mountainous areas that characterize most parts of Honduras (cf. Fakundiny, 1970; Manton, 1996) (Fig. 2). Other basement rocks are buried beneath late Cenozoic sedimentary rocks of the coastal plain of eastern Honduras (Mills and Barton, 1996), lie beneath carbonate banks of the offshore Nicaraguan Rise (Arden, 1975), and are obscured by extensive volcanic and tuffaceous rocks related to the Miocene Central American volcanic arc in western Honduras (Jordan et al., this volume).

Our objective is to use this information to better constrain the trends of basement rocks making up the Chortis block, which has been inferred by most previous workers to be a single, monolithic basement block (cf. Dengo, 1985; Case et al., 1990; Pindell and Barrett, 1990).

We have defined tectonic terranes on the Chortis block following the criteria of Howell et al. (1985, p. 4)[1]:

A tectonostratigraphic terrane is a fault-bounded package of rocks of regional extent characterized by a geologic history which differs from that of neighboring terranes. Terranes may be characterized internally by a distinctive stratigraphy, but in some cases a metamorphic or tectonic overprint is the most distinctive characteristic. In cases where juxtaposed terranes possess coeval strata, one must demonstrate different and unrelated geologic histories as well as the absence of intermediate lithofacies that might link the two terranes. In general, the basic characteristic of terranes is that the present spatial relations are not compatible with the inferred geologic histories. As additional geologic, paleontologic, and geophysical data are accumulated, terrane boundaries and classification can be modified and their nomenclature revised.

Our focus is the Chortis block of northern Central America, previously described by Case et al. (1984) as a "superterrane," or amalgamation of crustal terranes, although these authors made no attempt to define the component terranes or establish their boundaries within the Chortis block. As pointed out by Williams and Hatcher (1982), tectonic terranes are assumed to be "guilty" or "suspect" until their "innocence" can be proven by establishing their origin, history, and relation to adjacent terranes. Due to the

---

[1]Howell et al. (1985) define a terrane as a fault-bounded geologic entity or fragment that is characterized by a distinctive geologic history that differs markedly from that of adjacent terranes. Our field observations across the Chortis block reveal regions with markedly distinct geologic histories in the north, east, and south, reflected in differing stratigraphy and in the nature and grade of exposed basement rock. The extent to which these regions of the Chortis block are fault-bounded is uncertain because <15% of the Chortis block has been mapped at a scale of 1:50,000 or smaller. Therefore, instead of using field methods to delineate the boundaries of these distinct regions, we have employed a geophysical method that reflects the magnetic property of the basement rocks in each region. The correlation of magnetic property to the distinct geology of each region as described in this paper suggests that this approach is sound. The aeromagnetic survey we use encompasses a large part of the Chortis block, thereby allowing us to infer the boundaries of regions remote, inaccessible, or buried beneath young volcanic or alluvial strata. Therefore, the terranes we propose for the Chortis block are regions characterized by a distinctive geologic history that differs markedly from that of adjacent terranes that are in part fault-bounded and whose boundaries have been delineated on the basis of the magnetic properties of the basement rock.

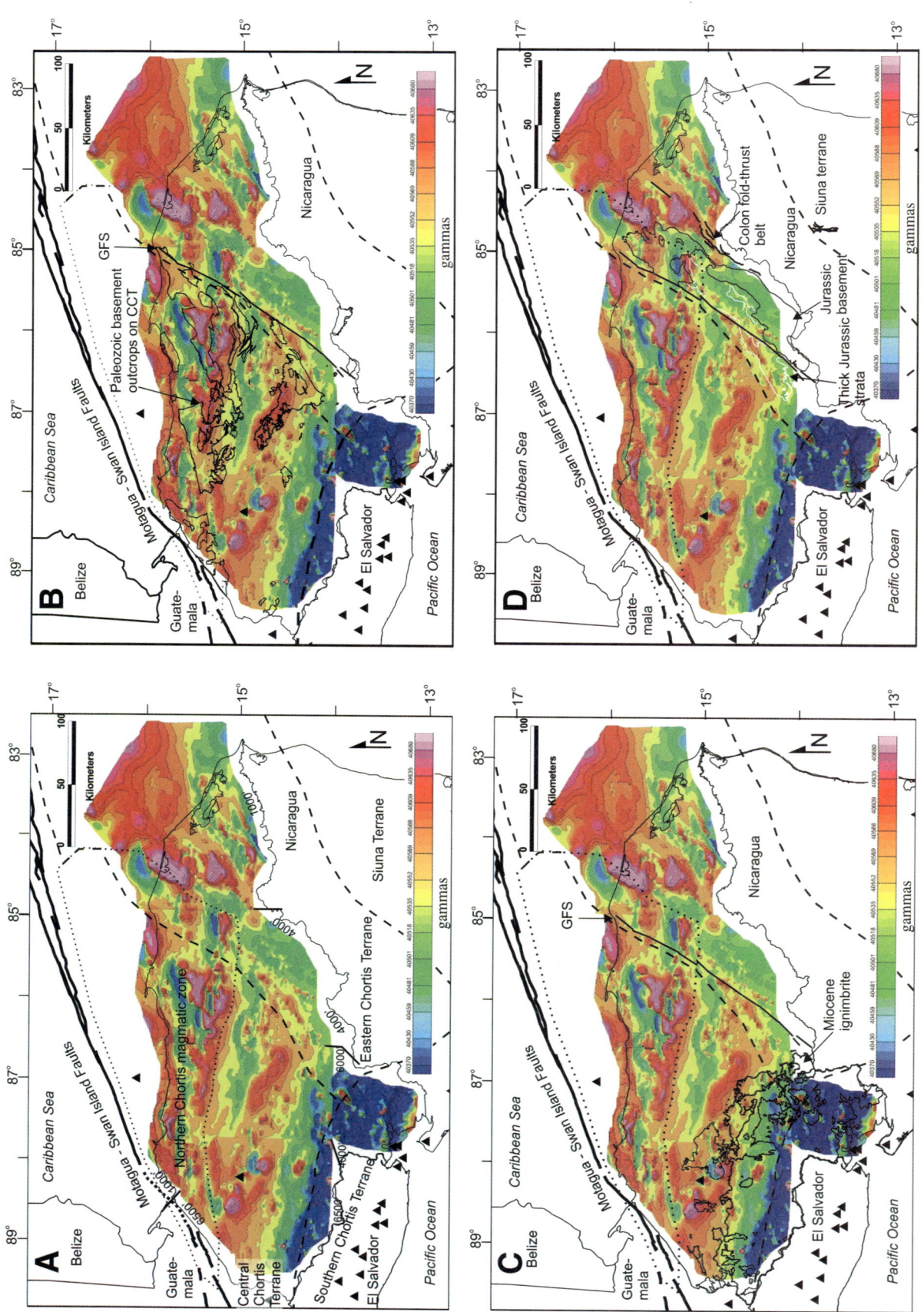

Figure 3. Composite total magnetic intensity anomaly map of Honduras (A) showing a higher but uniformly west-northwest–trending magnetic fabric across central Honduras (Central Chortis terrane); (B) a low-intensity region in southwestern Honduras (Southern Chortis terrane); (C) and a northeast-trending magnetic fabric in eastern Honduras (Eastern Chortis terrane) (D). The highly variable magnetic anomalies in northern Honduras (northern part of the Central Chortis terrane) are indicated by dotted outline. Quaternary volcanoes of the Central America volcanic arc are shown as black triangles. Key outcrop areas from the geologic map in Figure 2 are highlighted to show the geologic origins of the aeromagnetic anomalies. CCT—Central Chortis terrane; GFZ—Guayape Fault system.

sparse mapping of Honduras, the terranes we propose are "suspect" and subject to revision with improved mapping and dating. We have split the Chortis block into three terranes on the basis of aeromagnetic, geologic, and age isotopic observations until second-order analyses make the identity and genetic history of the pieces sufficiently well known to begin the "lumping" process. These proposed terranes and their boundaries should provide future workers a framework to plan more detailed mapping and dating studies.

Our objective is to characterize these suspect terranes and to identify the boundaries between these regions. To accomplish this, we utilize the regional aeromagnetic survey—a single geophysical data set that spans the regions of observed differing geology. We combine the aeromagnetic data with other geologic data compiled in this paper to (1) determine if this magnetic data can be correlated to the distribution of geologic features, and (2) determine if the magnetic data along with geologic data can be use to delineate the boundaries of these regions. Following these first two steps, using aeromagnetic and geologic data, we place the Chortis terranes in a more regional tectonic scheme by matching them with comparable tectonic terranes proposed by previous workers in southwestern Mexico (Fig. 1A).

GEOLOGIC AND TECTONIC SETTING OF THE CHORTIS BLOCK

The continental crust of the Chortis block forms the present-day northwest corner of the Caribbean plate, a plate otherwise composed of oceanic, oceanic plateau, or arc crust (Case et al., 1990) (Fig. 1). Subduction of the Cocos plate and its pre-Miocene predecessor, the Farallon plate, beneath the Chortis block (Caribbean plate) has produced the Middle America trench, a forearc basin, and the modern Central America volcanic arc (von Huene et al., 1980; Ranero et al., 2000) (Fig. 1).

Continental crustal rocks of northern Central America straddle the North America–Caribbean left-lateral, strike-slip plate margin, manifested by the prominent, arcuate-shaped fault valleys of the Motagua-Polochic fault strike-slip fault system in Guatemala (Figs. 1 and 2). Internal deformation of the northern and western parts of the Chortis block adjacent to these plate boundary faults occurs at rates of a few millimeters per year (DeMets et al., 2000; DeMets et al., this volume) has led to the formation of an ~east-west belt of about a dozen small, north-trending rifts of late Miocene to Holocene age (Burkart and Self, 1985; Guzman-Speziale, 2001; Rogers and Mann, this volume) (Figs. 1A and 1B). North of the strike-slip plate boundary, left-lateral motion of the North America–Caribbean plate is taken up by post–middle Miocene shortening, strike-slip faulting, and transpression in northern Guatemala (Guzman-Speziale and Meneses-Rocha, 2000). The southern and eastern boundaries of the Chortis block are loosely defined to be parallel to the Nicaragua–Costa Rica border and north of the Hess Escarpment in the Caribbean Sea (Dengo, 1985) (Fig. 2). GPS studies in these areas show that eastern Honduras is a stable region moving as part of the larger Caribbean plate (DeMets et al., this volume).

Geology of Central Honduras

Paleozoic continental crust (Cacaguapa Group of Fakundiny, 1970) is usually cited as the most common exposed basement type of the central Chortis block (cf. Horne et al., 1976a; Donnelly et al., 1990; Sedlock et al., 1993, Gordon 1993; Burkart, 1994) (Fig. 2). However, the oldest dated basement of Grenville age (1.0 Ga orthogneiss) has been found in several small inliers within larger areas of known or presumed Paleozoic crust in central and northern Honduras (Manton, 1996) (Fig. 2). In central Honduras, the Paleozoic gneiss and schist basement is overlain by mixed carbonate and clastic Cretaceous basins (Rogers et al., Chapter 5, this volume) (Fig 4B). Both basement and overlying Cretaceous strata are deformed by folding and thrusting into elongate outcrops representing inverted Albian to Aptian-age intra-arc rifts (Rogers et al., Chapter 5, this volume) (Fig. 2).

Geology of Northern Honduras

In northern Honduras, basement rocks consist of highly sheared, high-grade gneiss of Paleozoic to Precambrian age, felsic batholiths of Late Cretaceous and Early Tertiary age, and Cretaceous age metasedimentary strata (Horne et al., 1976b; Manton, 1987; Manton and Manton, 1999; Avé Lallemant and Gordon, 1999) (Fig. 2). The age and degree of deformation of these rocks contrasts with the low-grade phyllite and schist of the Cacaguapa Group of central Honduras (Fakundiny, 1970) (Fig. 2).

Geology of Eastern Honduras and Nicaragua

Exposures of crystalline basement of the Chortis block have not been reported from eastern Honduras (Fig. 2). Metasedimentary basement outcrops in the region southeast of the Guayape fault system have been correlated with unmetamorphosed lithologies of the Jurassic Agua Fria Formation found northeast of the Guayape fault system (Viland et al., 1996; Rogers et al., Chapter 6, this volume). Folds and southeast-dipping reverse faults affecting Cretaceous carbonate and clastic rocks of the northeast-trending Colon fold-thrust belt parallel the northeast-trending fabric that is prominent on topographic maps and satellite imagery (Rogers et al., Chapter 6, this volume) (Figs. 2 and 4C). Most of our proposed boundary between the Central and Eastern Honduras follows the Guayape strike-slip fault that partially forms the western extent of the Jurassic Agua Fria metasedimentary basement type of eastern Honduras (Gordon and Muehlberger, 1994) (Fig. 2).

In northern Nicaragua, Venable (1994) used field, isotopic, and petrologic data to show that igneous basement rocks south of the mining area of Siuna, Nicaragua, are a Cretaceous island arc assemblage (including ultramafic cumulates and serpentinites) that formed on oceanic crust (Rogers et al., Chapter 6, this volume) (Fig. 2). She proposed that these rocks formed the Siuna terrane and the basement for much of Nicaragua.

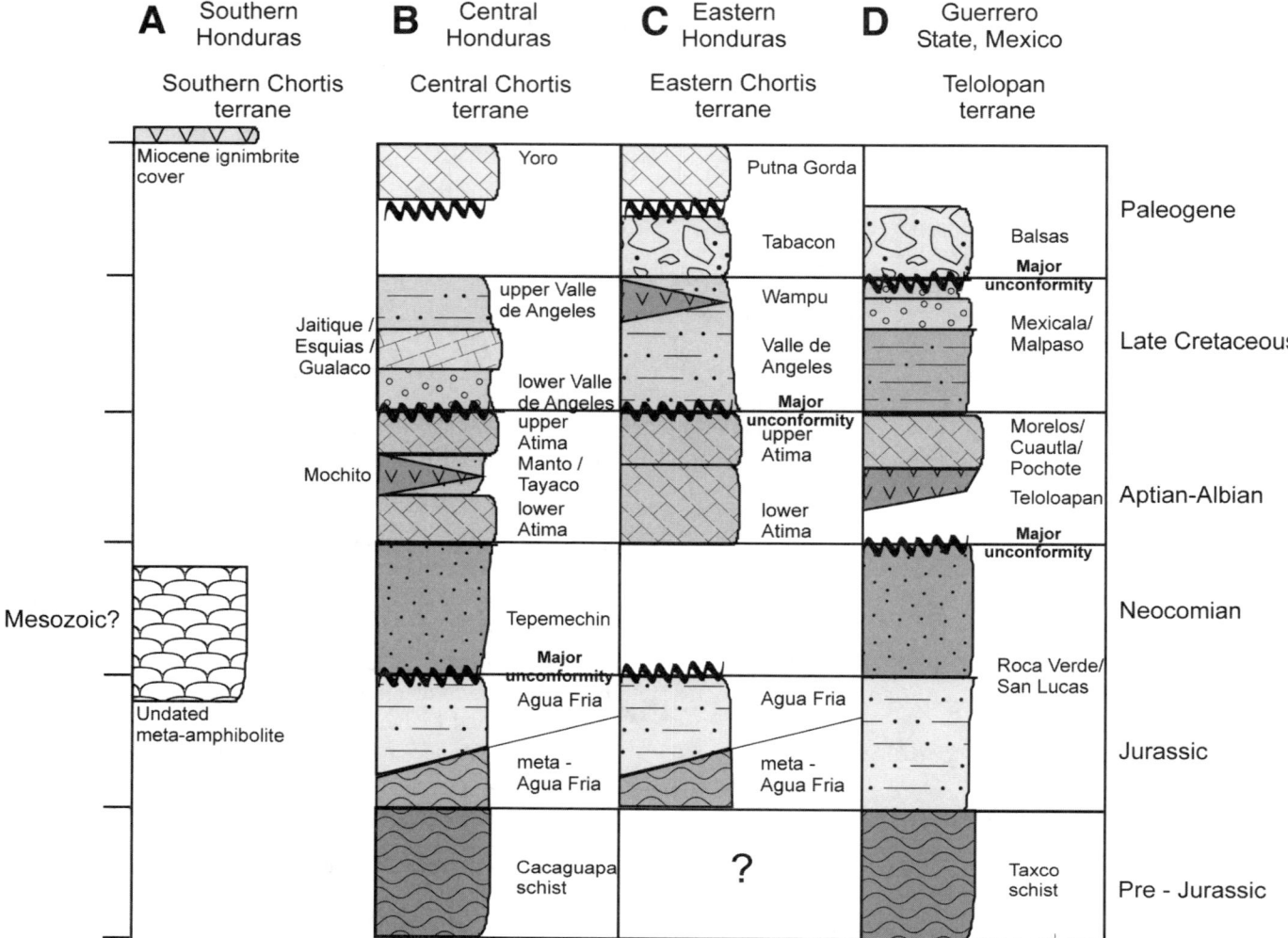

Figure 4. Regional correlation of late Mesozoic and Paleogene stratigraphy from Honduras and from the Guerrero state of southern Mexico. Data sources: Central Honduras terrane—Scott and Finch (1999); Kozuch (1991); and Rogers et al. (this volume, Chapter 5); eastern Honduras—Rogers et al. (this volume, Chapter 6); Guerrero state of Mexico—Cabral-Cano et al. (2000), Johnson et al. (1991), Cerca et al. (2006), and Lang et al. (1996). See text for discussion.

Geology of Southern Honduras

The distribution of Paleozoic rocks in southern and western Honduras is unknown because this region is blanketed by a thick cover, up to two kilometers, of Miocene volcanic strata produced by the Middle America volcanic arc (Williams and McBirney, 1969; Jordan et al., this volume) (Fig. 2). Geochemical data from modern lavas of the Central American volcanic arc show an absence of arc magma contamination or mixing with underlying Paleozoic continental crust of the type exposed in northern and Central Honduras (Carr et al., 2003). For this reason, Carr et al. (2003) infer that a post-Paleozoic, arc-type basement underlies the modern Middle America arc and areas extending an unknown distance north of the modern volcanic line. A single, isolated exposure of highly deformed, metavolcanic amphibolite is exposed in southern Honduras along the gorge of the Rio Choluteca (Markey, 1995; Harwood, 1993) (Figs. 2 and 4A). This single occurrence of older, higher-grade rocks provides our only window into the underlying basement geology of our proposed Southern Chortis terrane.

AEROMAGNETIC DATA FROM HONDURAS AND THE EASTERN NICARAGUAN RISE

The high-sensitivity aeromagnetic survey comprises ~41,400 line kilometers distributed across six survey altitudes zones (1000, 4000, 6000, and 6500 barometric feet), evident by discontinuities on merged data (downward continuation)

(Fig. 3). Traverse spacing was 3 km and 5 km with tie-line spacing of 12 km and 20 km. The survey utilized a Varian High-Sensitivity Magnetometer and data was processed to develop total magnetic intensity contour maps that were corrected for topography at time of production in 1985. The survey did not cover strips along international borders of Honduras with neighboring countries but included an area of the Caribbean Sea near northeastern Honduras (Fig. 3).

The aeromagnetic data were provided to us by the government of Honduras in the form of a large paper map with contoured values of magnetic data at a scale of 1:500,000. We scanned the map on a large-format scanner and digitized the total magnetic intensity values at 20 gamma (nanotesla) intervals (i.e., we digitized every fifth contour of those that were shown on the original paper map). These data were compiled on a computer to produce the color contoured map shown in Figure 3. This contour interval was judged sufficiently dense to accurately convey the regional-scale patterns that are described as follows. The aeromagnetic data and surface geologic map data were overlain and precisely geo-referenced using a geographic information system (GIS).

The aeromagnetic survey provides an ideal tool for examining the distribution and nature of the basement of the Chortis block because the source of the observed magnetic anomalies lies within the basement rocks and the influence of the sedimentary cover signature is insignificant. The aeromagnetic survey provides one of the best possible sources of regional data to constrain the type and structure of the basement character underlying the volcanic regions of southern Honduras and the alluvial or water-covered areas of eastern Honduras and the Nicaragua Rise (Fig. 2).

REGIONAL MAGNETIC TRENDS VISIBLE IN HONDURAN AEROMAGNETIC DATA

Four Distinctive Magnetic Provinces

Four magnetic provinces are apparent in Figure 3: (1) an extremely low, northwest-trending magnetic intensity region in southwest and southern Honduras adjacent to the Central American volcanic arc; (2) a higher-intensity area of west-northwest–trending magnetic lineations across central Honduras; (3) a high-intensity, northeast-trending, lineated magnetic fabric in eastern Honduras; and (4) and an area of high-intensity but highly variable speckled magnetic anomalies in northern Honduras.

Correlation of three of these magnetic provinces of Honduras with three geologic provinces of distinctive basement rocks, crustal properties, geochemical variations, and tectonic histories forms the basis of dividing the Chortis superterrane of Case et al. (1990) into three component terranes (Fig. 3A; Tables 1 and 2). In the following, each of the proposed Chortis terranes is defined based on correlation of known surface geology compiled in Figure 2 with the aeromagnetic nature of the terrane shown in Figure 3.

Central Chortis Terrane

Terrane Definition

The Central Chortis terrane contains broad, lineated magnetic gradients trending west-northwest across central Honduras (Fig. 3) that parallel the belts of deformed basement and overlying Cretaceous strata (Fig. 2). Topographic ridges of elevated basement rocks of mainly Paleozoic and Precambrian age form magnetic highs, while regions of thick accumulations of overlying Cretaceous strata produce bands of lower magnetic intensity (Rogers et al., Chapter 5, this volume) (Fig. 3B). Age dating of the Central Chortis terrane is compiled on Table 1 from many previous authors and plotted on the magnetic map in Figure 5 from 1 Ga (Grenville) to 55 Ma (Eocene) south of the Northern Chortis magmatic zone.

A prominent oroclinal bend in these east-west belts of older basement rocks is observed in the topography, the geology, and on the aeromagnetic map (Figs. 2 and 3). In Chapter 5 of this volume, Rogers et al., propose that oroclinal bending accompanied a phase of left-lateral shear long the Guayape fault during the Late Cretaceous.

The four structural belts of Honduras (Fig. 2), representing inverted Aptian-Albian rifts (Rogers et al., Chapter 5, this volume), align well with the west-northwest–trending magnetic fabric in the Central Chortis terrane (Fig. 3). The late Neogene rifts of western Honduras are not discernable from magnetic intensity despite their prominent basin and range topographic expression (Figs. 1B and 2). The abrupt southern boundary of the Central Chortis terrane with the adjacent Southern Chortis terrane is easily discernible beneath the thick Miocene volcanic cover of western Honduras (Fig. 3).

We infer the Central Chortis terrane represents the core of the continental Chortis superterrane (Case et al., 1990) with presently northwest-trending Cretaceous intra-arc basins deforming Paleozoic basement (Figs. 2 and 4). Structural highs are now occupied by outcrops of Paleozoic and Precambrian rocks as seen on the outcrop overlay in Figure 3B. This Paleozoic and Precambrian basement is the continental basement type of the Chortis block most commonly described by previous authors, such as Fakundiny (1970) and Simonson (1977), and summarized on Table 2.

Northern Chortis Magmatic Zone

The aeromagnetic character of the northern part of the Central Chortis terrane maintains the same west-northwest trend observed in the Central Chortis terrane but the trend is more variable and punctuated by pairs of magnetic highs and lows (dipoles) that align with mapped, subcircular exposures of felsic batholiths of Late Cretaceous and early Tertiary age (Horne et al., 1976a, 1976b; Manton, 1996) (Figs. 2 and 3, Table 1). A belt of Late Cretaceous to early Tertiary magmatism occurs in

TABLE 1. COMPILATION OF CHORTIS BLOCK RADIOMETRIC DATA (cf. FIG. 5)

Sample	System	Mineral	Age (Ma)	Source and comments
Northern Chortis metamorphic zone—basement rocks				
Banaderos metaigneous rock	Rb/Sr	wr	720 ± 260	Horne et al., 1976a
Quebrada Seca metaigneous rock	Rb/Sr	wr	305 ± 12	Horne et al., 1976a
Adamellite gneiss	Rb/Sr	wr	230–203	Sedlock et al., 1993
Adamellite gneiss	Rb/Sr	wr	150–125	Sedlock et al., 1993
Chamelecon granodiorite	K/Ar	p	224.0 ± 17	MMAJ, 1980a—metaigneous?
Northern Chortis metamorphic zone—igneous rocks				
San Marcos granodiorite	Rb/Sr	wr	150.0 ± 13	Horne et al., 1976a
Tela augite tonalite	K/Ar	h	93.3 ± 1.9	Horne et al., 1976b
Rio Jalan granite	U/Pb	z	92.7	Manton and Manton, 1984
Granodiorite	K/Ar		89.0 ± 4	MMAJ, 1980a
Chamelecon diorite	K/Ar		86.3 ± 3.7	MMAJ, 1980a
Rio Cangrejal gneiss	Rb/Sr	wr, b	83.5	Manton and Manton, 1984—sheared pluton
Rio Cangrejal gneiss	U/Pb	z	79.8	Manton and Manton, 1984—crystallization age
El Carbon granodiorite	U/Pb	z	81.0 ± 0.1	Manton and Manton, 1984—crystallization age
El Carbon granodiorite	U/Pb	z	76.3	Manton and Manton, 1984—metamorphic age
El Carbon granodiorite	Rb/Sr	k, p	68.9	Manton and Manton, 1984—metamorphic age
Tela augite tonalite	K/Ar	b	80.5 ± 1.5	Horne et al., 1976b
Trujillo granodiorite	Rb/Sr	wr	80.0	Manton and Manton, 1984
Tela augite tonalite	K/Ar	b	73.9 ± 1.5	Horne et al., 1976b
Piedras Negras tonalite	K/Ar	h	72.2 ± 1.5	Horne et al., 1976b
Mezapa tonalite	K/Ar	b	71.8 ± 1.4	Horne et al., 1976b
Las Mangas tonalite	K/Ar	m	57.3 ± 1.1	Horne et al., 1976b
Piedras Negras tonalite	K/Ar	b	56.8 ± 1.1	Horne et al., 1976b
Rio Cangrejal granodiorite	Rb/Sr	b	54.4	Manton and Manton, 1984
Sula Valley granodiorite	Rb/Sr	b	38.5	Manton and Manton, 1984
San Pedro Sula granodiorite	K/Ar	b	35.9 ± 0.7	Horne et al., 1976b
Confadia granite	Rb/Sr	b	35.0	Manton and Manton, 1984*
Banderos granodiorite	Rb/Sr	wr, b	30.8	Manton and Manton, 1984
Banderos granodiorite	U/Pb	z	28.9	Manton and Manton, 1984
Central Chortis terrane—basement rocks				
Orthogneiss	U/Pb	z	1000	Manton, 1996
Mica schist La Arada	K/Ar	b	222.0 ± 8	MMAJ, 1980b
Central Chortis terrane—igneous rocks				
San Andres dacite	U/Pb	z	124.0 ± 2	Drobe and Cann, 2000
San Ignacio granodiorite	K/Ar	h	122.7 ± 2.5	Emmet, 1983
San Ignacio granodiorite		b	117.0 ± 1.9	Emmet, 1983
San Ignacio adamellite	K/Ar	b	114.4 ± 1.7	Horne et al., 1976b
HN-149, basalt	K/Ar	wr	86.30 ± 3.8	ENEE, 1987
Gabbro dike	K/Ar	h	79.6 ± 1.6	Emmet, 1983
HN-205, diorite/basalt	K/Ar	wr	62.20 ± 2.6	ENEE, 1987
Minas de Oro granodiorite	K/Ar	b	60.6 ± 1.3	Horne et al., 1976b
Minas de Oro granodiorite	K/Ar	b	59.30 ± 0.9	Emmet, 1983
Minas de Oro granodiorite		h	55.00 ± 1.1	Emmet, 1983
San Francisco dacite pluton	K/Ar	b	58.6 ± 0.7	Horne et al., 1976b
Eastern Chortis terrane—igneous rocks				
Dipilto granite	Rb/Sr	wr	140.0 ± 15	Donnelly, et al., 1990
Carrizal diorite	Rb/Sr	k, b	118.0 ± 2	Manton and Manton, 1984
Wampu basaltic andesite	K/Ar	p	80.7 ± 4.3	Weiland et al., 1992
Wampu basaltic andesite	K/Ar	p	70.4 ± 3.4	Weiland et al., 1992
Siuna terrane—igneous rocks				
Andesite	Ar/Ar	h	75.62 ± 1.3	Venable, 1994
Diorite	Ar/Ar	b	59.88 ± 0.47	Venable, 1994

Note: Abbreviations: b—biotite; h—hornblende; z—zircon; p—plagioclase; k—potassium feldspar; wr—whole rock; ENEE—Empressa Nacional de Energia Electrica; MMAJ—Mineral Mining Agency of Japan.
*Initial $^{87}Sr/^{86}Sr$ = 0.7040 assumed.

TABLE 2. SUMMARY OF GEOLOGIC CHARACTERISTICS OF THE CHORTIS TERRANES AND REGIONS (cf. FIGS. 3 AND 4)

Name	Type	Crustal Component	Defining characteristics	Exposed boundaries	Age	Origin
Central Chortis terrane (this study)	Continental fragment	Crystalline continental crust	Paleozoic basement, Cretaceous overlap sequence	Motagua-Polochic fault	Grenville-Paleozoic	Dispersed from nuclear Mexico
Eastern Chortis terrane (this study)	Continental margin fragment	Attenuated continental crust?	Meta-Jurassic basement, Jurassic strata, Cretaceous overlap sequence	Guayape fault, extent of Jurassic strata	Exposed Jurassic, inferred Paleozoic	Attenuated Central Chortis, dispersed from Mexico
Southern Chortis terrane (this study)	Volcanic arc fragment	Island arc built on oceanic crust	Magnetic, Tertiary volcanic overlap, inferred island arc basement	Inferred buried suture beneath Miocene ignimbrites	Inferred Jurassic–early Cretaceous	Accreted to Central Chortis, dispersed from Mexico
Siuna terrane (Venable, 1994)	Volcanic arc fragment	Island arc built on oceanic crust	Cretaceous volcanic basement, noncontinental isotopic signature, Tertiary volcanic overlap	Siuna suture	Early Cretaceous	Accreted to Eastern Chortis terranes
Northern Chortis metamorphic zone (this study)	Metamorphic	Overprinted continent and continental margin	Magnetic, high-grade metamorphic basement, Late Cretaceous–early Triassic plutons, Cretaceous overlap sequence metamorphosed	Shear zones, generally unmapped	Overprinted Grenville-Paleozoic	Metamorphic overprint of Central and Eastern Chortis terranes

a broad, 100-km-wide swath across northern Honduras (Horne et al., 1976a, 1976b; Kozuch, 1991) and is associated with a basement of sheared, high-grade gneiss (Manton, 1996; Manton and Manton, 1999) (Fig. 5, Tables 1 and 2). The concentrated belt of Late Cretaceous to early Tertiary magmatic activity on the northern part of the Central Chortis terrane suggests that the variability of the observed magnetic signature results from superimposed magmatic activity that we refer to here as the Northern Chortis magmatic zone. We do not distinguish this region as a separate terrane because its basement crustal types and isotopic ages are similar to the Central Honduras terrane to the south (Horne et al., 1976a, 1976b; Manton, 1996) (Fig. 5). The distinctive speckled aeromagnetic signature seen on the aeromagnetic map in Figure 3 is clearly related to intrusions within the Northern Chortis magmatic zone and therefore a local thermal and intrusive event affecting only the northern part of the Central Chortis terrane.

The variable character of the magnetic field of the Northern Chortis magmatic zone continues into the Eastern Chortis terrane where it is characterized by large high-low magnetic dipoles occurring beneath the northwestern part of the Mosquitia Plain and the Nicaragua Rise, suggesting buried intrusive bodies in the subsurface. Arden (1975) describes oil industry wells from the Nicaraguan Rise that encountered plutonic rocks of Late Cretaceous and early Cenozoic age that are unconformably overlain by Cenozoic carbonate banks of the Nicaraguan Rise.

The uncertain boundaries of the overprinting effect of the magmatic province across the northern margin of the Central Chortis block is indicated by the dotted lines in Figure 3. As drawn, the Northern Chortis magmatic zone also overprints the Guayape fault system shown in Figure 3 in a manner that indicates the Late Cretaceous pulse of magmatic overprinting post-dates significant lateral offset along the fault (Gordon and Muehlberger, 1994). (Figs. 2 and 5).

Faulting Related to the Strike-Slip Plate Margin

Some of the variability of the magnetic field of the northern part of the Central Chortis terrane appears to have been produced by late Neogene transtensional deformation of the Honduras borderlands related to the North America–Caribbean strike-slip plate boundary (Rogers and Mann, this volume) (Fig. 2). This late tectonic control on the aeromagnetic signature is shown by the prominent magnetic low aligned with the active, alluvial-filled Aguan rift basin south of the northern coastal mountain range of Honduras (Figs. 2 and 3).

Eastern Chortis Terrane

Magnetic intensity of the Eastern Chortis terrane trends to the northeast (Fig. 3) and is confined mainly to the area east of the Guayape fault zone (Fig. 2). The regional magnetic fabric of the Eastern Chortis terrane is roughly perpendicular to the prominent west-northwest magnetic trends of the Central Chortis terrane (Fig. 3). The boundary between the Central and Eastern Chortis terranes is placed at the eastern extent of the crystalline basement exposure and coincides with the northern part of the Guayape fault. In this area, the pronounced northeast-trending topographic fabric of the Eastern Chortis terrane is well-reflected by the northeast orientation of outcrops and trends in the Colon fold-thrust belt of Late Cretaceous age (Rogers et al., Chapter 6, this volume) (Figs. 2 and 3D).

-In the area of the southern Guayape fault zone south of the town of Catacamas (Fig. 2), we propose that the boundary separating the Central and Eastern terranes does not coincide with

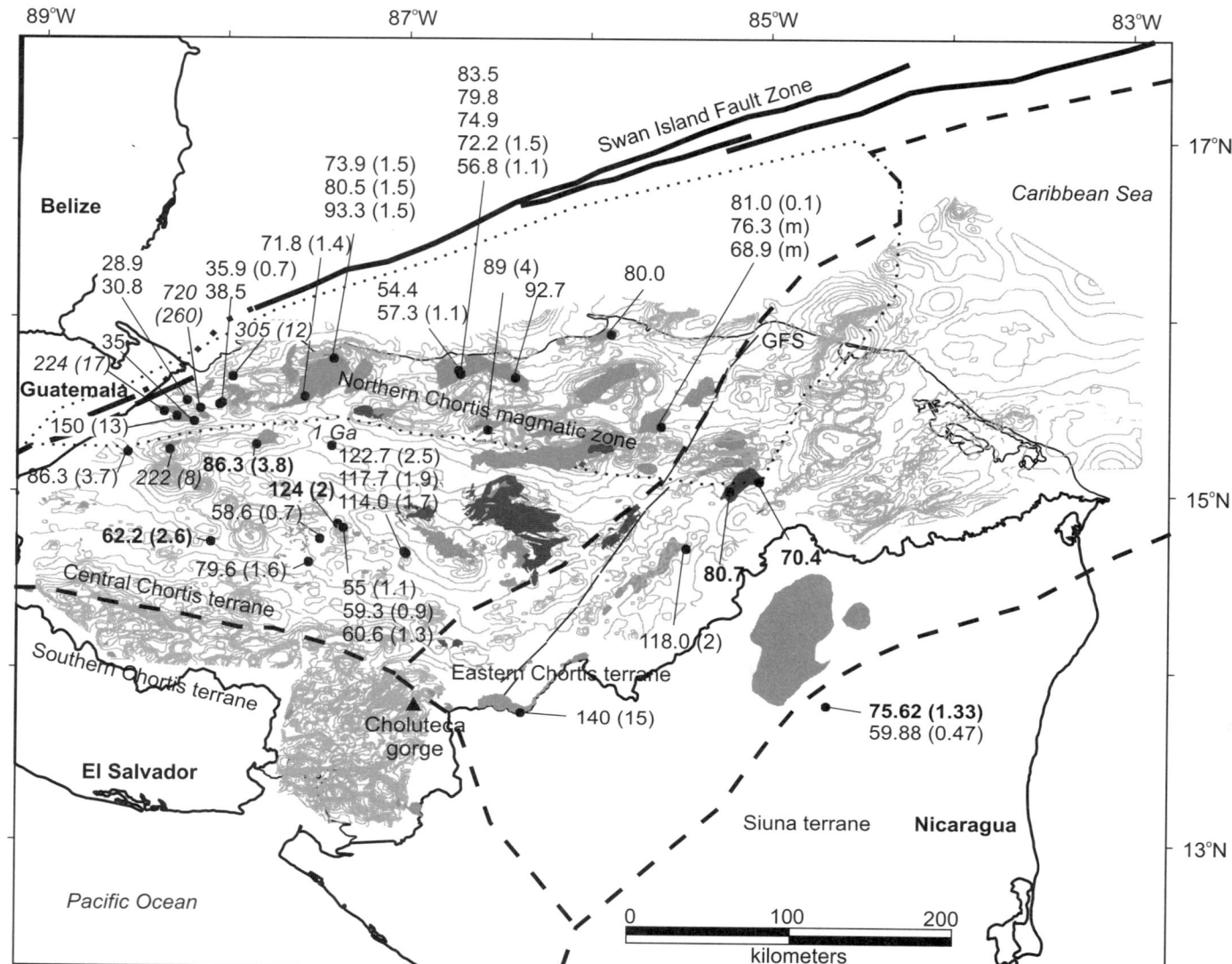

Figure 5. Compilation map of pre-Neogene reported radiometric ages of igneous rocks in Honduras and Nicaragua, in m.y.; error shown in parentheses (also in m.y.). Outcrop exposures of all plutonic rocks from the geologic map in Figure 2 are shown by light shading; outcrop exposures of all volcanic rocks are shown in dark shading (for simplicity, Neogene volcanic rocks are omitted). Note the concentration of Late Cretaceous to early Tertiary magmatic activity that coincides with the northern edge of the Central Chortis terrane. Bold numbers indicate reported radiometric dates for volcanic rocks; italics indicate dates from metamorphic rocks; (m) denotes radiometric dates for the deformation age of sheared igneous rocks. Gray background contours are magnetic intensity map of Honduras shown in Figure 3. Sources for all data are given in Table 1. Triangle—location of the Choluteca gorge exposure of basement rock in the Southern Chortis terrane; GFS—Guayape Fault system.

the Guayape fault but instead follows a prominent magnetic lineament west of the Guayape fault (Fig. 3). This magnetic lineament roughly coincides with the western margin of the Jurassic sedimentary basin (Rogers et al., Chapter 6, this volume). A small exposure of crystalline basement rock occurs adjacent to the southernmost segment of the Guayape fault zone, although it is not clear by what mechanism this area was exposed (Rogers, 1995).

We infer that the East Chortis terrane represents the rifted, southern margin of Chortis (North American plate) developed during the Jurassic opening of the North and South America plates (Table 2). The original Jurassic orientation of these rift-related faults was likely to be more east-west than their present strike because a large, Cenozoic counterclockwise rotation needs to be taken into account (Gose and Swartz, 1977; Gose, 1985). This implies that attenuated crystalline continental basement may exist beneath the exposed Jurassic metasedimentary rocks of the Eastern Chortis terrane.

We have no direct evidence for Jurassic age extensional faults, especially on the regional scale depicted by James (2006). Thickening of the Jurassic sedimentary units from a few tens of meters on the Central Chortis terrane to >2 km on the Eastern Chortis terrane lends some support to the interpretation by James (2006) that at least part of the Guayape fault may have originated as a Late Jurassic normal fault. Exposed basement of the Eastern Chortis terrane is limited to greenschist-grade phyllite and schist of the metamorphosed Jurassic-age Agua Fria Formation (Rogers et al., Chapter 6, this volume). The northeast magnetic trends of the Eastern Chortis terrane align well with the distribution of folded Jurassic rocks. These pronounced magnetic trends indicate that folds and faults are likely thick-skinned features that may also affect inferred crystalline basement rocks at depth beneath the outcrops.

Southern Chortis Terrane

The boundary between the Southern and the Central Chortis terranes appears on the magnetic map of Honduras as 200-km-long lineament across which an abrupt southward decrease in total magnetic intensity is exhibited (Fig. 3). The low but variable magnetic field in the Southern Chortis terrane results in an undulating pattern of magnetic highs and lows that do not correlate with Quaternary volcanic centers in the area (Fig. 2). The Southern Chortis terrane is overlain by up to 2 km of mid-Miocene pyroclastic strata of the Chortis block (Williams and McBirney, 1969; Jordan et al., this volume) (Fig. 2). A single known exposure of basement of the Southern Chortis terrane is a metavolcanic amphibolite mapped by Markey (1995) (Fig. 5).

The absence of the Paleozoic basement underlying the Southern Chortis terrane is indicated by the geochemistry of Quaternary volcanic lavas (Carr et al., 2003). We propose that post-Paleozoic arc or oceanic-type basement of the Southern Chortis terrane may extend to the southeast and underlie the offshore Cretaceous forearc basins (von Huene et al., 1980; Ranero et al., 2000) (Fig. 1).

Because the Southern Chortis terrane sharply abuts the southwestern edge of the continental Central Chortis terrane, the Southern Chortis terrane is inferred to be an accreted oceanic element to more inboard continental terranes (as is the case for the Guerrero terrane to nuclear Mexico) (Centeno-Garcia et al., 1993). Future work may identify more outcrops of the basement rock of the Southern Chortis terrane, which is now known only from one locality (Fig. 5).

Siuna Terrane

Adjacent to the Chortis terranes is the Siuna terrane (Venable 1994) in the area of the Siuna mining district in northern Nicaragua (Figs. 2 and 5). Because the area is in Nicaragua, it is not covered by the Honduras aeromagnetic map shown in Figure 3. At Siuna, exposures of thrusted serpentinite and associated ultramafic cumulates exhibit isotopic characteristics of oceanic crust (Rogers et al., Chapter 6, this volume) and is consistent with the interpretation that the Siuna terrane consists of an Early Cretaceous oceanic island arc developed on oceanic basement and accreted to the Eastern Chortis terrane in the Late Cretaceous (Venable, 1994). The Colon fold-thrust belt in eastern Honduras and the northern Nicaragua Rise developed in response to the suturing of the Siuna terrane to the eastern Chortis terrane in the Late Cretaceous (Rogers et al., Chapter 6, this volume).

This terrane exposed in northern Nicaragua is an exotic oceanic crustal element accreted to the Eastern Chortis terrane in the Late Cretaceous (Table 2). We infer that the close association of the Siuna terrane island arc with the Colon fold-thrust belt of the Eastern Chortis terrane indicates that both were deformed in the arc-continent collision between the Guerrero-Caribbean arc and the continental margin of the southern Chortis block (Rogers et al., Chapter 6, this volume).

Kesler et al. (1990) show that the continental blocks of Central America and Mexico display distinct clustering of lead isotopic ratios from samples taken from basement and overlying volcanic rocks. These lead values therefore provide a useful basis to distinguish among the complexly amalgamated terranes of Central America and Mexico. A plot of lead values is shown in Figure 6 for basement or volcanic samples from the following terranes: Central Chortis, Siuna, Eastern Chortis, Tertiary volcanic rocks in Nicaragua, the Maya block, and the Caribbean large igneous province (all sources of data are given in the figure caption and are discussed in detail by Venable (1994) and Rogers et al. (Chapter 6, this volume). The lead values of the Siuna terrane cluster outside the Caribbean large igneous province and indicate that the Siuna arc is not underlain by crust of the Caribbean large igneous province (Fig. 6). Instead, it is likely that the Caribbean arc system developed at the edge of the Caribbean large igneous province rather than directly above it. Other interesting correlations are that lead values from the Central and Eastern Chortis areas form distinct clusters and support our proposal that both areas occupy distinctive terranes.

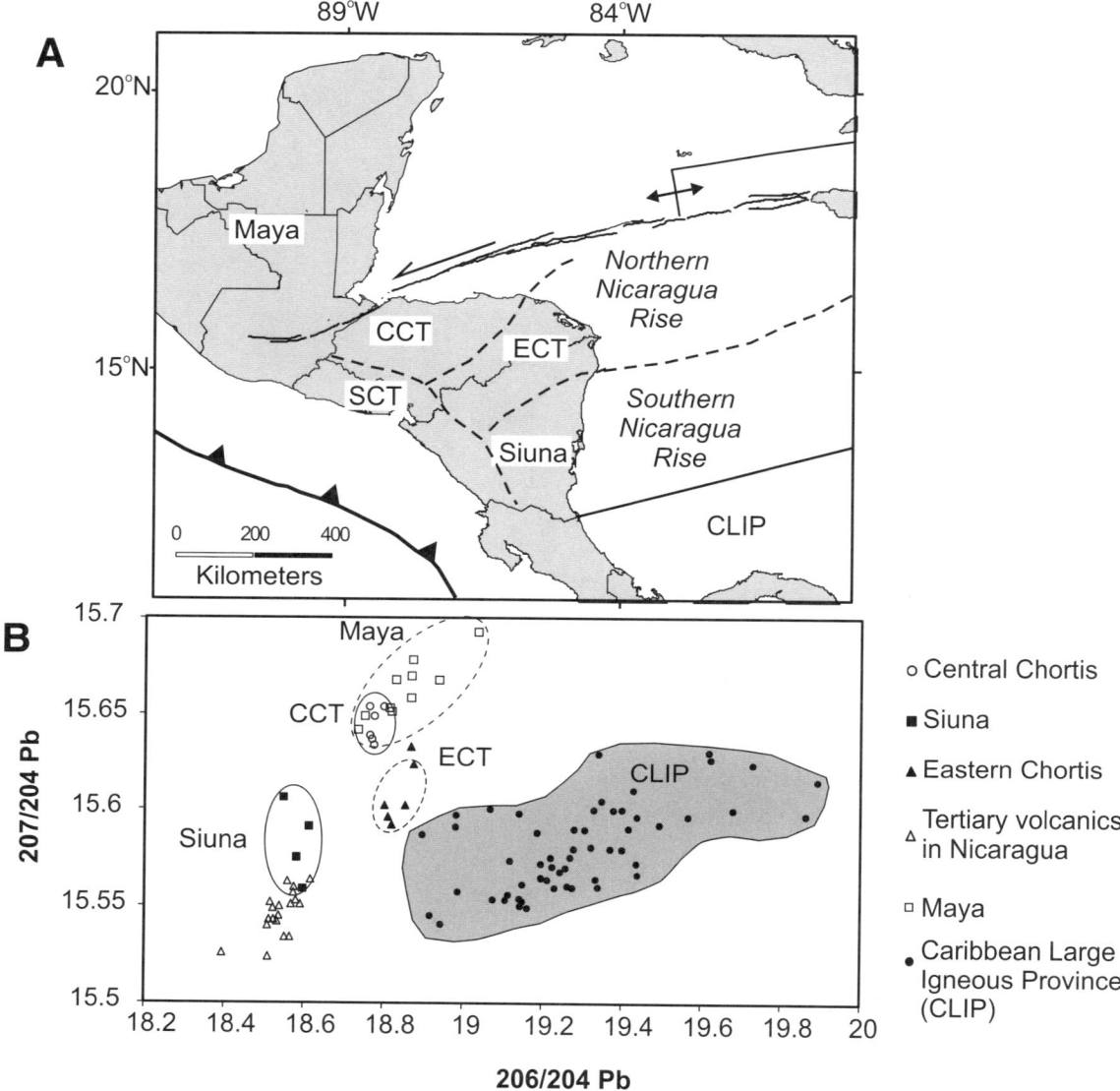

Figure 6. (A) Map showing Maya and Chortis blocks and proposed terranes of the Chortis block: CCT—Central Chortis terrane; ECT—Eastern Chortis terrane; SCT—Southern Chortis terrane; Siuna—Siuna terrane. Summary of lead isotopic data from Maya and Chortis continental blocks. (B) Comparison of common lead isotope data for Chortis block (including the Siuna terrane) with surrounding terranes and the Maya block. Note that lead data distinguishes the accreted Siuna terrane inferred to have originated as part of the Great Arc of the Caribbean. No lead data currently exists for the magmatic rocks along the northern part of the Central Chortis terrane. Lead data compiled from previous work by (Cumming and Kesler, 1976; Cumming et al. 1981); Sunblad et al. (1979); Sinton et al. (1997); Hauff et al. (2000); and Hoernle et al. (2002).

DISCUSSION

Defining Terranes on the Chortis Block

In order to define the Chortis terranes, we have applied the definition of a tectonic terrane as a "fault-bounded geologic entity or fragment that is characterized by a distinctive geologic history that differs markedly from that of adjacent terranes" (Howell et al., 1985, p. 4). Our terrane definitions are preliminary because of the large tracts of Honduras that remained unstudied or are known only at a reconnaissance level. However, the use of the aeromagnetic data supports our interpretations and shows that the Chortis block is a monolithic "superterrane" as envisioned by Case et al. (1984) and appears to be a composite of at least three terranes that have become welded together by Jurassic normal faulting or later strike-slip and shortening events along relatively narrow fault zones. Due to the sparse mapping of Honduras, the terranes we propose are "suspect" and subject to revision with improved mapping and dating. However, these proposed terranes and their boundaries should provide future workers a framework by which to plan more detailed mapping and dating studies.

PIERCING LINES AND OTHER FEATURES COMMON TO SOUTHERN MEXICO AND THE CHORTIS BLOCK

Many previous workers have speculated that the Chortis block originated along the southwestern margin of Mexico prior to Cenozoic translation from the North America plate to the Caribbean plate (Dengo, 1985; Ross and Scotese, 1988; Pindell and Barrett, 1990; Riller et al., 1992; Herrmann et al., 1994; Schaaf et al., 1995; Dickinson and Lawton, 2001). Placement of the Chortis along the southern Mexico margin by these workers has remained speculative because these reconstructions lacked piercing points or lines that could reconnect the rocks in Honduras with those in southwestern Mexico.

In this discussion, we introduce six geological and geophysical features to constrain the Late Cretaceous position of the Chortis block along the southwestern Mexican margin. Two features and three piercing lines are common to southern Mexico and the Chortis block: (1) Precambrian basement, (2) similar Mesozoic cover, (3) a north-trending mid-Cretaceous arc and geochemical trends, (4) north-trending Late Cretaceous structural belts, and (5) a north-trending common magnetic signature. (6) A fourth piercing line is the east-trending alignment of the eastern Honduras Colon fold belt with Late Cretaceous east-trending fold belt of southeastern Guatemala (Rogers et al., Chapter 6, this volume). Our best fit alignment of these features and piercing lines places the Chortis block along the southern margin of Mexico as displayed on the latest Cretaceous reconstruction in Figure 7A.

Common Feature 1 of Chortis and Southern Mexico

The Precambrian basement of Mexico is documented in the Oaxaca region of Mexico and is believed to represent the southward continuation of North American Gondwana elements into Mexico (Ortega-Gutierrez et al., 1995, Dickinson and Lawton, 2001). The distribution of Precambrian basement exposures (Oaxaca terrane of Ortega-Gutierrez et al., 1995) in Mexico is shown on Figures 1A and 6. On the Chortis block, Manton (1996) reports Precambrian basement of 1.0 Ga in the Yoro region of Honduras (Fig. 7A, near the circled "3"). The reconstruction we propose Figure 7A juxtaposes these two areas of Precambrian crust in pre-Eocene times.

Common Feature 2 of Chortis and Southern Mexico

A remarkably similar sequence of Late Cretaceous clastic, marine sandstone and shale overlying an Early Cretaceous shallow-water carbonate platform rocks occurs both on the Chortis block and in the area of the Teloloapan terrane of Guerrero state in southern Mexico (Figs. 4 and 7A). In Honduras, Late Cretaceous marine sandstone and shale is represented by the Valle de Angeles Formation (Figs. 4B and 4C) (Mills et al., 1967; Wilson, 1974; Horne et al., 1974; Finch, 1981; Scott and Finch, 1999; Rogers et al., this volume, Chapters 5 and 6) and in Mexico by the Mexicala Formation (Johnson et al., 1991; Lang et al., 1996; Cabral-Cano et al., 2000; Cerca et al., 2004) (Fig. 4D). Mid-Cretaceous carbonate deposition resulted in deposition of the Morelos Formation in Mexico (e.g., Cabral-Cano et al., 2000) and the Atima Formation on the Chortis block (e.g., Finch, 1981). The continuity of Cretaceous strata across the Eastern and Central Chortis terranes and into Mexico constitutes an overlap assembly above the older Paleozoic and Precambrian terranes. The reconstruction in Figure 7A restores both areas of similar Cretaceous stratigraphy.

North-South Piercing Line 1 of Chortis and Southern Mexico

The Early Cretaceous and arc affinity of the Manto Formation on the Chortis block correlates with Early Cretaceous arc activity in southern Mexico (Rogers, et al., Chapter 5, this volume) (Figs. 2 and 7A). In Mexico, three separate Early Cretaceous arcs have been recognized and include the Arteaga, Arcelia, and Teloloapan (Mendoza and Suastegui, 2000). Rogers et al., (this volume, Chapter 5) compared the multi-elemental geochemical patterns of the Manto Formation volcanic rocks of Honduras with the reported values for the three Early Cretaceous volcanic arcs in Mexico and concluded that the geochemical data supports a direct correlation between the Teloloapan arc of Mexico and the Manto Formation of Honduras. The reconstruction in Figure 7 restores both areas of similar Early Cretaceous arc activity.

North-South Piercing Line 2 of Chortis and Southern Mexico

Late Cretaceous shortening, attributed to Laramide-age deformation in Mexico, is observed both in southern Mexico (Campa, 1985; Burkart et al., 1987; Lang et al., 1996; Cerca et al., 2004) and on the Chortis block (Horne et al., 1974; Donnelly et al., 1990; Rogers et al., Chapter 5, 2006). Our best-fit positioning of the Chortis block along the southern margin Mexico aligns the Late Cretaceous north-trending fold thrust belts of Mexico with the four Late Cretaceous structural belts in Honduras (Fig. 7).

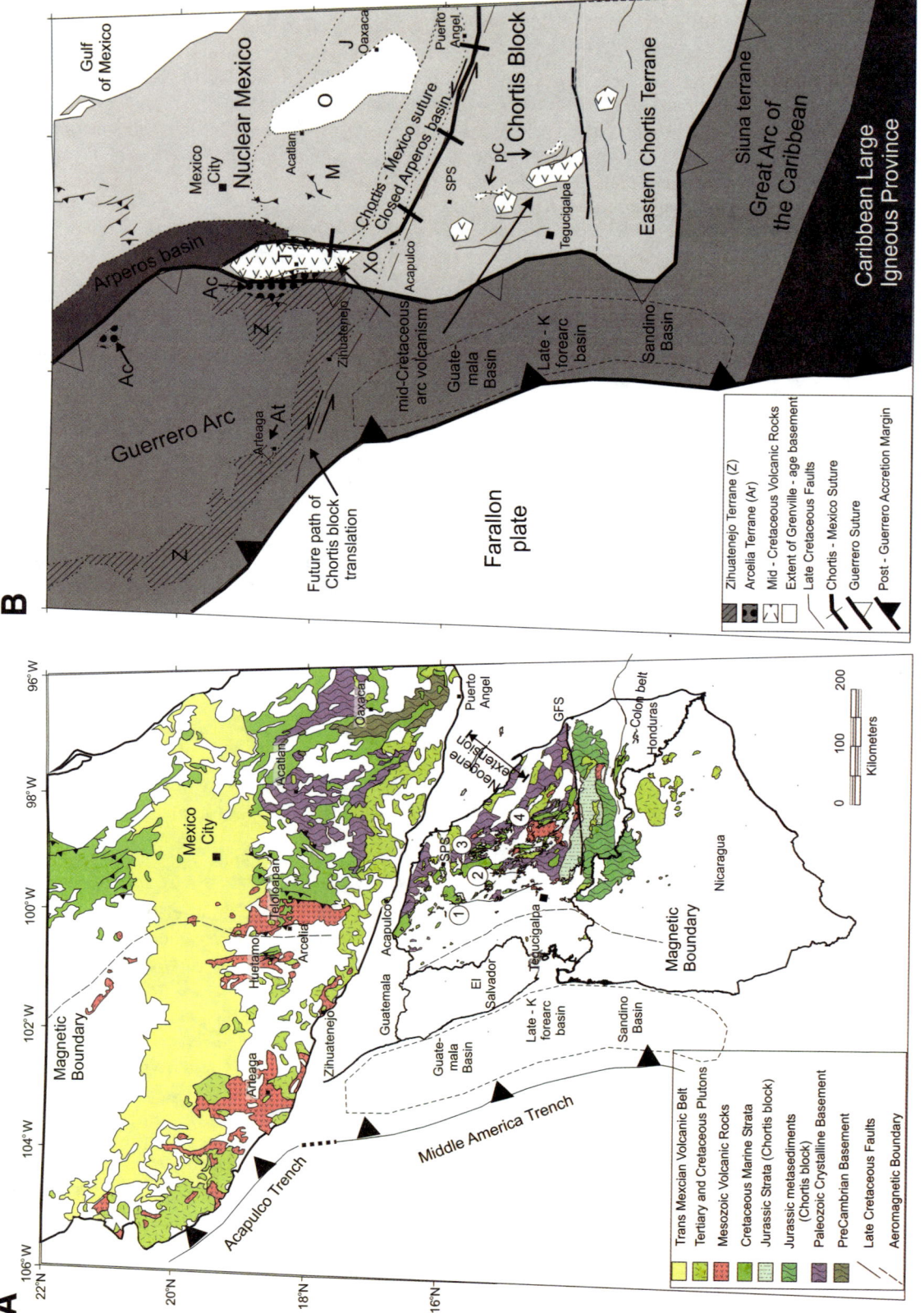

Figure 7. (A) Restoration of the Chortis block to a position adjacent to southwest Mexico in order to realign similar, pre-Tertiary structures and geologic units on the Chortis block and in southwestern Mexico. Key features in Honduras and southern Mexico: (1) a Cretaceous volcanic arc (cf. Rogers et al., Chapter 5, this volume); (2) a late Mesozoic stratigraphy (cf. stratigraphic columns in Fig. 4); (3) a Grenville-age basement exposure of Central Honduras terrane; (4) aligned rifts formed in late Cretaceous times (Rogers et al., Chapter 5, this volume); and (5) a distinct magnetic boundary between the accreted Guerrero terrane and nuclear Mexico and between the Southern and Central Chortis terranes boundaries (Fig. 8). Structural belts on the Chortis block are noted by circled numbers: 1—Comayagua belt; 2—Minas de Oro belt; 3—Montaña de la Flor belt; 4—Frey Pedro belt; abbreviations: SPS—San Pedro Sula; GFS—Guayape fault system. (B) Latest Cretaceous tectonic setting of the southwestern corner of the North American continent prior to the proposed eastward displacement of the Chortis block as part of the Caribbean plate. Following previous workers, we assume that the Guerrero arc and the Great arc of the Caribbean form a continuous, west and north-facing arc system that wraps around the continental promontory of southwestern North America. These large scale lithologic trends are supported by the aeromagnetic data shown in Figure 3. Ac—Arcelia; At—Arteaga; J—Jurassic; K—Cretaceous; M—Miocene; O—Oligocene; pC—pre-Cambrian; SPS—San Pedro Sula; T—Tertiary; Xo—Xolopa.

North-South Piercing Line 3 of Chortis and Southern Mexico

Prominent magnetic expression boundaries exist on both the Chortis block (Fig. 3) and in western Mexico (Fig. 8) with the location of the main terrane boundaries shown on Figure 7. In Mexico, the prominent magnetic boundary separates the accreted oceanic and arc Guerrero terrane from the autochthonous continental terranes of nuclear Mexico (Campa and Coney, 1983; Dickinson and Lawton, 2001). A similar relationship is present on the Chortis block where a major terrane boundary separates the Southern Chortis terrane known from a single basement to be underlain metaigneous (perhaps oceanic) rocks and the Central Chortis terranes known from outcrop studies to be underlain by continental rocks of Precambrian and Paleozoic age (Fig. 4).

East-West Piercing Line of Chortis and Southern Mexico

The Colon foldbelt of eastern Honduras represents the collisional response to the suturing of an island arc to the southern margin of North America in the late Cretaceous (Rogers et al., this volume, Chapter 6). We interpret this as the eastern extent of the same event that emplaced the Santa Cruz ophiolites and deformed the southern Guatemala in the late Cretaceous (Rosenfeld, 1981; Donnelly et al., 1990). Cenozoic translation of the Chortis to the southeast places the Colon belt to the east of its Guatemala counterpart.

Restoring the Chortis block along the southern margin of Mexico based on the best-fit alignment of features common to both regions results in the following configuration of the southwestern corner of North America prior to the eastward Cenozoic translation of the Chortis block that is summarized on the map in Figure 7B:

1. Continuity of the Chortis block and the autochthonous terranes of nuclear Mexico provides a common basement and platform for the deposition of late Mesozoic strata and its subsequent Late Cretaceous shortening (Fig. 7).
2. Continuity of the geochemically similar Cretaceous Teloloapan and Manto arcs links a north-south–trending arc now offset left-laterally (Fig. 7).
3. An open western and southern margin along the southwest corner of North America provide a backstop for the Guerrero arc and the Caribbean arc to accrete. Caribbean arc components include the Guerrero terrane, the southern Chortis terrane, and the Siuna terrane (Fig. 7).

Correlation between Chortis Terranes and Terranes in Southwestern Mexico

Using the common features and piercing lines between the Chortis block and southwestern Mexico, the late Cretaceous alignment of the Chortis block along the southern Mexico margin corresponds with the alignment of similar Chortis and Mexican terranes (Fig. 7). In this position we propose a correlation between the Chortis terranes and the terranes in Mexico based on common type and age of basement and correlative events (i.e., stratigraphic, igneous and deformation). We utilize the terrane names of Campa and Coney (1983) rather the different set of terms introduced by Sedlock et al., (1993). We take into account the recent terrane revision of Keppie (2004) where his revision is relevant to our proposed correlations.

To facilitate discussion of the various crustal elements of the southern Mexican Cordillera, Central America and the Caribbean studied by many previous authors, Table 3 summarizes key crustal elements of all proposed terranes. This table lists the terrane nomenclature used in the discussion below in the left hand column and is consistent with labeling of terranes shown on the regional map in Figure 1A. The terranes are organized by affinity to crustal elements following the approach of Dickinson and Lawton (2001) and include some of the modifications by Keppie (2004).

Central Chortis Terrane—Del Sur Block (Mixteco-Oaxaca Terranes)

The Central Chortis terrane with its basement of Paleozoic and Grenville-age metamorphic rocks and thick continental crust is correlated with the Del Sur block (Dickinson and Lawton, 2001) composed of elements of neighboring Mixteca and Oaxaca terranes (Campa and Coney, 1983; Sedlock et al., 1993) and the southern Oaxaquia terrane of Keppie (2004) (Fig. 1). Similarities between Cretaceous sequences of the central Chortis terrane and those of the Mixteco terranes and the Teloloapan subterrane of the Guerrero terrane include: (1) early Cretaceous rifting, platform carbonate and terrestrially derived redbed deposition; (2) development of intra-arc volcanism; and (3) Laramide-age inversion of Mesozoic sequences (Lang et al., 1996; Mendoza and Suastegui, 2000; Cabral-Cano, et al., 2000; Cerca et al., 2004; Figure 4 in Rogers et al., Chapter 5, this volume).

The Eastern Chortis terrane with a Jurassic metasedimentary basement was controlled by opening and rifting between the Americas in early Jurassic time (cf. Pindell and Barrett, 1990). The little studied Juarez or Cuicateco terrane of Mexico with its Jurassic metasedimentary and metavolcanic basement (Sedlock et al. 1993) appears a likely genetic equivalent to the Eastern Chortis terrane (Fig. 1). Dickinson and Lawton (2001) note that the Juarez terrane likely formed during proto-Caribbean opening and prior to the early Cretaceous arrival of the Maya (Yucatan) block from the northern Gulf of Mexico (Marton and Buffler, 1994).

The Southern Chortis Terrane—Guerrero Terrane

We correlate the Southern Chortis terrane to the Mexican Guerrero composite terrane. The Mexican Guerrero terrane is a Jurassic-Cretaceous island arc built on oceanic crust accreted to the western margin of Mexico (Centeno-Garcia et al., 1993; Sedlock et al., 1993). Guerrero accretion to western Mexico is interpreted as a diachronous collision from north to south spanning much of Cretaceous time (Tardy et al., 1994; Dickinson and Lawton, 2001).

The aeromagnetic map of Mexico (North American Magnetic Anomaly Group, 2002) displays a strong contrast between nuclear

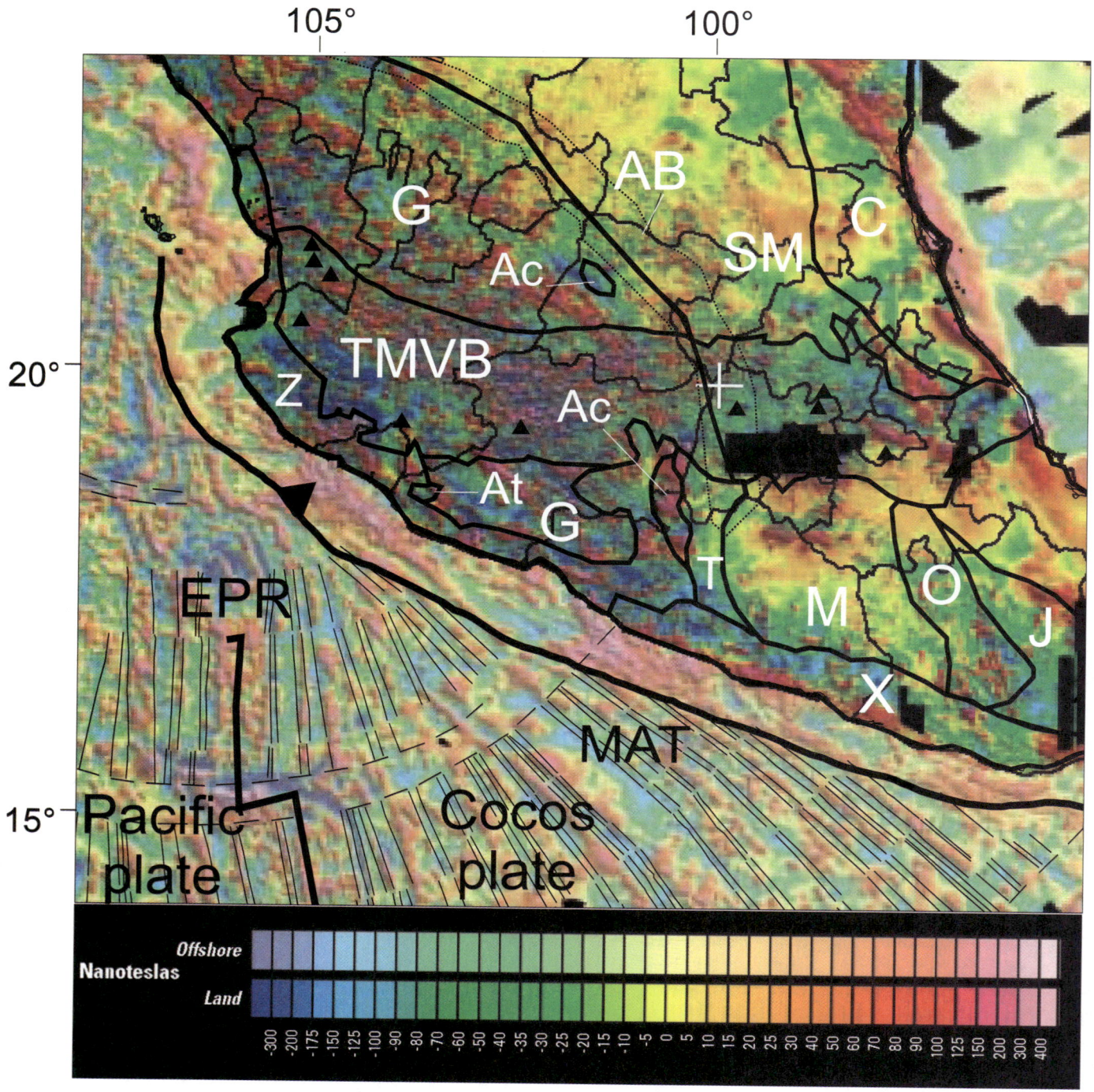

Figure 8. Magnetic intensity map of southern Mexico (data from the North American Magnetic Anomaly Group, 2002). The magnetic map is overlain by terrane boundaries (heavy black lines) from Campa and Coney (1983). Note that an abrupt magnetic contrast marks the suture zone between the allochthonous Guerrero arc terrane to the west with autochthonous, continental terranes of nuclear Mexico to the east. The outline of the Late Cretaceous Arperos basin (AB) formed along the suture zone is shown by the dotted line. Note that the Teloloapan terrane (assumed by some workers to be a Guerrero subterrane) actually lies to the east of the magnetic boundary–suture zone, suggesting that it is distinct from the Guerrero terrane. There is a similar, strong magnetic intensity contrast between the Guerrero arc terrane and nuclear, continental Mexico as that between the arc-related Southern Chortis terrane and the continental Central Chortis terrane (Fig. 3). Mexican terrane abbreviations: C—Coahuila; G—Guerrero; J—Juarez; M—Mixteca; O—Oaxaca; SM—Sierra Madre; X—Xolopa. Guerrero subterrane (Mexico) abbreviations: Ac—Argelia; At—Arteaga; T—Teloloapan; Z—Zihuatanejo. AB—Arperos basin, Mexico; TMVB—Trans-Mexican Volcanic Belt. Thin lines in Mexico are state borders. EPR—East Pacific Rise; MAT—Middle America trench.

TABLE 3. CRUSTAL ELEMENTS OF SOUTHERN MEXICO, CENTRAL AMERICA, AND THE CARIBBEAN (cf. FIGS. 1 AND 6)

Crustal element	Basement age	Names from previous workers
Gondwana elements (continental crust)		
Oaxaca terrane (1)	Grenville	Del Sur (2); Zapoteca (3) Oaxaquia (10)
Mixteca terrane (1)	Paleozoic	Del Sur (2)
Xolopa terrane (1)	Paleozoic	Del Sur (2); Chatino (3)
Central Chortis terrane*	Grenville	Undifferentiated Chortis (2, 3); includes El Tambor Western Chortis (4); and Motagua (10)
Northern Chortis metamorphic zone*	Paleozoic	Undifferentiated Chortis (2, 3); northern Chortis, western Honduras borderlands and Explorer province (4)
Tampico block (SE Maya) (2)	Paleozoic	eastern Sierra Madre and northwest Maya (1, 3); Guachichili (3); Chiapas-Peten, Chiapas massif, Yucatan platform, (4)
Teloloapan subterrane (6)	Pre-Mesozoic?	Tierra Caliente complex (7); Nahuatl (3)
Proto-Caribbean elements (transitional to oceanic crust)		
Eastern Chortis terrane*	Jurassic	Undifferentiated Chortis (2, 3); Southern Chortis, northern Mosquitia-Nicaragua Rise basins (4)
Juarez terrane (1)	Jurassic	Cuicateco (3)
Subduction complexes and suture belts		
Arperos basin (2) (closed)	Late Jurassic– Early Cretaceous	Parral, Sombrerete, Toliman (1); western Tepehuano (3); overlaps eastern Guerrero (1, 2) and Nahuatl, Tahue (3)
Motagua (4)	Cretaceous	southernmost Maya (1, 2, 3)
Siuna terrane (5)	Cretaceous	Undifferentiated Chortis (2, 3); Yolaina, southern Mosquitia-Nicaragua Rise basins (4)
Intraoceanic (paleo-Pacific) island arc complex		
Guerrero terrane (1, 2)	Late Jurassic– Early Cretaceous	Nahuatl, Tahue (3)
Arteaga subterrane (6)	Late Jurassic– Early Cretaceous	part of Nahuatl (3)
Zihuatanejo subterrane (6)	Late Jurassic– Early Cretaceous	part of Nahuatl (3)
Arcelia subterrane (6)	Late Jurassic– Early Cretaceous	part of Nahuatl (3)
Southern Chortis terrane*	Mesozoic?	Middle America volcanic province, forearc basin and northern forearc ridge (4); Guerrero (2, 8)
Siuna terrane (5)	Late Jurassic– Early Cretaceous	Undifferentiated Chortis (2, 3); Yolaina, southern Mosquitia-Nicaragua Rise basins (4)
Cuba (amalgamated)	mainly Late Jurassic– Early Cretaceous	Zaza, Pural, Sierra Maestra, Escambray, Las Villas, Cifuentes-Plancetas, Cascarajicara, Organos-Rosario, Isla de Pinos (4)
Cayman Ridge	Tertiary?	Pickle (4)
Oceanic crust		
Cayman Trough	post-Paleocene	
Yucatan Basin	Eocene	
Intraoceanic (paleo-Pacific) plateau		
Caribbean large igneous province	Cretaceous	Columbia basin (4)
Chorotega block (9) Built on CLIP	Cretaceous	Talamanca-Gatun, southern Middle America forearc ridge, San Blas-Darian, (4)

Note: References: (1)–Campa and Coney (1983); (2)–Dickinson and Lawton (2001); (3)–Sedlock et al. (1993); (4)–Case et al. (1984, 1990); (5)–Venable (1994); (6)–Mendoza and Suastegui (2000); (7)–Cabal-Cano et al. (2000); (8)–Tardy et al. (1994); (9)–Dengo (1985); (10)–Keppie (2004). Abbreviations: CLIP—Caribbean Large Igneous Province.
*Defined in this paper.

Mexico and the accreted Guerrero arc terrane (Fig. 8) that is similar to the strong contrast between the Southern Chortis terrane of proposed arc affinity and the Central Chortis terrane of known continental affinity (Fig. 3) (Table 2). Restoring the Chortis block to its pre-Tertiary position along the southwest margin of Mexico places what is now the southern Chortis block of Honduras near Acapulco, Mexico, where the Guerrero terrane intersects the Mexican margin (Fig. 7) (Rogers et al., Chapter 5, this volume). This suggests the Southern Chortis terrane is most similar to the Arcelia or Zihuatanejo terranes of the composite Guerrero terrane.

Siuna Terrane—Caribbean Arc

Venable (1994) recognized and defined the Siuna terrane of the Chortis superterrane as an accreted island arc built on oceanic crust and accreted to the southern margin of the Chortis block in the late Cretaceous. Refinements of southern Cordillera and Caribbean evolution by various authors (Tardy et al., 1994; Moores, 1998; Mann, 1999) suggest that the Siuna terrane was the westernmost accreted terrane produced by the collision between the Caribbean arc and the continental margin of Chortis (Rogers et al., Chapter 6, this volume).

Northern Chortis Magmatic Zone—Xolopa Terrane

The Northern Chortis magmatic zone and the Xolopa terrane of Mexico appear to reflect early Tertiary magmatic and metamorphic overprinting (Fig. 1). We assume that the observed overprinting on both terranes suggests that both blocks shared a common history.

In the Northern Chortis magmatic zone the high grade and sheared gneiss with east-west trending left-slip indicators suggests exposure of much deeper crustal levels than observed to the south in the Central Chortis terrane (Southernwood, 1986; Manton, 1987) (Fig. 2). Late Cretaceous and early Tertiary felsic intrusives are common (Fig. 5) in the magmatic zone along with metamorphosed Mesozoic strata including marble containing Cretaceous rudists (Wilson, 1974) (Fig. 2).

In Mexico, the Xolopa terrane displays: (1) progressively increasing metamorphic grade toward the truncated Mexican coast of the Mixteco terrane (Oretega-Gutierrez and Elías-Herrera, 2003); (2) intrusion by plutons of decreasing age to the southeast (Herrmann et al., 1994; Schaaf et al., 1995); and (3) near-vertical southeast-striking mylonitic shear zones with left-slip kinematic indicators (Riller et al., 1992). Keppie (2004) eliminates Xolopa entirely as a terrane. We emphasize the strong similarities remain between this region of Mexico and the Northern Chortis magmatic zone and realign them on the reconstruction shown in Figure 7.

We suggest that both the Northern Chortis magmatic zone and the area of the Xolopa terrane share a genetic origin with the development of a magmatic arc in the early Tertiary oblique to the Farallon-Chortis trench, in a manner similar to the present-day Trans-Mexican Volcanic Belt (Cerca et al., 2004). The obliquity of the early Tertiary arc relative to the trench combined with oblique convergence of Farallon-North America (Engebretson et al., 1985; Schaaf et al., 1995) provides the tectonic setting to translate the Chortis block as a large forearc sliver detached along shear zones developed in the magmatic arc (Riller et al., 1992). This interpretation involving the breaking of a shear zone through the continental arc margin contrasts with the model by Keppie (2004) and Keppie and Morán-Zenteno (2005) for prolonged subduction erosion of a hundreds of kilometers wide swath of the Mexican continental margin

Tectonic Evolution of the Chortis Block

Restoring the Chortis block to its pre-translation (pre-Eocene) position adjacent to the truncated margin of southwestern Mexico re-aligns the Colon fold-thrust belt of the Eastern Chortis terrane with the fold-thrust belts and ophiolites north of the Motagua suture in Guatemala (Donnelly et al., 1990; Burkart et al., 1994) (Fig. 9A). The present day configuration of these elements is shown in Figure 9B. This reconstruction also provides a best fit of the following elements common to the Chortis block and southwestern Mexico that are shown in Figure 7 and include (1) **influx of late Cretaceous terrigenous sandstone and shale** over early Cretaceous shallow marine platform limestone of both southern Mexico and Chortis, (2) **Grenville-age basement** common to both areas, (3) **Late Cretaceous shortening** structures common to both areas, and (4) **mid-Cretaceous arc volcanism** that is geochemically similar in both areas.

Three independent lines of evidence support our proposed interpretation shown in Figure 9A that the eastern Chortis block records the collision of the Caribbean arc with the southern margin of North America in the late Cretaceous. The first is the 350 km-long Colon fold belt with Campanian-age, northwest-directed shortening described by Rogers et al. in Chapter 6 of this volume. The second is the spatial association and inferred accretion of the intraoceanic Siuna island arc complex on the southern margin of the Chortis block in the late Cretaceous (Venable, 1994). The third is the Pacific origin of the Caribbean arc and its position at the leading edge of the Caribbean large igneous province (Pindell and Barrett, 1990; Pindell et al., 2006). The entry of this arc-plateau feature into the area of the proto-Caribbean Sea shown in Figure 9A led to the subduction of proto-Caribbean oceanic crust and partial accretion of the Caribbean arc and the oceanic plateau at the "gateways" to the Caribbean in Colombia in South America and in southern Central America where large areas of the crust appear to have been built on oceanic plateau material (Kerr et al., 1998). The Chortis block and southwestern Mexico "corner" shares many common elements to the corner of northwestern South America: in both areas a core of continental rocks is surrounded by collided arc rocks (Fig. 7B and 9A). In the case of Chortis and southwestern Mexico, the arc rocks are not yet as well understood as those in northwestern South America.

CONCLUSIONS

Correlation of a 1985 aeromagnetic survey of Honduras with mapped basement outcrop exposures in Honduras and Nicaragua forms the basis to define three terranes on the Chortis block: (1) **Central Chortis terrane** with exposed Paleozoic basement of continental origin; the northern edge of this terrane is modified by early Tertiary metamorphism and intrusions that overprints areas of the Central and Eastern Chortis terranes; (2) **Eastern Chortis terrane** with a Jurassic metasedimentary basement; and (3) **Southern Chortis terrane** of low magnetic intensity covered by Miocene pyroclastic strata and limited basement exposure. The **Siuna terrane** of Nicaragua of oceanic island origin that accreted to the Chortis block in the late Cretaceous; and the **Northern Chortis metamorphic zone** is defined by early Tertiary metamorphism and intrusions that overprints areas of the Central and Eastern Chortis terranes. These proposed terranes are preliminary but are based on the most up to date compilation of data that is available including the aeromagnetic data. Future mapping and sampling will likely modify the terrane definitions and precisely locate their boundaries with adjacent terranes.

Shared geologic and magnetic characteristics between Chortis and southwestern Mexican terranes—previously described by Campa and Coney (1983), Sedlock et al. (1993) Dickinson

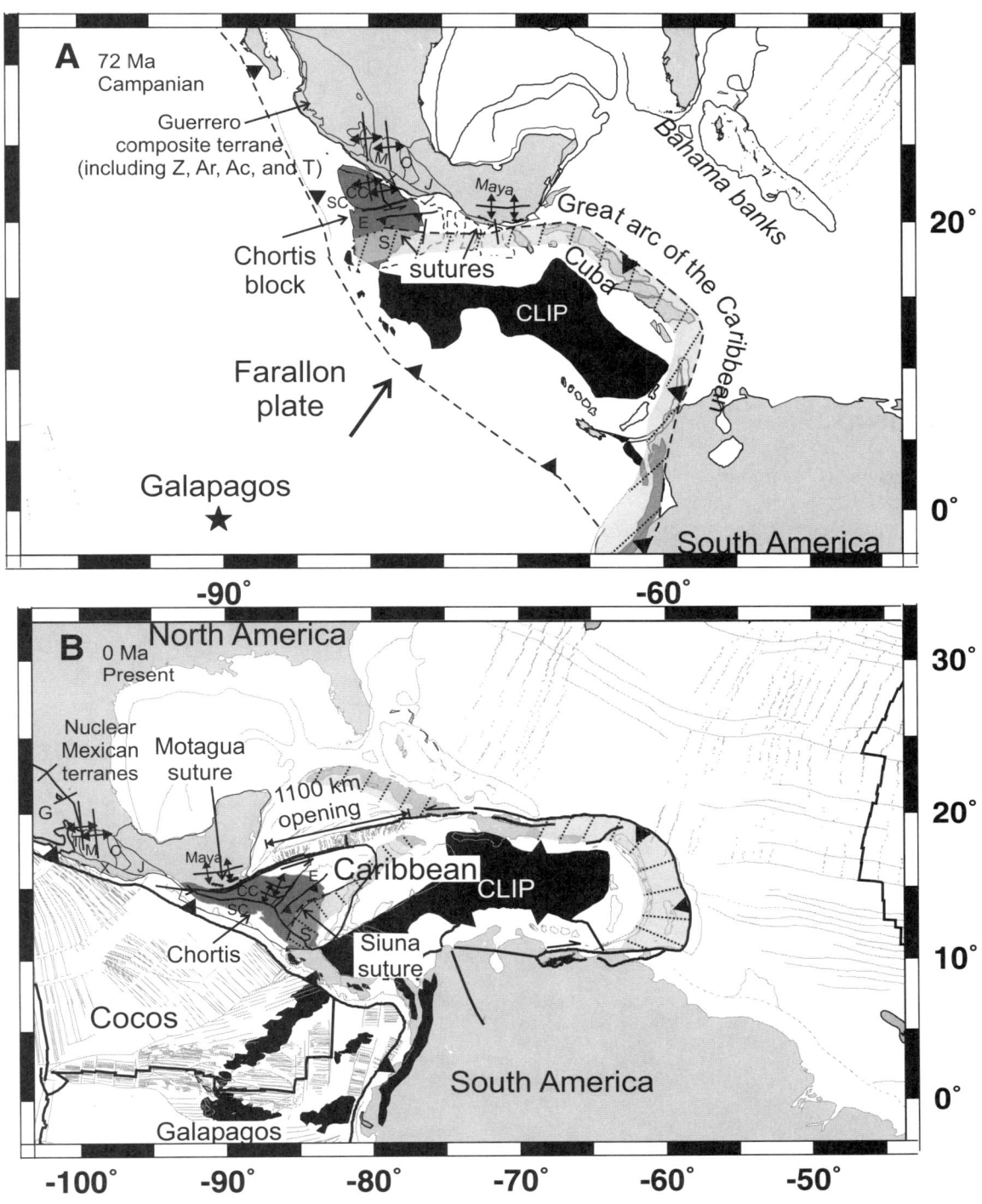

Figure 9. (A) Plate reconstruction of the Chortis block during Campanian time (ca. 72 Ma). This reconstruction uses a mantle reference frame, and the Galapagos hotspot is held fixed. At this time, the Chortis block (medium-gray area) is proposed to form the southwestern continental corner of the North America plate. The Caribbean large igneous province (CLIP; dark area) formed ca. 88 Ma in the eastern Pacific Ocean as part of the Farallon plate and moved northeastward with the Caribbean arc at its leading edge. The Caribbean arc and trailing Caribbean large igneous province collided with the edges of North and South America on either side of the gap (proto-Caribbean Sea) formed by Late Jurassic–Early Cretaceous rifting and oceanic spreading between the two Americas. Accreted parts of the Caribbean large igneous province in South America are shown by the darker areas. Ac—Arcelia; Ar—Arteaga; CC—Central Chortis terrane; E—Eastern Chortis terrane; J—Juarez terrane; M—Maya block; O—Oaxaca terrane; S—Siuna terrane; SC—Southern Chortis terrane; T—Teloloapan; Z—Zihuatanejo. (B) Present-day position of the same tectonic elements shown in A.

and Lawton (2001) and Keppie (2004)—were used to correlate: (1) the Central Chortis terrane to the Mixteca and Oaxaca terranes of Mexico, (2) the Eastern Chortis terrane with the Juarez terrane of Mexico (3) the Southern Chortis terrane to the Guerrero terrane of Mexico, (4) the Northern Chortis metamorphic zone to the Xolopa terrane of Mexico and (5) the Siuna terrane to the circum-Caribbean volcanic arc, or the "Great Arc of the Caribbean" (Pindell and Barrett, 1990; Mann, 1999).

Linking Chortis and Mexican terranes in pre-Tertiary time provides a starting point for improved understanding of the complex geology of terranes in both areas. We present plate reconstructions using all regional plate kinematic and geologic constraints to provide insights into the main tectonic events affecting the Chortis block.

ACKNOWLEDGMENTS

Funding for R. Rogers and P. Mann was provided by the Petroleum Research Fund of the American Chemical Society (grant 33935-AC2 to P. Mann). We thank the Dirección de Energia of Honduras and the Japanese Petroleum Exploration Co. Geoscience Institute for releasing aeromagnetic data used in the study. Bill Dickinson, Ginny Sisson, Randy Marrett, Bill Muehlberger, Ric Finch, Mark Gordon, and Wulf Gose provided many useful discussions on the geology and tectonics of the Chortis block. Special thanks to Lisa Bingham for her help with the text and figures. We thank two anonymous reviews for helpful comments. The authors acknowledge the financial support for publication costs provided by the University of Texas at Austin's Geology Foundation and the Jackson School of Geosciences. This is University of Texas Institute for Geophysics contribution no. 1869.

REFERENCES CITED

Arden, D., 1975, Geology of Jamaica and the Nicaraguan Rise, *in* Nairn, A., and Stehli, F., eds., The ocean basins and margins: New York, Plenum Press, v. 3, p. 616–661.

Avé Lallemant, H., and Gordon, M., 1999, Deformation history of Roatan Island: Implications for the origin of the Tela basin (Honduras), *in* Mann, P., ed., Caribbean basins: Sedimentary basins of the world series: Amsterdam, Elsevier, v. 4, p. 197–218.

Burkart, B., 1994, Northern Central America, *in* Donovan, S., and Jackson, T., eds., Caribbean geology: An introduction: Jamaica, University of the West Indies Publisher's Association, p. 265–284.

Burkart, B., and Self, S., 1985, Extension and rotation of crustal blocks in northern Central America and effect on the volcanic arc: Geology, v. 13, p. 22–26, doi: 10.1130/0091-7613(1985)13<22:EAROCB>2.0.CO;2.

Burkart, B., Denton, B., Dengo, C., and Moreno, G., 1987, Tectonic wedges and offset Laramide structures along the Polochic fault of Guatemala and Chimpas, Mexico: Reaffirmation of large Neogene displacement: Tectonics, v. 9, p. 411–422.

Campa, M., 1985, The Mexican thrust belt, *in* Howell, D.G., ed., Tectonostratigraphic terranes of the Circum-Pacific Region: Houston, Texas, Circum-Pacific Council on Energy and Mineral Resources, Earth Science Series no. 1, p. 299–313.

Campa, M., and Coney, P., 1983, Tectono-stratigraphic terranes and mineral resource distributions in Mexico: Canadian Journal of Earth Sciences, v. 20, p. 1040–1051.

Cabral-Cano, E., Lang, H., and Harrison, C., 2000, Stratigraphic assessment of the Arcelia-Teloloapan area, southern Mexico: Implications for southern Mexico's post-Neocomian tectonic evolution: Journal of South American Earth Sciences, v. 13, p. 443–457, doi: 10.1016/S0895-9811(00)00035-3.

Carr, M., Feigenson, M., Patino, L., and Walker, J., 2003, Volcanism and geochemistry in Central America: Progress and problems, *in* Eiler, J., and Abers, G., eds., The subduction factory: American Geophysical Union Monograph Series, v. 138, p. 153–179.

Case, J., Holcombe, T., and Martin, R., 1984, Map of geological provinces in the Caribbean region: The Geological Society of America Memoir 162, p. 1–30.

Case, J., MacDonald, W., and Fox, P., 1990, Caribbean crustal provinces; seismic and gravity evidence, *in* Dengo, G., and Case, J.E., eds., The Caribbean Region: Boulder, Colorado, The Geological Society of America, The Geology of North America, v. H, p. 1536.

Centeno-Garcia, E., Ruiz, J., Coney, P., Patchett, P., and Ortega-Gutierrez, F., 1993, Guerrero Terrane of Mexico; its role in the Southern Cordillera from new geochemical data: Geology, v. 21, p. 419–422, doi: 10.1130/0091-7613(1993)021<0419:GTOMIR>2.3.CO;2.

Cerca, M., Manetti, P., Ferrari, L., Bonini, M., and Corti, G., 2004, The role of crustal heterogeneity in controlling vertical coupling during Laramide shortening and the development of the Caribbean–North America transform boundary in southern Mexico: Insights from analogue models: Geological Society [London] Special Publication 227, p. 117–140.

Cumming, G., Kesler, S., and Krstic, D., 1981, Source of lead in Central America and Caribbean mineralization: Earth and Planetary Science Letters, v. 31, p. 262–268.

DeMets, C., 2001, A new estimate for present-day Cocos-Caribbean place motion: Implications for slip along the Central American volcanic arc: Geophysical Research Letters, v. 28, p. 4043–4046/

DeMets, C., Jansma, P., Mattioli, G., Dixon, T., Farina, F., Bilham, R., Calais, E., and Mann, P., 2000, GPS geodetic constraints on Caribbean–North American plate motion: Geophysical Research Letters, v. 27, p. 437–440, doi: 10.1029/1999GL005436.

DeMets, C., Mattioli, G., Jansma, P., Rogers, R., Tenorios, C., and Turner, H.L., 2007, this volume, Present motion and deformation of the Caribbean plate: Constraints from new GPS geodetic measurements from Honduras and Nicaragua, *in* Mann, P., ed., Geologic and tectonic development of the Caribbean plate boundary in northern Central America: The Geological Society of America Special Paper 428, doi: 10.1130/2007.2428(02).

Dengo, G., 1985, Mid America: Tectonic setting for the Pacific margin from southern Mexico to northwestern Columbia, *in* Nairn, A., and Stehli, F., eds., The ocean basins and margins: New York, Plenum Press, v. 7, p. 123–180.

Dickinson, W., and Lawton, T., 2001, Carboniferous to Cretaceous assembly and fragmentation of Mexico: The Geological Society of America Bulletin, v. 113, p. 1142–1160, doi: 10.1130/0016-7606(2001)113<1142:CTCAAF>2.0.CO;2.

Dirección General de Minas e Hidrocarburos, 1985, Aeromagnetic survey of Honduras: Tegucigalpa, Honduras, Dirección General de Minas e Hidrocarburos, scale 1:500,000, 1 sheet.

Donnelly, T., Horne, G., Finch, R., and López-Ramos, E., 1990, Northern Central America: The Maya and Chortis blocks, *in* Dengo, G., and Case J., eds., The Caribbean Region: Boulder, Colorado, The Geological Society of America, The Geology of North America, v. H, p. 37–76.

Drobe, J., and Cann, R., 2000, Cu-Au skarn mineralization, Minas de Oro District, Honduras, Central America: Exploration and Mining Geology, v. 9, p. 51–63, doi: 10.2113/0090051.

Emmet, P., 1983, Geology of the Agalteca Quadrangle, Honduras, Central America [M.S. thesis]: Austin, Texas, University of Texas at Austin, 201 p.

ENEE (Empressa Nacional de Energia Electrica), 1987, Geologic investigations of proposed hydrothermal sites, Honduras: Tegucigalpa, Honduras, Honduras Electrical Company, 26 p.

Engebretson, D., Cox, A., and Gordon, R., 1985, Relative motions between oceanic and continental plates in the Pacific Basin: The Geological Society of America Special Paper 206, 59 p.

Fakundiny, R.H., 1970, Geology of the El Rosario Quadrangle, Honduras, Central America [PhD dissertation]: Austin, Texas, University of Texas at Austin, 234 p.

Finch, R., 1981, Mesozoic stratigraphy of central Honduras: AAPG Bulletin, v. 65, p. 1320–1333.

Gordon, M.B., 1993, Revised Jurassic and Early Cretaceous (Pre-Yojoa Group) stratigraphy of the Chortis block: Paleogeographic and tectonic implications, in Pindell, J.L., and Perkins, R.F., eds., Mesozoic and Early Cenozoic development of the Gulf of Mexico and Caribbean region: A context for hydrocarbon exploration: Austin, Texas, Gulf Coast Section Society of Economic Paleontologists and Mineralogists Foundation, p. 143–154.

Gordon, M., and Muehlberger, W., 1994, Rotation of the Chortís block causes dextral slip on the Guayape fault: Tectonics, v. 13, p. 858–872, doi: 10.1029/94TC00923.

Gose, W., 1985, Paleomagnetic results from Honduras and their bearing on Caribbean tectonics: Tectonics, v. 4, p. 565–585.

Gose, W., and Swartz, D., 1977, Paleomagnetic results from Cretaceous sediments in Honduras: Tectonic implications: Geology, v. 5, p. 505–508, doi: 10.1130/0091-7613(1977)5<505:PRFCSI>2.0.CO;2.

Guzman-Speziale, M., 2001, Active seismic deformation in the grabens of northern Central America and its relationship to the relative motion of the North America–Caribbean plate boundary: Tectonophysics, v. 337, p. 39–51, doi: 10.1016/S0040-1951(01)00110-X.

Guzman-Speziale, M., and Meneses-Rocha, J.J., 2000, The North America–Caribbean plate boundary west of the Motagua-Polochic fault system, A jog in southeastern Mexico: Journal of South American Earth Sciences, v. 13, p. 459–468, doi: 10.1016/S0895-9811(00)00036-5.

Harwood, R., 1993, Mapa Geológica de Honduras, Hoja de Yuscaran: Tegucigalpa, Honduras, Instituto Geográfico Nacional, scale 1:50,000.

Herrmann, U., Nelson, B., and Rathschbacher, L., 1994, The origin of a terrane: U/Pb zircon geochronology and tectonic evolution of the Xolapa complex (southern Mexico): Tectonics, v. 13, p. 455–474, doi: 10.1029/93TC02465.

Horne, G., Atwood, M., and King, A., 1974, Stratigraphy, sedimentology, and paleoenvironment of Esquias Formation of Honduras: AAPG Bulletin, v. 58, p. 176–188.

Horne, G., Clark, G., and Pushkar, P., 1976a, Pre-Cretaceous rocks of northwestern Honduras: Basement terrane in Sierra de Omoa: AAPG Bulletin, v. 60, p. 566–583.

Horne, G., Pushkar, P., and Shafiqullah, M., 1976b, Laramide plutons on the landward continuation of the Bonacca ridge, northern Honduras: Guatemala, Publicaciones Geológicas del ICAITI, v. 5, p. 84–90.

Howell, D.G., Jones, D.L., and Schermer, E.R., 1985, Tectonostratigraphic terranes of the Circum-Pacific region: Principles of terrane analysis, in Howell, D.G., ed., Tectonostratigraphic terranes of the Circum-Pacific region: Circum-Pacific Council for Energy and Mineral Resources, Earth Science Series, v. 1, p. 3–31.

INETER (Instituto Nicaragüense de Estudios Territoriales), 1995, Mapa Geológica Minero de la Republica de Nicaragua: Managua, Nicaragua, escala 1:500,000.

James, K., 2006, Arguments for and against the Pacific origin of the Caribbean Plate: Discussion, finding for an inter-American origin: Geological Acta, v. 4, p. 279–302.

Johnson, C., Lang, H., Cabral-Cano, E., Draper, G., Harrison, C., and Barros, J., 1991, Preliminary assessment of stratigraphy and structure, San Lucas region, Michoacan and Guerrero states, SW Mexico: Mountain Geologist, v. 28, p. 121–135.

Jordan, B.R., Sigurdsson, H., Carey, S., Lundin, S., Rogers, R., Singer, B., and Barquero-Molina, M., 2007, this volume, Petrogenesis of Central American Tertiary ignimbrites and associated Caribbean Sea tephra, in Mann, P., ed., Geologic and tectonic development of the Caribbean plate boundary in northern Central America: The Geological Society of America Special Paper 428, doi: 10.1130/2007.2428(07).

Karig, D.E., Cardwell, R.K., Moore, G.F., and Moore, D.G., 1978, Late Cenozoic subduction and continental margin truncation along the northern Middle America Trench: The Geological Society of America Bulletin, v. 89, p. 265–276, doi: 10.1130/0016-7606(1978)89<265: LCSACM>2.0.CO;2.

Keppie, J., 2004, Terranes of Mexico revisited: A 1.3 billion year odyssey: International Geology Review, v. 46, p. 765–794.

Keppie, J., and Morán-Zenteno, D., 2005, Tectonic implications of alternative Cenozoic reconstructions for southern Mexico and the Chortis block: International Geology Review, v. 47, p. 473–491.

Kerr, A., Tarney, J., Nivia, A., Marriner, G., and Saunders, A., 1998, The internal structure of oceanic plateaus: Inferences from obducted Cretaceous terranes in western Colombia and the Caribbean: Tectonophysics, v. 292, p. 173–188, doi: 10.1016/S0040-1951(98)00067-5.

Kesler, S., Levy, E., and Martín, F., 1990, Metallogenic evolution of the Caribbean region, in Dengo, G., and Case, J.E., eds., The Caribbean Region: Boulder, Colorado, The Geological Society of America, The Geology of North America, v. H, p. 459–482.

Kozuch, M., 1991, Mapa Geológica de Honduras: Tegucigalpa, Honduras, Instituto Geográfico Nacional, escala 1:500,000.

Lang, H., Barros, J., Cabral-Cano, E., Draper, G., Harrison, C., Jansma, P., and Johnson, C., 1996, Terrane deletion in northern Guerrero State: Geofísica Internacional, v. 35, p. 349–359.

Leroy, S., Mauffret, A., Patriat, P., and Mercier de Lepinay, B., 2000, An alternative interpretation of the Cayman trough evolution from a reidentification of magnetic anomalies: Geophysical Journal International, v. 141, p. 539–557, doi: 10.1046/j.1365-246x.2000.00059.x.

MacDonald, W.D., 1976, Cretaceous-Tertiary evolution of the Caribbean, in Causso, R., ed., Transactions of the 7th Caribbean Geologic Conference, Guadaloupe: Paris, Bureau de Recherches Geologiques et Minières Publication 1524, p. 69–78.

Mann, P., 1999, Caribbean sedimentary basins: Classification and tectonic setting from Jurassic to Present, in Mann, P., ed., Caribbean basin, Sedimentary basins of the world series: Amsterdam, Elsevier, v. 4, p. 3–31.

Manton, W., 1987, Tectonic interpretation of the morphology of Honduras: Tectonics, v. 6, p. 633–651.

Manton, W., 1996, The Grenville of Honduras: The Geological Society of America Abstracts with Programs, v. 28, no. 7, p. A-493.

Manton, W., and Manton, R., 1984, Geochronology and Late Cretaceous–Tertiary tectonism of Honduras: Tegucigalpa, Honduras, Dirección General de Minas e Hidrocarburos, 55 p.

Manton, W., and Manton, R., 1999, The southern flank of the Tela basin, Republic of Honduras, in Mann, P., ed., Caribbean basins: Sedimentary basins of the world series: Amsterdam, Elsevier, v. 4, p. 219–236.

Markey, R., 1995, Mapa Geológica de Honduras, Hoja de Moroceli (Geologic Map of Honduras, Morocelli sheet): Tegucigalpa, Honduras, Instituto Geográfico Nacional, escala 1:50,000.

Marton, G., and Buffler, R., 1994, Jurassic reconstruction of the Gulf of Mexico basin: International Geology Review, v. 36, p. 545–586.

Mendoza, T.O., and Suastegui, M., 2000, Geochemistry and isotopic composition of the Guerrero terrane (western Mexico): Implications for the tectono-magmatic evolution of southwestern North America during the Late Mesozoic: Journal of South American Earth Sciences, v. 13, p. 297–324, doi: 10.1016/S0895-9811(00)00026-2.

Metal Mining Agency of Japan (MMAJ), 1980a, Report on geology survey of the western area (Olancho): Japan International Cooperation Agency, Government of Japan, v. 5, 138 p.

Metal Mining Agency of Japan (MMAJ), 1980b, Report on geology survey of the western area (consolidated report): Japan International Cooperation Agency, Government of Japan, v. 6, 177 p.

Mills, R., and Barton, R., 1996, Geology of the Ahuas area in the Mosquitia Basin of Honduras: Preliminary report: AAPG Bulletin, v. 58, p. 189–207.

Mills, R., Hugh, K., Feray, D., and Swolfs, H., 1967, Mesozoic stratigraphy of Honduras: AAPG Bulletin, v. 51, p. 1711–1786.

Moores, E., 1998, Ophiolites, the Sierra Nevada, "Cordillera," and orogeny along the Pacific and Caribbean margins of North and South America: International Geology Review, v. 40, p. 40–54.

North American Magnetic Anomaly Group, 2002, Magnetic anomaly map of North America: U.S. Geological Survey Special Map, scale 1:10,000,000, http://pubs.usgs.gov/sm/mag_map/.

Oretega-Gutierrez, F., and Elías-Herrera, M., 2003, Wholesale melting of the southern Mixteco terrane and origin of the Xolapa complex: The Geological Society of America Abstracts with Program, v. 35, no. 4, p. 66.

Oretega-Gutierrez, F., Ruiz, J., and Centeno-Garcia, E., 1995, Oaxaquia, a Proterozoic microcontinent accreted to North America during the late Paleozoic: Geology, v. 23, p. 1127–1130, doi: 10.1130/0091-7613(1995)023<1127:OAPMAT>2.3.CO;2.

Oretega-Gutierrez, F., Solari, L., Sole, J., Martens, U., Gomez-Tuena, A., Moran-Ical, S., Reyes-Salas, M., and Ortega-Obregon, C., 2004, Polyphase, high temperature eclogite-facies metamorphism in the Chuacus Complex, central Guatemala: Petrology, geochronology, and tectonic implications: International Geology Review, v. 46, p. 445–470.

Pindell, J., and Barrett, S., 1990, Geological evolution of the Caribbean region; a plate-tectonic perspective, *in* Dengo, G., and Case, J.E., eds., The Caribbean Region: Boulder, Colorado, The Geological Society of America, The Geology of North America, v. H, p. 405–432.

Pindell, J., and Dewey, J., 1982, Permo-Triassic reconstruction of western Pangea and the evolution of the Gulf of Mexico/Caribbean region: Tectonics, v. 1, p. 179–212.

Pindell, J., Kennan, L., Stanek, K., Maresch, W., and Draper, G., 2006, Foundations of Gulf of Mexico and Caribbean evolution: Eight controversies resolved: Geologica Acta, v. 4, p. 303–341.

Ranero, C., von Huene, R., Flueh, E., Duarte, M., Baca, D., and McIntosh, K., 2000, A cross section of the convergent Pacific margin of Nicaragua: Tectonics, v. 19, p. 335–357, doi: 10.1029/1999TC900045.

Riller, U., Ratschbacher, L., and Frisch, W., 1992, Left-lateral transtension along the Tierra Colorada deformation zone, northern margin of the Xolapa magmatic arc of southern Mexico: Journal of South American Earth Sciences, v. 5, p. 237–249, doi: 10.1016/0895-9811(92)90023-R.

Rogers, R.., 1995, Mapa geológica de Honduras, Hoja de Valle de Jamastran: Tegulcigalpa, Honduras Instituto Geográfico Nacional, escala 1:50,000.

Rogers, R.D., and Mann, P., 2007, this volume, Transtensional deformation of the western Caribbean–North America plate boundary zone, *in* Mann, P., ed., Geologic and tectonic development of the Caribbean plate boundary in northern Central America: The Geological Society of America Special Paper 428, doi: 10.1130/2007.2428(03).

Rogers, R.D., Mann, P., Emmet, P.A., and Venable, M.E., 2007, this volume, Colon fold belt of Honduras: Evidence for Late Cretaceous collision between the continental Chortis block and intraoceanic Caribbean arc, *in* Mann, P., ed., Geologic and tectonic development of the Caribbean plate boundary in northern Central America: The Geological Society of America Special Paper 428, doi: 10.1130/2007.2428(06).

Rogers, R.D., Mann, P., Scott, R.W., and Patino, L., 2007, this volume, Cretaceous intra-arc rifting, sedimentation, and basin inversion in east-central Honduras, *in* Mann, P., ed., Geologic and tectonic development of the Caribbean plate boundary in northern Central America: The Geological Society of America Special Paper 428, doi: 10.1130/2007.2428(05).

Rosencrantz, E., Ross, M., and Sclater, J., 1988, Age and spreading history of the Cayman trough as determined from depth, heat flow, and magnetic anomalies: Journal of Geophysical Research, v. 93, p. 2141–2157.

Rosenfeld, J.H., 1981, Geology of the Western Sierra de Santa Cruz, Guatemala, Central America: An ophiolite sequence [Ph.D. thesis]: Binghamton, New York, State University of New York at Binghamton, 313 p.

Ross, M., and Scotese, C., 1988, A hierarchical tectonic model of the Gulf of Mexico and Caribbean region: Tectonophysics, v. 155, p. 139–168, doi: 10.1016/0040-1951(88)90263-6.

Schaaf, P., Morán-Zenteno, D., Hernandez-Bernal, M., Solis-Pichardo, G., Tolson, G., and Kohler, H., 1995, Paleogene continental margin truncation in southwestern Mexico: Geochronological evidence: Tectonics, v. 14, p. 1339–1350, doi: 10.1029/95TC01928.

Scott, R., and Finch, R., 1999, Cretaceous carbonate biostratigraphy and environments in Honduras, *in* Mann, P., ed., Caribbean basins: Sedimentary basins of the world series: Amsterdam, Elsevier, v. 4, p. 151–166.

Sedlock, R., Oretega-Gutierrez, F., and Speed, R., 1993, Tectonostratigraphic terranes and tectonic evolution of Mexico: The Geological Society of America Special Paper 278, 153 p.

Simonson, B., 1977, Mapa geológica de Honduras, El Porvenir: Tegulcigalpa, Honduras, Instituto Geográfico Nacional, escala 1:50,000.

Solari, L., Keppie, J., Ortega-Gutierrez, F., Cameron, K., Lopez, R., and Hames, W., 2003, 990 and 1100 Grenvillian tectonothermal events in the northern Oaxaca complex, southern Mexico: Roots of an orogen: Tectonophysics, v. 365, p. 257–282, doi: 10.1016/S0040-1951(03)00025-8.

Southernwood, R., 1986, Late Cretaceous limestone clast conglomerates of Honduras [M.S. thesis]: Dallas, Texas, University of Texas at Dallas, 300 p.

Sundblad, K., Cumming, G., and Krstic, D., 1991, Lead isotope evidence for the formation of epithermal gold quartz veins in the Chortis block, Nicaragua: Economic Geology and the Bulletin of the Society of Economic Geologists, v. 86, p. 944–959.

Talavera-Mendoza, O., Ruiz, J., Gehrels, G., Meza-Figueroa, D., Vega-Granillo, R., and Campa-Uranga, M., 2005, U-Pb geochronology of the Acatlan Complex and implications for the Paleozoic paleogeography and tectonic evolution of southern Mexico: Earth and Planetary Science Letters, v. 235, p. 682–699, doi: 10.1016/j.epsl.2005.04.013.

Tardy, M., Lapierre, H., Freydier, C., Coulon, C., Gill, J., Mercier de Lepinay, B., and Beck, C., Martinez-R., Talavera-M., Ortiz-H., E., Stein, G., Bourdier, J., and Yta, M., 1994, The Guerrero suspect terrane (western Mexico) and coeval arc terranes (the Greater Antilles and the Western Cordillera of Colombia): A late Mesozoic intra-oceanic arc accreted to cratonal America during the Cretaceous: Tectonophysics, v. 230, p. 49–73.

Venable, M., 1994, A geological, tectonic, and metallogenetic evaluation of the Siuna terrane (Nicaragua) [Ph.D. dissertation]: Tucson, Arizona, University of Arizona, 154 p.

Viland, J., Henry, B., Calix, R., and Diaz, C., 1996, Late Jurassic deformation in Honduras: Proposals for a revised regional stratigraphy: Journal of South American Earth Sciences, v. 9, p. 153–160, doi: 10.1016/0895-9811(96)00002-8.

von Huene, R., Aubouin, J., Azema, J., Blackinton, G., Carter, J.A., Colbourn, W., Cowan, D.S., Curiale, J.A., Dengo, C.A., Faas, R.W., Harrison, W., Hesse, R., Hussong, D.M., Laad, J.W., Muzylov, N., Shiki, T., Thompson, P.R., and Westberg, J., 1980, Leg 67: The Deep Sea Drilling Project Mid-America Trench transect off Guatemala: The Geological Society of America Bulletin, v. 91, p. 421–432, doi: 10.1130/0016-7606(1980)91<421:LTDSDP>2.0.CO;2.

Weiland, T., Suayah, W., and Finch, R., 1992, Petrologic and tectonic significance of Mesozoic volcanic rocks in the Río Wampú area, eastern Honduras: Journal of South American Earth Sciences, v. 6, p. 309–325, doi: 10.1016/0895-9811(92)90049-5.

Williams, H., and Hatcher, R., 1982, Suspect terranes and accretionary history of the Appalachian orogeny: Geology, v. 10, p. 530–536, doi: 10.1130/0091-7613(1982)10<530:STAAHO>2.0.CO;2.

Williams, H., and McBirney, A., 1969, Volcanic history of Honduras: Berkeley, University of California Publications in Geological Sciences no. 85, 101 p.

Wilson, H., 1974, Cretaceous sedimentation and orogeny in nuclear Central America: AAPG Bulletin, v. 58, p. 1348–1396.

Manuscript Accepted by the Society 22 December 2006

Cretaceous intra-arc rifting, sedimentation, and basin inversion in east-central Honduras

Robert D. Rogers*
Paul Mann

Institute for Geophysics, Jackson School of Geosciences, University of Texas at Austin, J.J. Pickle Research Campus, Bldg. 196 (ROC), 10100 Burnet Road (R2200), Austin, Texas 78758-4445, USA

Robert W. Scott

Precision Stratigraphy Associates and Geosciences Department, Tulsa University, RR3 Box 103-3, Cleveland, Oklahoma, 74020, USA

Lina Patino

Department of Geology, Michigan State University, 206 Natural Science Building, East Lansing, Michigan, 48824-1115, USA

ABSTRACT

This study describes the geology of a well-exposed but previously unmapped section of Paleozoic–early Cenomanian metamorphic, sedimentary, and igneous rocks in the Frey Pedro study area of the Agalta Range of east-central Honduras. The objective of the study is to use these new structural, stratigraphic, biostratigraphic, and geochemical data to better constrain the geologic and tectonic history of this part of the Chortis block during the period of time from Aptian to early Cenomanian. The study revealed that the topographic Agalta Range exposes a thick stratigraphic section (3.5 km) deposited in an Albian-Aptian intra-arc rift and on the rift shoulders. This rift feature, named here the Agua Blanca rift, presently trends northwest and is parallel to three other belts of deformed Cretaceous rocks in Honduras (the Comayagua, Minas de Oro, and Montaña de la Flor belts) that also may correspond to Cretaceous intra-arc rifts produced during the same phase of intra-arc extension. These other three deformed belts are west of the Agalta Range and also form topographically elevated mountain ranges.

Albian-Aptian calc-alkaline volcanic rocks and pyroclastic flows of the Manto Formation record arc affinity, while the volcaniclastic wedges of the Tayaco Formation record syn-rift deposition. These rift and arc-related units occupy the stratigraphic position between two major, extensive, shallow-water carbonate units, the lower and upper Atima Formations, also of middle Cretaceous age. Thickening of volcaniclastic rocks of the Tayaco Formation strata in the Agua Blanca rift accompanied erosion of the adjacent rift shoulders and eruption of Manto calc-alkaline volcanic rocks both within and adjacent to the Agua Blanca rift. The Agua Blanca

---

*Present address: Department of Geology, California State University Stanislaus, 801 West Monte Vista Avenue, Turlock, California 95382, USA; rrogers@geology.csustan.edu.

Rogers, R.D., Mann, P., Scott, R.W., and Patino, L., 2007, Cretaceous intra-arc rifting, sedimentation, and basin inversion in east-central Honduras, in Mann, P., ed., Geologic and tectonic development of the Caribbean plate boundary in northern Central America: Geological Society of America Special Paper 428, p. 89–128, doi: 10.1130/2007.2428(05). For permission to copy, contact editing@geosociety.org. ©2007 The Geological Society of America. All rights reserved.

intra-arc rift was inverted by a regional shortening event presumably in the Late Cretaceous. Rocks within the rift were intensely shortened, while rocks on the rift shoulders were shortened less because they are underlain by more competent metamorphic basement rocks.

In order to better understand the Aptian–early Cenomanian tectonic setting for intra-arc rifting and subsequent rift inversion on the Chortis block, we reconstructed the Chortis block relative to the terranes studied by previous workers in southern Mexico. Five geologic and tectonic features were selected for realigning the two now widely separated areas: (1) areas of Precambrian basement outcrops, (2) areas of similar Mesozoic stratigraphy, (3) aligned trends of Mesozoic volcanic arc rocks exhibiting a similar arc geochemistry, (4) aligned trends of Late Cretaceous folds and thrusts, and (5) alignment of magnetic boundary.

Keywords: Caribbean, Honduras, Cretaceous, Chortis block.

INTRODUCTION

The Cretaceous southwestern margin of North America has been truncated by Cenozoic plate movement as North America migrated to the west while the Chortis block remained behind to become incorporated as part of the Caribbean plate (Fig. 1A) (Riller et al., 1992; Tardy et al., 1994; Herrmann et al., 1994). This hypothesis predicts that elements of the southwestern North American Cordillera occur in Honduras (Fig. 1B) and that these elements were sheltered from Cenozoic overprinting associated with the arc-subduction systems along southwestern Mexico. In the under-studied regions of the stable interior of east-central Honduras, extensive exposure of Cretaceous strata provide a record of events occurring along the southernmost North American Cordillera during Cretaceous times.

The Chortis block of northern Central America forms the only continental part of the present-day Caribbean plate and is therefore an important constraint on the origin and Cretaceous-Cenozoic displacement history of the Caribbean plate (Fig. 1A). Closure of the 1100-km-long Eocene-Recent strip of oceanic crust in the Cayman trough pull-apart basin (Rosencrantz et al., 1988; Leroy et al., 2000) places the Chortis block adjacent to an area of mainly continental terranes in southern Mexico (Pindell and Barrett, 1990; Sedlock et al., 1993; Dickinson and Lawton, 2001). A problem with previous reconstructions of Chortis–southern Mexico is that the geology of the Chortis area in Honduras is much less studied and understood than correlative terranes in Mexico. Previous reconstructions of the Chortis block and southern Mexico that have relied almost entirely on data from southern Mexico include Azéma et al. (1985), Pindell and Barrett (1990), Riller et al. (1992), Herrmann et al. (1994), Tardy et al. (1994), and Schaaf et al. (1995). A consequence of the sparse information from the Chortis has been models such as Keppie and Moran-Zenteno (2005) that abandon connections between the Chortis and Mexico.

The objective of this study is to present new structural, stratigraphic, biostratigraphic, and geochemical data from east-central Honduras to determine the Aptian–early Cenomanian tectonic

Figure 1. (A) Present-day plate structure, oceanic magnetic anomaly, and tectonic terrane map of southern Mexico, Central America, and adjacent ocean basins showing terrane nomenclature used in this chapter. Boxes show the Frey Pedro study area in east-central Honduras and the Teloloapan region of Guerrero State, southern Mexico, which exhibits similar stratigraphic, tectonic, and geochemical history to the Frey Pedro study area. Definitions of Caribbean terranes from Case et al. (1990) for the Caribbean Sea area, Marshall et al. (2000) for Costa Rica and Panama, and Rogers et al. (Chapter 4, this volume) for Honduras, Nicaragua, and El Salvador: NCMZ—Northern Chortis magmatic zone; CCT—Central Chortis terrane; ECT—Eastern Chortis terrane; SCT—Southern Chortis terrane; ST—Siuna terrane. Definitions of southern Mexican terranes from Campa and Coney (1983) and Dickinson and Lawton (2001): C—Cohuila; G—Guerrero; J—Juarez; M—Mixteca; MA—Maya; O—Oaxaca; SM—Sierra Madre; X—Xolopa. Abbreviations of Guerrero subterranes: At—Arteaga; Ac—Arcelia; T—Teloloapan; Z—Zihuatanejo. Other abbreviations: AB—Arperos basin (modified from Tardy et al., 1994); CT—Cayman trough; EPR—East Pacific Rise; ES—El Salvador; LIP—large igneous province; MAT—Middle America trench; TMVB—Trans-Mexican volcanic belt. Magnetic anomalies shown in the Pacific Ocean are compiled from Wilson (1996), Klitgord and Mammerickx (1982), and Barckhausen et al. (2001), and Cayman trough anomalies are from Rosencrantz (1994). Plate motions relative to a fixed Caribbean plate are from DeMets et al. (2000) and DeMets (2001). Triangles represent Quaternary arc and intraplate volcanoes. (B) Topographic map of northern Central America showing physiographic expression of terranes shown in A and boxed location of Frey Pedro study area. Note that structural grain and prominent faults like the Guayape fault system (GFS) are at right angles to the trend of the present-day Middle America trench and related volcanic arc (black triangles). CB—Colon fold belt; CCT—Central Chortis terrane; ECT—Eastern Chortis terrane; HB—Honduran borderlands; HRS—Honduran rift system; MCSC—Mid Cayman Spreading Center; MFZ—Motagua fault zone; NCMZ—Northern Chortis magmatic zone; ND—Nicaraguan depression; NNR—Northern Nicaragua Rise; PFZ—Polochic fault zone; SCT—Southern Chortis terrane; SIFZ—Swan Island transform fault; SNR—Southern Nicaragua Rise; ST—Siuna terrane. Heavy dotted line offshore is 100 m bathymetric contour.

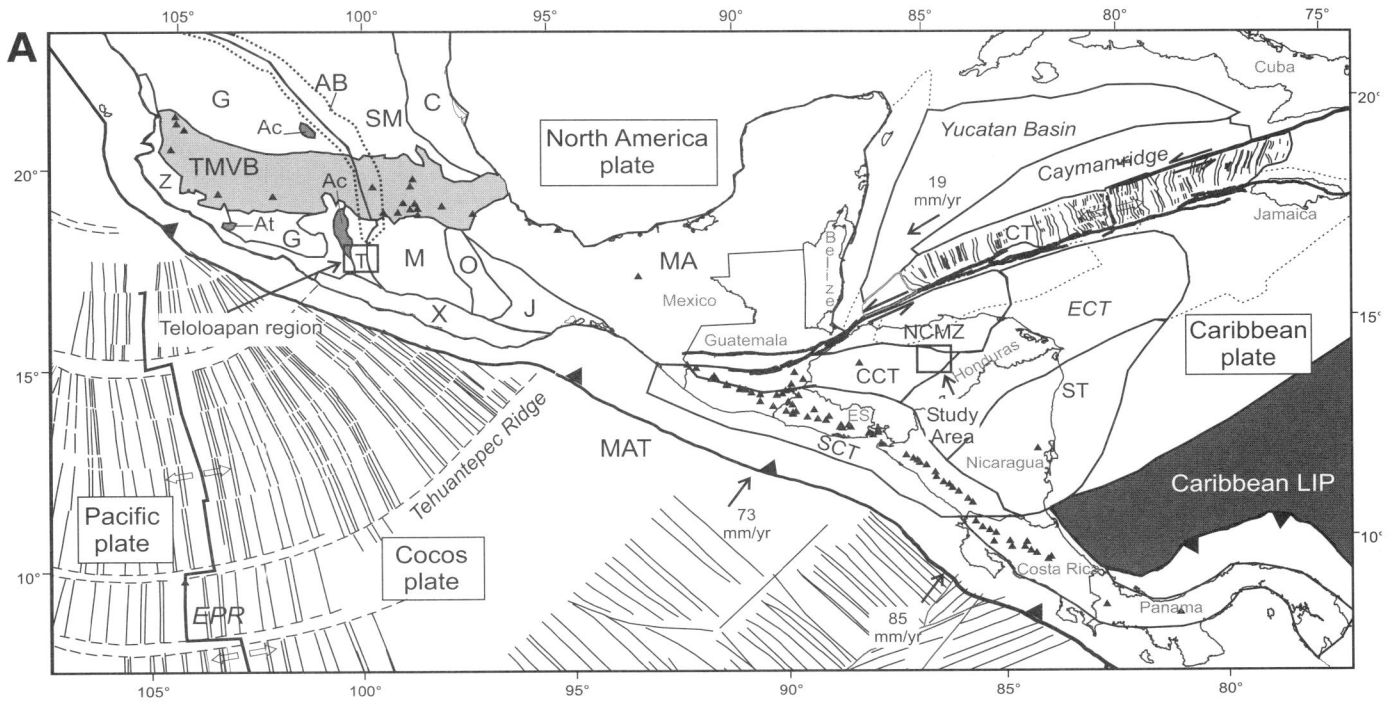

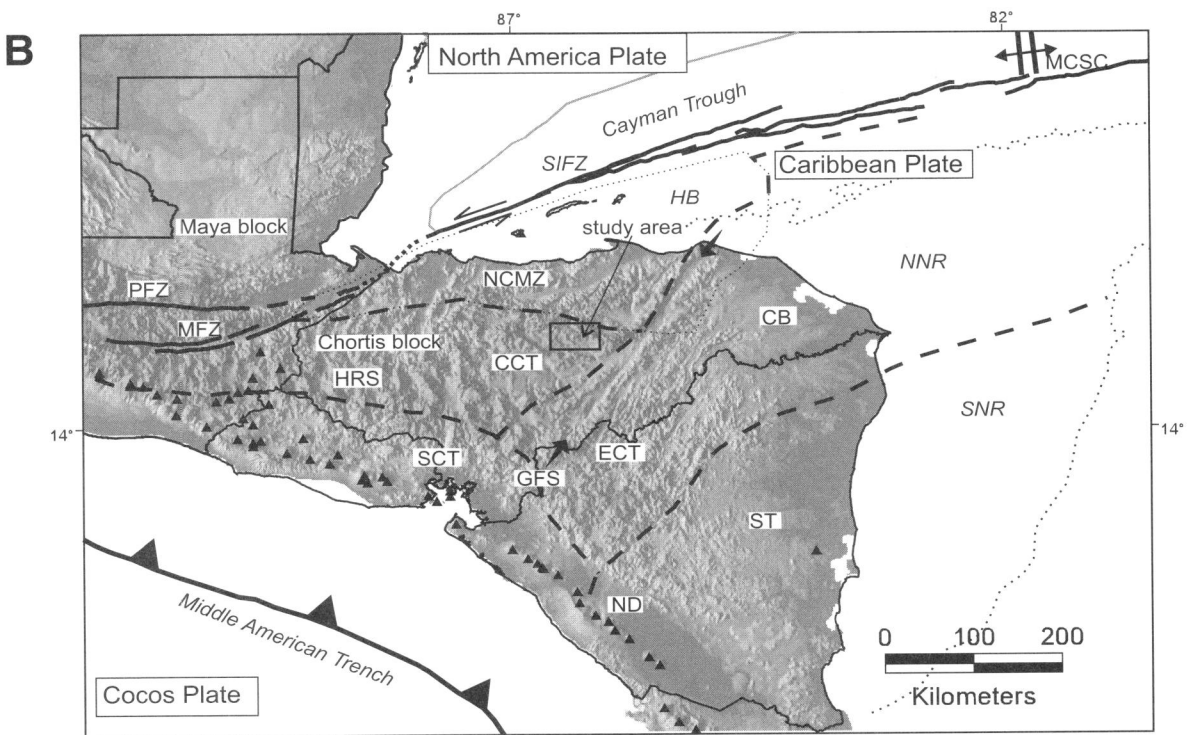

setting for the Chortis block when previous workers have proposed that the Chortis block formed a southern extension to the continental and arc terranes of southern Mexico. We then use these new data to compare to previously published results from southern Mexico to identify features and piercing lines on the Chortis block and southern Mexico that best constrain the restoration of the two areas.

This field study describes the geology from a previously unmapped, 2000 km^2 region in the interior of east-central Honduras (Frey Pedro area of Agalta Range). This region was selected for detailed study because (1) it has not been covered by Tertiary volcanic strata as found in western Honduras (Kozuch, 1991), and (2) it has not been subjected to Cenozoic plate boundary deformation related to the North America–Caribbean plate boundary zone deformation (DeMets et al., 2000; Rogers and Mann, this volume). For these reasons, the area provides a valuable record of Cretaceous structural and stratigraphic events that have not been overprinted by younger, Cenozoic events.

Methods included standard geologic mapping on base maps of 1:50,000 scale topographic sheets, measured sections, and measurements and analysis of structural information. Eight weeks were spent in the field in February through April of 2000, and three weeks in January of 2001.

Sedimentary samples were collected for micropaleontological dating; all results of age and environmental interpretations by R.W. Scott are presented in this study. Samples of volcanic rocks were collected; major element chemistry was analyzed by T. Vogel using X-ray fluorescence (XRF), and trace element chemistry was analyzed by L. Patino using an inductively-coupled plasma mass spectrometer (ICP-MS) (Hannah et al., 2002).

GEOLOGY OF THE CHORTIS BLOCK

The Chortis block is a polycomponent block because it includes five distinct terranes whose boundaries are defined principally by their contrasting geologic and magnetic basements (Rogers, 2003; Rogers et al., Chapter 4, this volume) (Fig. 1B). The study area is on the Central Chortis terrane and is adjacent to the Eastern Chortis terrane. The basement of the Central Chortis terrane consists of Paleozoic schist and gneiss of the Cacaguapa Group (Fakundiny, 1970; Horne et al., 1976; Metal Mining Agency of Japan (MMAJ), 1980; Kozuch, 1991) and Grenville-age orthogneiss (Manton, 1996) (Fig. 2). Basement of the Eastern Chortis terrane consists of Jurassic unmetamorphosed Agua Fria Formation strata overlying metamorphosed Agua Fria strata (Gordon, 1993a; Rogers, 1994; Rogers et al., Chapter 6, this volume).

A series of four northwest-trending structural belts and topographic mountain ranges record regional, Late Cretaceous shortening of the central Chortis terrane (Fig. 2). The westernmost Comayagua and Minas de Oro belts are partially buried by Miocene-age ignimbrites and overprinted by active north-trending rifts of western Honduras (Rogers and Mann, this volume). The eastern Montaña de Flor and Frey Pedro belts occur in the stable interior of the Chortis block, east of the rifts and south of the actively deforming Honduran borderlands (Rogers and Mann, this volume) (Fig. 2).

The northwest-trending belts are nearly perpendicular to the northeast-trending Guayape fault system, active in the Neogene as a right-lateral strike-slip fault (Fig. 2). The series of normal-fault bounded valleys along the Guayape fault system includes the Catacamas rift, formed in response to this phase of dextral motion on the Guayape fault (Gordon and Muehlberger, 1994) (Fig. 2). The easternmost part of the Montaña de la Flor belt is truncated by one of these normal fault splays related to shear on the Guayape fault system. The easternmost part of the Frey Pedro belt is oroclinally bent in a counter-clockwise sense (Fig. 2). We infer that this bending occurred as a result of an earlier period of left-lateral strike-slip motion on the Guayape fault that accompanied formation of the Late Cretaceous Colon fold-thrust belt (Rogers et al., Chapter 6, this volume).

THE FREY PEDRO STUDY AREA, EAST-CENTRAL HONDURAS

The study area is centered on the eastern part of the Frey Pedro structural belt and Agalta Range north of the Catacamas valley (Fig. 3). The study area is in a dry climatic zone within the dissected, Central American plateau of Honduras (Rogers et al., 2002). Deep plateau dissection and the tectonic stability of the area results in the preservation and excellent exposure of a thick Cretaceous section in the Agalta Range. Car access to this region of Honduras is via the Juticalpa-Trujillo unpaved highway. The main access crossing the study area is a 24-km-long gravel road constructed in 1988 that crosses the western Agalta Range and Montaña Frey Pedro, Montaña Agua Blanca, and Montaña Jacaleapa at a right angle and connects the towns of San Francisco de la Paz and Gualaco (Fig. 3). Outcrops along the road provide near-continuous exposure of the structure and stratigraphy of the Frey Pedro belt (Fig. 3). This is the only major road in Honduras that crosses one of the four northwest-trending structural belts shown on Figure 2 at a high angle. Car and foot access within the study area shown in Figure 3 is along a network of numerous unimproved logging roads and trails.

Previous geologic studies of the region were limited to mapping in an area to the west of the study area by the Metal Mapping Agency of Japan (MMAJ, 1980) and quadrangle mapping to the south along the Guayape fault system (Kozuch, 1990; Gordon, 1993b; Gordon and Muehlberger, 1994) (Fig. 3). Southernwood (1986) examined exposures of a Late Cretaceous limestone clast conglomerate along part of the San Francisco de la Paz–Gualaco road prior to its improvement and straightening in 1988.

The Agalta Range splits into three parallel and smaller ranges or "mountains" in the study area: the Montaña Frey Pedro is an antiformal block cored by schist and gneiss, as is the Jacaleapa Range to its north (Fig. 4). A foliated orthogneiss underlies the northern flank of the Jacaleapa Range. Between these ranges, steeply dipping northwest-striking Cretaceous carbonate and clastic strata form a series of northeast- and southwest-dipping

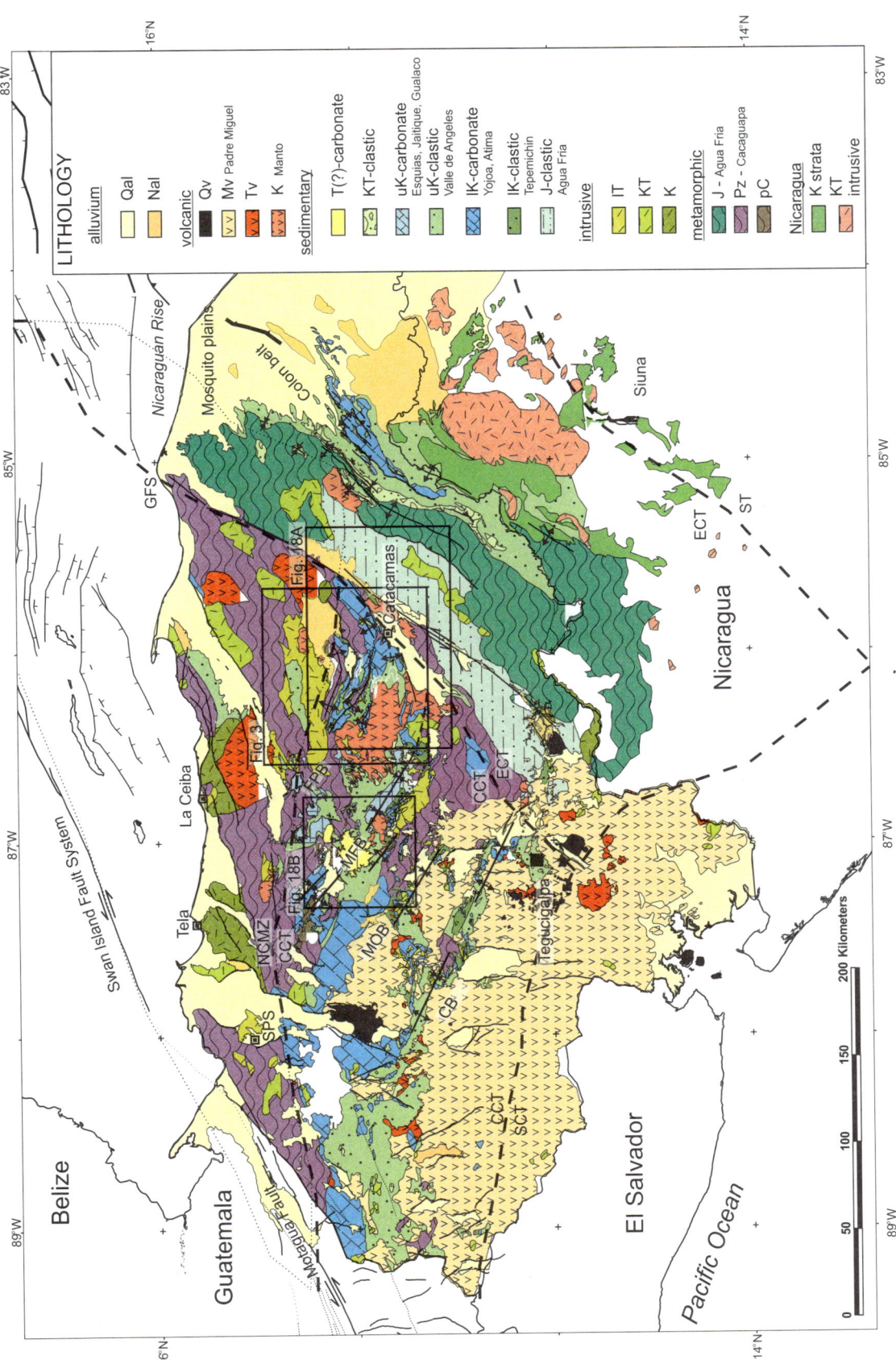

Figure 2. Geologic map of Honduras modified from national geologic compilation map by Kozuch (1991) and northeastern Nicaragua modified from national geologic compilation map by INETER (1995) showing superimposed terrane boundaries (dashed lines) from Figure 1. Boxed Frey Pedro study area, shown in detail in Figure 3, is located in stable, intraplate area of the Caribbean plate on the Central Chortis terrane (CCT) of east-central Honduras. Northwest-trending alignment of deformed Mesozoic strata and faults on the CCT west of the Guayape fault system (GFS) are the Frey Pedro belt (FPB), Montaña de la Flor belt (MFB), Minas de Oro belt (MOB), and Comayagua belt (CB). NCMZ—Northern Chortis magmatic zone; ECT—Eastern Chortis terrane; SCT—Southern Chortis terrane; SPS—San Pedro Sula; ST—Siuna terrane. East of GFS, mapping and reconnaissance reveals an extensive area of Jurassic Agua Fria strata (Gordon, 1993; Rogers, 1995; Rogers et al., 2003) and the northeast-trending Colon fold-thrust belt (Rogers et al., this volume, Chapter 6). North-trending faults actively rift western Honduras, while transtensional deformation of the Honduras borderlands extends onshore in northern Honduras (Rogers and Mann, this volume, Chapter 3). Honduras geology modified from Kozuch (1991), and Nicaragua geology (pre-Tertiary rocks only) modified from INETER (1995). Lithology abbreviations: lK—lower Cretaceous; J—Jurassic; K—Cretaceous; KT—Cretaceous-Tertiary; Mv—Miocene volcanic rocks; Nal—Neogene alluvium; Pz—Paleozoic; pC—Precambrian; Qal—Quaternary alluvium; Qv—Quaternary volcanic rocks; T—Tertiary; Tv—Tertiary volcanic rocks; uK—Upper Cretaceous.

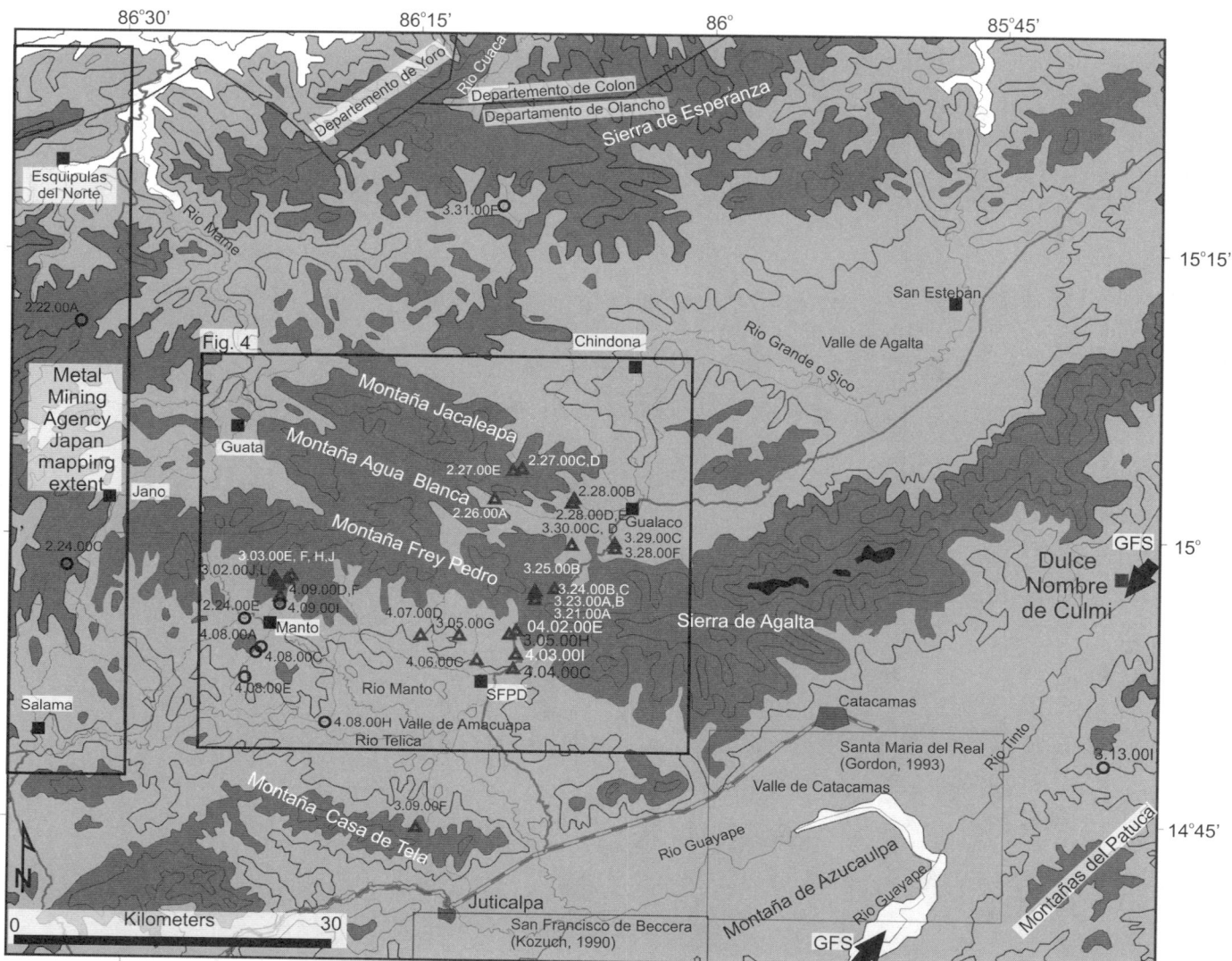

Figure 3. The Frey Pedro study area lies in a previously unmapped region of the Departamento of Olancho, east-central Honduras (topographic contour interval is 250 feet; shading reflects topography). Boxes indicate previous studies in the region surrounding the Frey Pedro study area. Left: the Metal Mining Agency of Japan (1980) mapped the area to the west at a scale of 1:50,000. Lower right: 1:50,000-scale mapping of Gordon (1993b) of the Santa Maria del Real quadrangle. Bottom center: 1:50,000-scale mapping of Kozuch (1990) of the San Francisco de Beccera quadrangle. The Gualaco–San Francisco de la Paz (SFDP) gravel road is the only access across Montaña Frey Pedro; outcrop exposures along this road and smaller, unimproved roads leading off of it are the main source of observations for this study. Samples for paleontological study are shown by numbered open triangles; samples for geochemical analysis are shown by open circles. Position of the Guayape fault system (GFS) is indicated by arrows. Box in center-left marks the study area detailed in Figure 4.

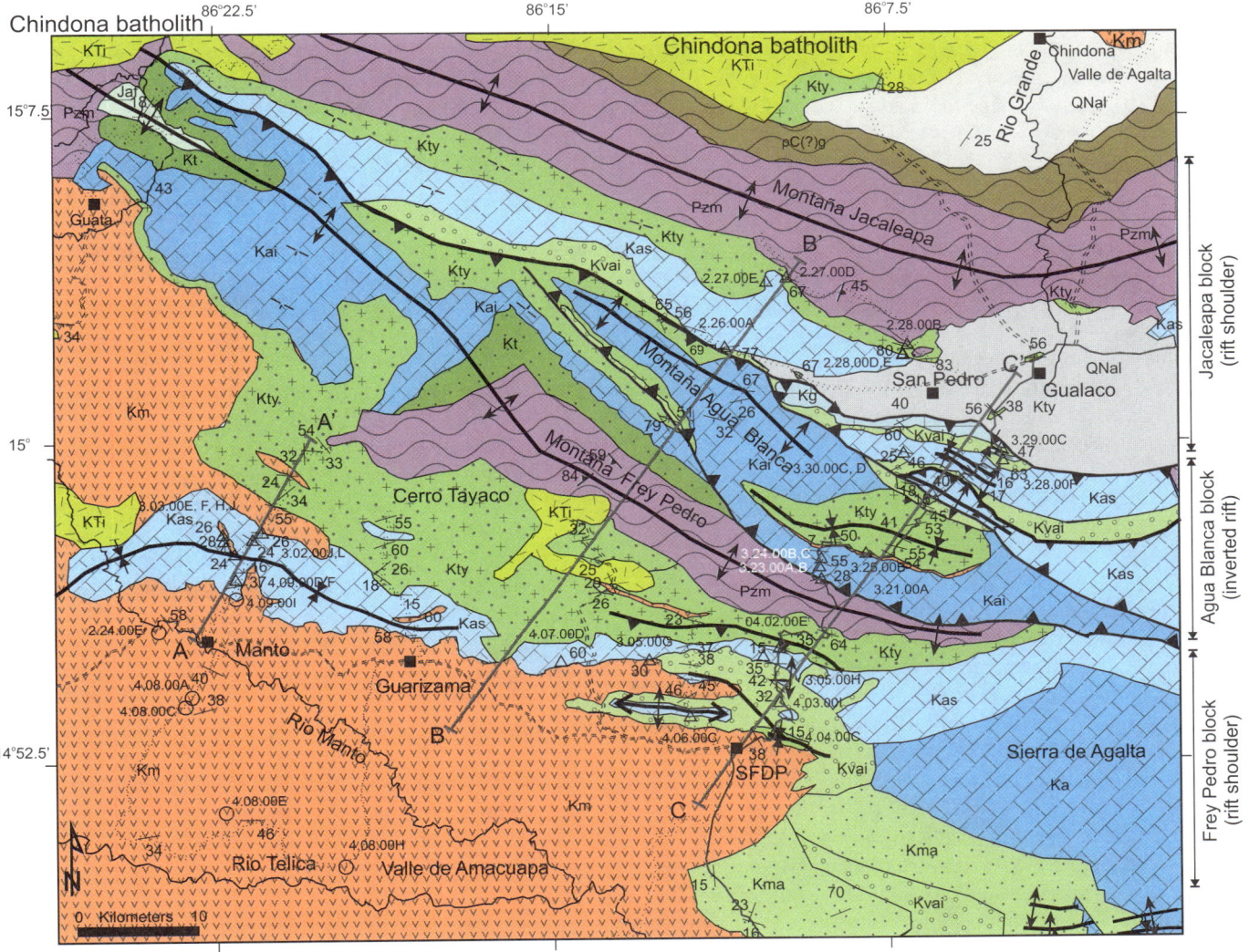

Figure 4. Map showing geologic setting of Montaña Frey Pedro, two northwest-trending antiforms cored by basement schist (Pzm) and orthogneiss (pC(?)g) and flanked by tilted and folded Cretaceous strata. Intense Late Cretaceous shortening of Mesozoic strata of the inverted Agua Blanca rift is more intense than shortening of the Frey Pedro and Jacaleapa basement blocks to NE and SW. Cretaceous volcanic rocks and volcaniclastic strata form the exposures in the Valle de Amacuapa. Numbered triangles indicate fossil samples; circles denote volcanic samples. Locations of structural cross sections A–A′, B–B′, and C–C′ shown in Figure 15 are indicated. Jaf—Agua Fria Formation; Ka—undifferentiated Atima Formation; Kai—lower Atima Formation; Kas—upper Atima Formation; Kg—Gualaco Formation; Km—Manto volcanic strata; Kt—Tepemechin Formation; KTi—felsic intrusive rock; Kty—Tayaco Formation; Kva—undifferentiated Valle de Angeles Formation; Kvai—lower Valle de Angeles Formation; pC(?)g—orthogneiss; Pzm—Paleozoic (?) basement schist and gneiss; QNal—valley floor alluvium. See Figure 5 for a stratigraphic column of all formations exposed in this area. SFDP—town of San Francisco de la Paz. The Frey Pedro composite stratigraphic section shown in Figure 5 is based on measured sections (Figs. 8–12) along the San Francisco de la Paz–Gualaco road.

hogbacks and domal hills. Reverse faults and folds attest to intense deformation of the Cretaceous strata between these ranges. South of the Frey Pedro Range, the Cretaceous strata are more gently dipping, with deformation expressed as open folds, in contrast to the intense deformation between the ranges. Along the lower southern flanks of the Frey Pedro Range, flows of andesite and dacitic pyroclastic deposits are exposed and occur within the carbonate strata. These volcanic rocks and associated volcaniclastic strata floor the Amacuapa valley and extend southward to form the Montaña Casa de Tela (Figs. 3 and 4).

North of the Jacaleapa Range is the Chindona batholith, a Late Cretaceous–early Tertiary felsic intrusive complex weathered to grus and eroded in the east to form the Valle de Agalta (Figs. 2–4). Intermediate volcanic rocks and metamorphosed strata similar to the Cretaceous sedimentary rock and Cretaceous volcanic rocks are exposed as resistant ridges within the Chindona batholith (Fig. 2). Similar, smaller felsic stocks, sills, and dikes intrude the basement rocks and the Cretaceous strata of the Frey Pedro Range, resulting in contact metamorphism of the Cretaceous strata (Fig. 4).

STRATIGRAPHY OF THE FREY PEDRO BELT, CENTRAL HONDURAS

The Frey Pedro region provides excellent exposures of the Cretaceous formations of the Central Chortis terrane along with their unfaulted stratigraphic contacts. Two features of the 3.5-km-thick Frey Pedro stratigraphic section are common to the Cretaceous stratigraphy of the Chortis block studied by previous workers, such as Finch (1981): (1) formation thickness varies abruptly over short distances, and (2) Cretaceous units of different ages can be found onlapping the same metamorphic basement. Unlike other parts of the Chortis block, the Frey Pedro region displays extensive evidence for calc-alkalic volcanic rocks interbedded in the middle to late Cretaceous carbonate section (Fig. 5).

The road between San Francisco de La Paz and Gualaco exposes most of the stratigraphy in the Frey Pedro region with the exception of the basal clastic rocks of the Tepemechin Formation (Fig. 4). Fifty-six measured sections of road outcrops from the north and south sides of Montaña Frey Pedro were combined to produce the composite stratigraphic column shown in Figure 5. Locations of numbered measured sections of continuous exposures along the San Francisco de La Paz and Gualaco road are keyed to the map in Figure 6. The lithologic patterns used in all the measured sections are displayed in Figure 7. Figures 8–12 show these measured sections. The combined thickness of the Cretaceous section is 3.5 km as measured from the basal Lower Cretaceous Tepemechin Formation through the upper Albian to lower Cenomanian Gualaco Formation (Fig. 5).

Age determination for the stratigraphic units in the Frey Pedro study area is based on paleontological identification of fossils from carbonate beds. The stratigraphic position of the samples is indicated on Figure 5, and the locations of samples are shown in Figures 3 and 4. Identification and biostratigraphic range of fauna identified by thin-section examination of the samples are provided in Table 1 along with paleoenvironmental interpretation based on fauna content and carbonate petrology by R.W. Scott.

Basement Rocks of the Frey Pedro and Jacaleapa Blocks

Biotite and chlorite schist and quartz augen gneiss form the basement of the Frey Pedro block (Fig. 4). The augen gneiss crops out only along the north side of the topographic crest of the Montaña Frey Pedro, adjacent to the large reverse fault separating the Frey Pedro and Agua Blanca blocks. Schist and gneiss exposures on the Frey Pedro block are similar to exposures of the schist and gneiss of the Cacaguapa Group of presumed Paleozoic age that has been described by previous workers in other parts of central Honduras (Fakundiny, 1970; Horne et al., 1976; MMAJ, 1980; Kozuch, 1991) (Fig. 2).

The basement of the Jacaleapa block contains, in addition to the schist and gneiss of the Cacaguapa group, an exposure of foliated orthogneiss extending along the northern side of Montaña Jacaleapa to the southern margin of the Agalta valley (Fig. 4). The orthogneiss is composed of recrystallized quartz and plagioclase feldspar with minor amounts of biotite and chlorite that display a foliated fabric. Associated with the orthogneiss are quartz boulders (up to 2 m in diameter) that form a distinctive lag on the weathered landscape underlain by the orthogneiss. Previous reports of foliated quartz feldspar orthogneiss in Honduras are limited to the Central Chortis terrane of the Yoro region of Honduras (Mohl, 1969; Williams and McBirney, 1969; Manton, 1987) and along the Montaña de la Flor belt (Fig. 2) where Manton (1996) reports a Grenville U/PB age of 1 Ga for the Yoro orthogneiss. The Jacaleapa orthogneiss may represent a new locality for Grenville-age rocks on the eastern edge of the Central Chortis terrane, but further field studies and radiometric dating are needed to confirm this inference.

Tepemechin Formation

The oldest unit in the study area is the Lower Cretaceous Tepemechin Formation composed of 230 m of conglomerate and sandstone. The Tepemechin Formation lies unconformably on the basement schist of the Cacaguapa Group with the intervening Upper Jurassic Agua Fria Formation absent except in the core of an anticline north of the village of Guata (Fig. 4).

Outcrop Distribution and General Stratigraphy

A 230-m-thick sequence of a red, quartz, and metamorphic pebble conglomerate in a coarse-grained, lithic sand matrix occurs locally along the northern side of Montaña Frey Pedro, west of the San Francisco de la Paz–Gualaco road (Fig. 4). This unit is in unconformable contact with the underlying basement schist and gneiss and in conformable contact with the overlying massive limestone of the lower Atima Formation.

The unit is assigned to the Tepemechin Formation based on its lithology and stratigraphic position beneath the lower

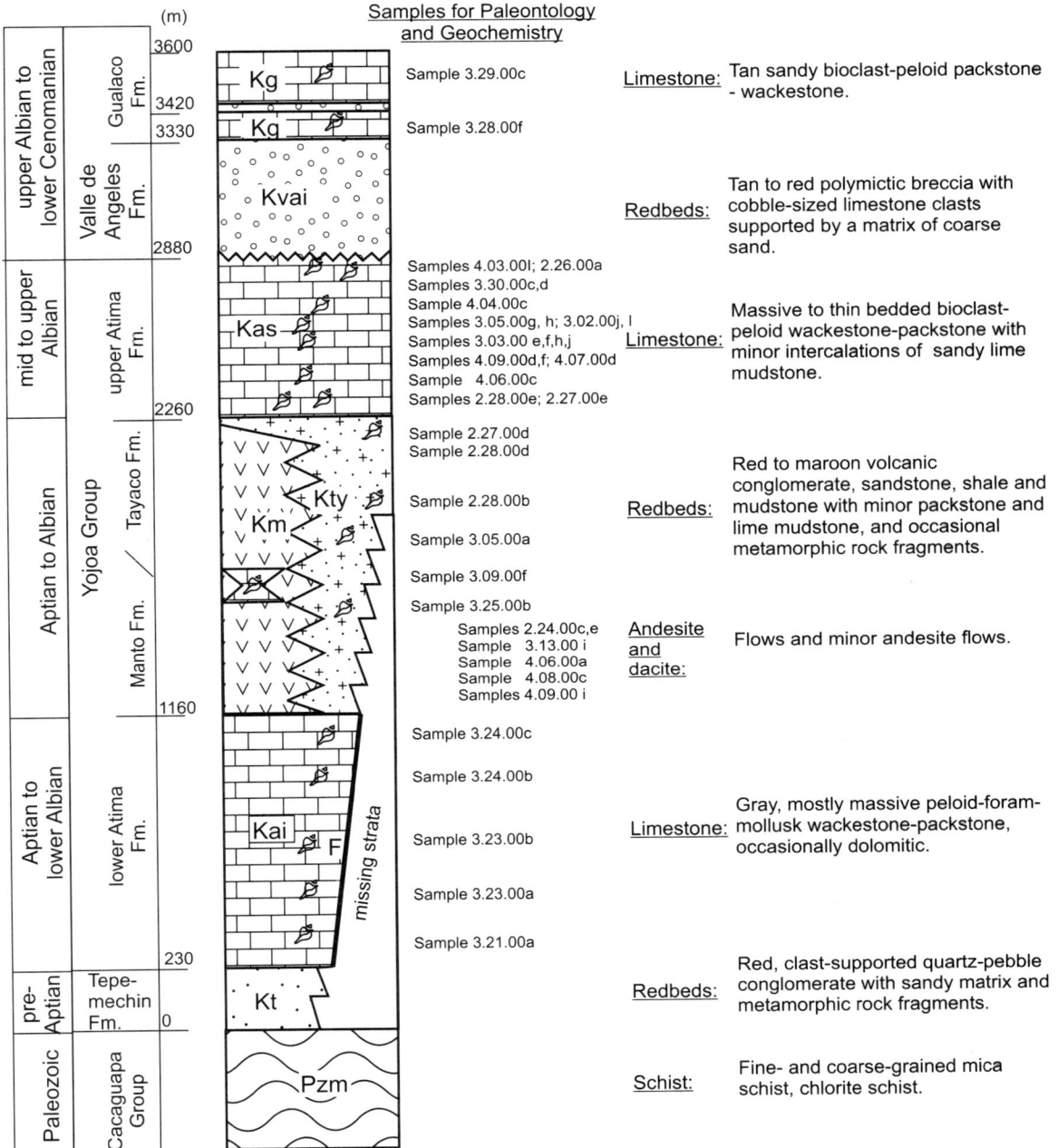

Figure 5. Composite stratigraphic column of Frey Pedro study area is based on 3.5 km of semi-continuous, fresh rock exposures along the San Francisco de la Paz–Gualaco road (location shown in Fig. 4). Fossil samples are tabulated in Table 1 and are shown here in approximate stratigraphic position. F—fault; Kai—lower Atima Formation; Kas—upper Atima Formation; Kg—Gualaco Formation; Km—Manto volcanic strata; Kt—Tepemechin Formation; Kty—Tayaco Formation; Kvai—lower Valle de Angeles Formation; Pzm—Paleozoic (?) basement schist and gneiss.

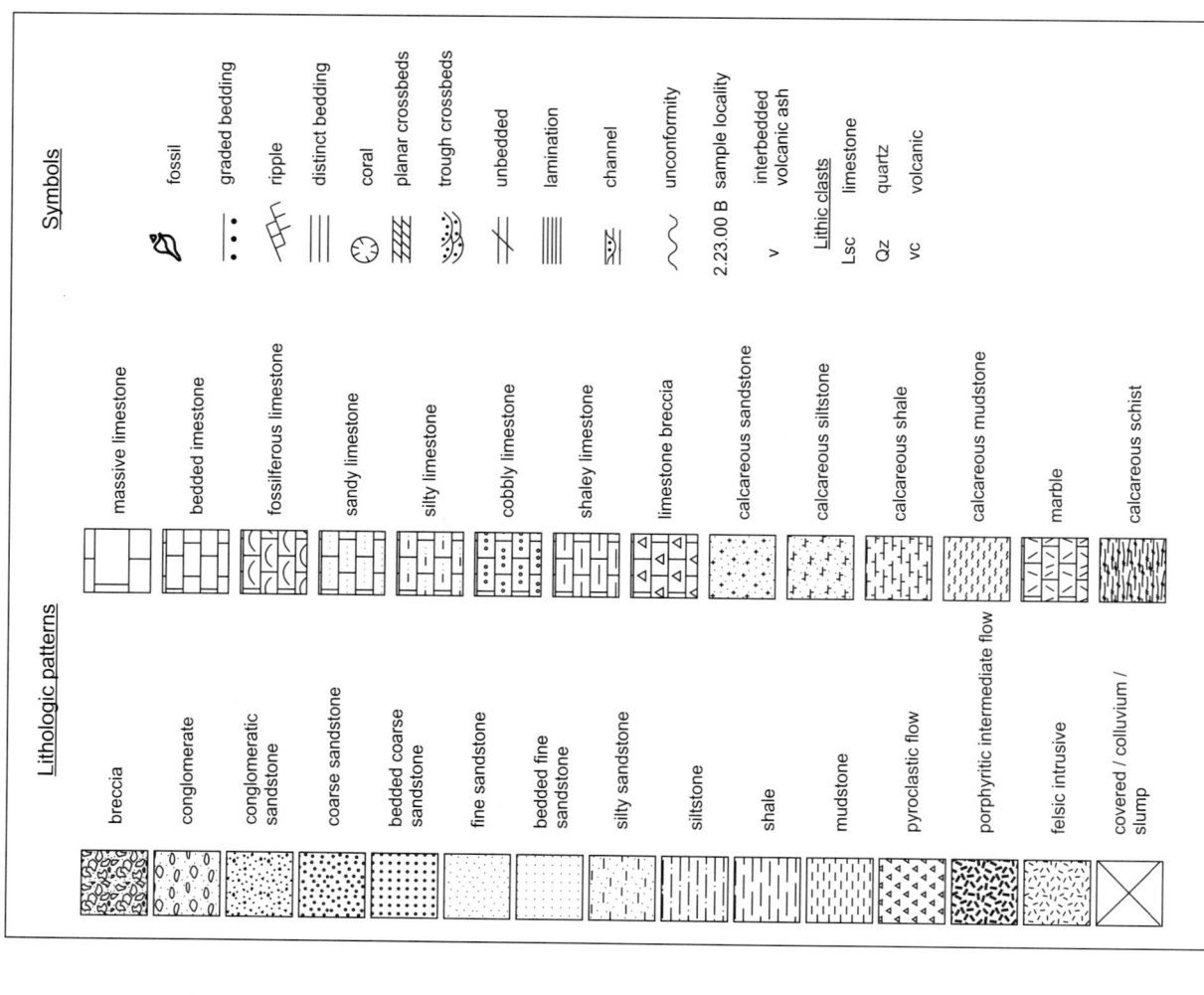

Figure 6. Map showing detailed geology along the San Francisco de la Paz (SFDP)–Gualaco road crossing Montaña Frey Pedro. Numbers and alternating black and white sections of road show locations of measured sections from roadside outcrops that are shown in detail in Figures 8–12. Kai—lower Atima Formation; Kas—upper Atima Formation; Kg—Gualaco Formation; Km—Manto volcanic strata; Kty—Tayaco Formation; Kvai—lower Valle de Angeles Formation; Qal—Quaternary alluvium.

Figure 7. Key to lithologic patterns and symbols used in Figures 8–12.

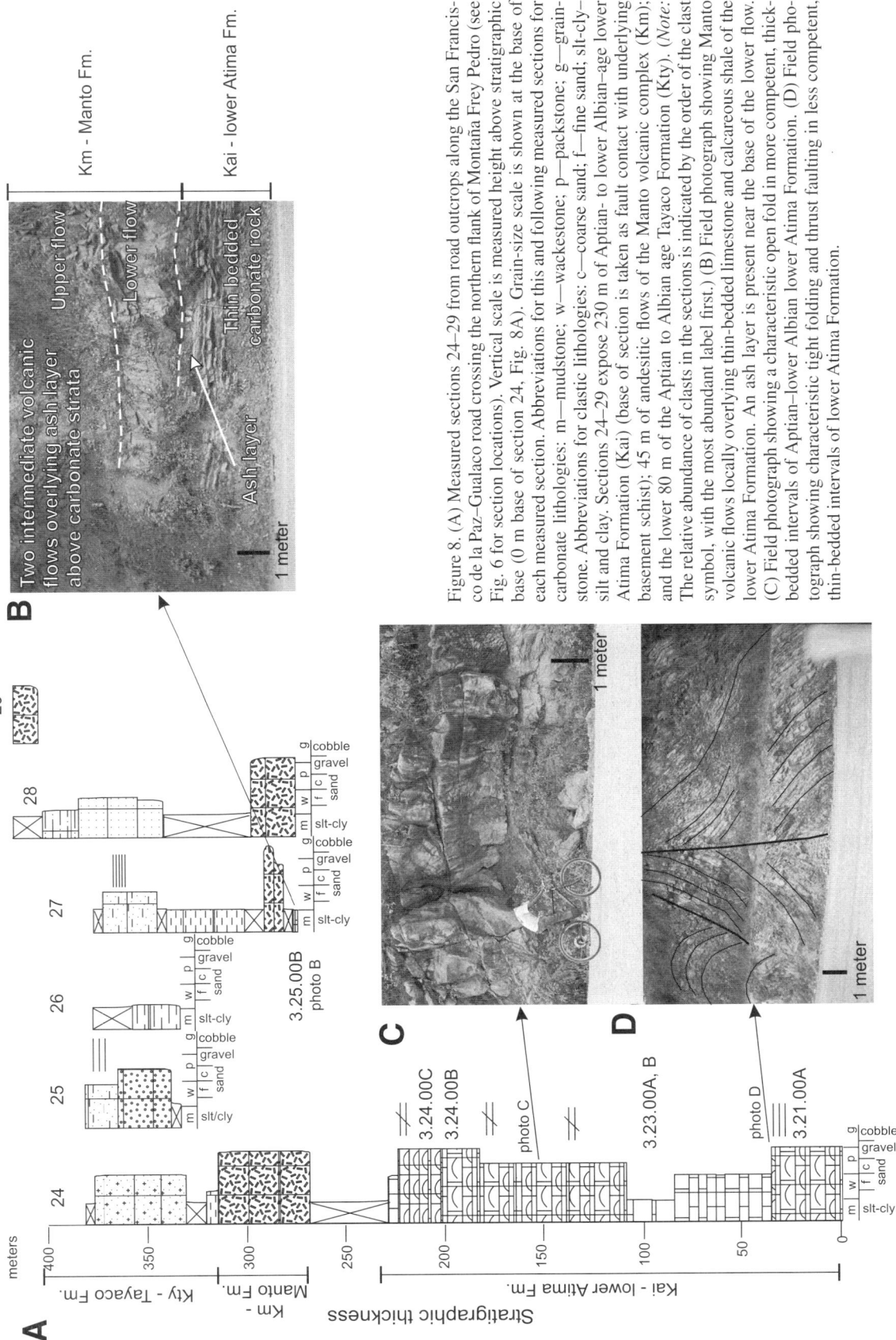

Figure 8. (A) Measured sections 24–29 from road outcrops along the San Francisco de la Paz–Gualaco road crossing the northern flank of Montaña Frey Pedro (see Fig. 6 for section locations). Vertical scale is measured height above stratigraphic base (0 m base of section 24, Fig. 8A). Grain-size scale is shown at the base of each measured section. Abbreviations for this and following measured sections for carbonate lithologies: m—mudstone; w—wackestone; p—packstone; g—grainstone. Abbreviations for clastic lithologies: c—coarse sand; f—fine sand; slt-cly—silt and clay. Sections 24–29 expose 230 m of Aptian- to lower Albian-age lower Atima Formation (Kai) (base of section is taken as fault contact with underlying basement schist); 45 m of andesitic flows of the Manto volcanic complex (Km); and the lower 80 m of the Aptian to Albian age Tayaco Formation (Kty). (Note: The relative abundance of clasts in the sections is indicated by the order of the clast symbol, with the most abundant label first.) (B) Field photograph showing Manto volcanic flows locally overlying thin-bedded limestone and calcareous shale of the lower Atima Formation. An ash layer is present near the base of the lower flow. (C) Field photograph showing a characteristic open fold in more competent, thick-bedded intervals of Aptian–lower Albian lower Atima Formation. (D) Field photograph showing characteristic tight folding and thrust faulting in less competent, thin-bedded intervals of lower Atima Formation.

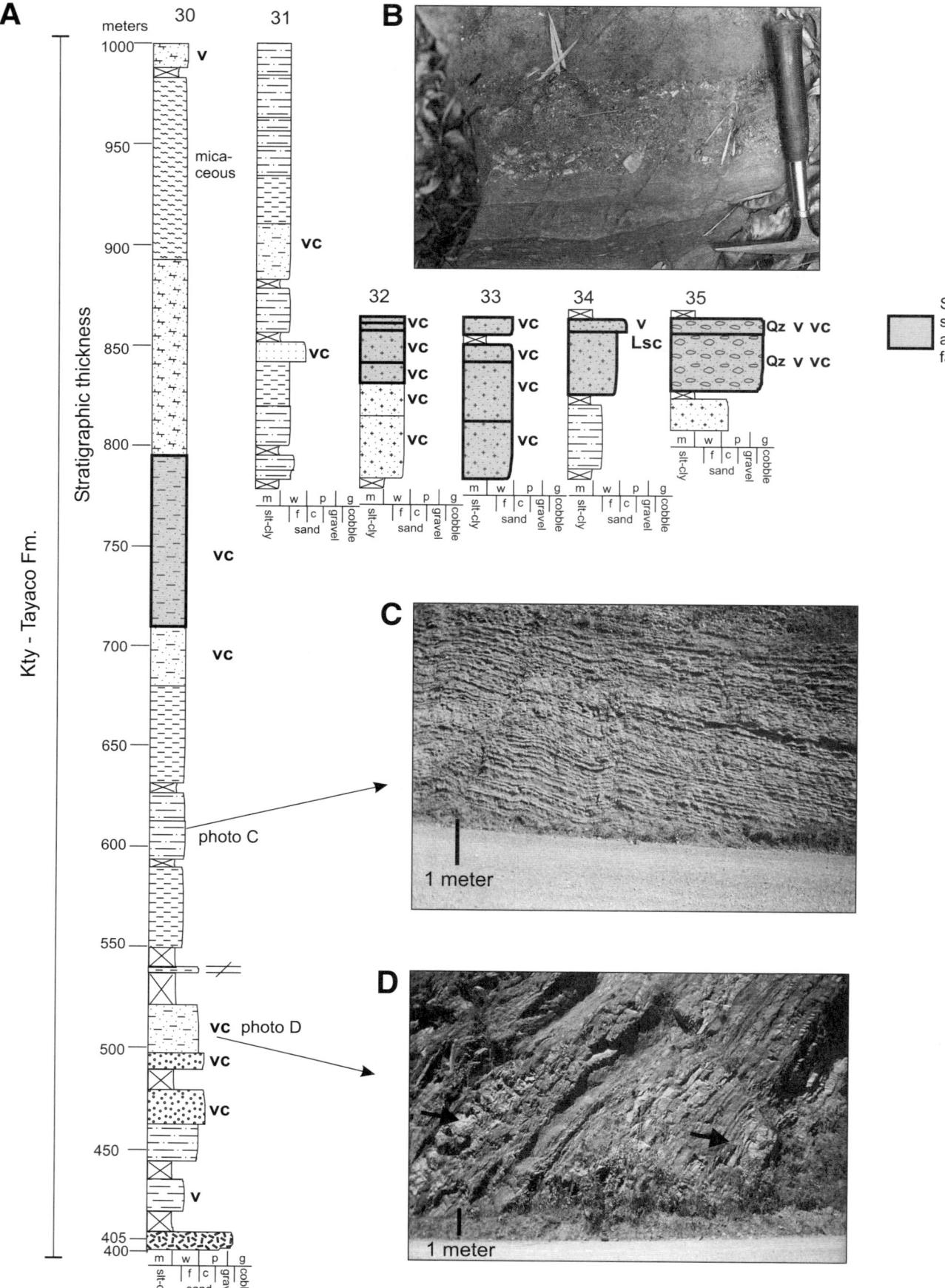

Figure 9. (A) Measured sections 30–35 from road outcrops along the San Francisco de la Paz–Gualaco road crossing the northern flank of the Montaña Frey Pedro (see Fig. 6 for locations of sections). Vertical scale is measured height above stratigraphic base (0 m base of section 24, Fig. 8A). Lsc—limestone; Qz—quartz; vc—volcanic; see Figure 8 for other lithologic abbreviation definitions. Sections 30–35 expose 600 m of the Aptian-Albian–age Tayaco Formation. Interpreted shoreface facies of Tayaco Formation are highlighted. (*Note:* The relative abundance of clasts in the sections is indicated by the order of the clast symbol, with the most abundant label first.) (B) Field photograph showing limestone pebble-clasts derived from the lower Atima Formation in calcareous redbeds of the Tayaco Formation. (C) Field photograph showing characteristic thinly bedded and gently folded siltstone of the Tayaco Formation. (D) Sandy beds of the Tayaco Formation containing volcanic rock fragments, indicated by arrows.

Atima Formation (Gordon, 1993a). An additional exposure of the Tepemechin Formation was mapped northeast of the town of Guata in the northeast part of the study area (Fig. 4) in the core of a large, breached anticline. In this locality, the clastic rocks of the Tepemechin Formation conformably overlie dark gray shale assigned to the Jurassic Agua Fria Formation.

Facies and Paleoenvironment of the Tepemechin Formation

Conglomerate of the Tepemechin Formation is clast-supported with mainly planar beds ranging between 10 and 50 cm in thickness. Metamorphic, lithic clasts within the conglomerate are predominantly subangular, while quartz clasts are subrounded. Metamorphic clast angularity is controlled by rock cleavage present in the schist and gneiss lithologies. Conglomerate clast sizes range from fine to large pebbles. The environment of deposition of the Tepemechin Formation is inferred to be fluvial and debris flow alluvial fans that were proximal to upland areas underlain by metamorphic basement. This interpretation is supported by rare but broad channels with trough crossbeds and normal grading in the more sandy intervals, subrounded quartz clasts, and very rare scoured bases of beds or channels. The conformable contact of the fluvial Tepemechin Formation with the overlying shallow-marine limestone of the lower Atima Formation represents a basal transgressive sequence in the Early Cretaceous as first noted by Simonson (1977).

Age of the Tepemechin Formation

Gordon (1993a) assigns an Early Cretaceous age to the Tepemechin Formation in central Honduras. The stratigraphic position of the Tepemechin Formation below the Aptian to lower Albian lower Atima Formation in the Frey Pedro belt is consistent with the Early Cretaceous age of the Tepemechin Formation elsewhere in Honduras.

Lower Atima Formation

The lower Aptian to Albian lower Atima Formation conformably overlies the Tepemechin Formation. The thickness of the mainly massive limestone of the lower Atima Formation is estimated to total 930 m based on the construction of cross sections.

Stratigraphic Nomenclature

The Atima Formation of the Yojoa Group is divided on the basis of a prominent clastic break called the Mochito shale in western Honduras (Finch, 1985) and in the Colon Mountains of eastern Honduras (Scott and Finch, 1999; Rogers et al., Chapter 6, this volume). In the Frey Pedro belt, an intervening clastic break, the Tayaco Formation, allows the Atima Formation to be divided into an upper Atima Formation and a lower Atima Formation (Fig. 5). This nomenclature is informal and provisional. Rather than introduce new names for recognized stratigraphic elements or propose elevating Atima to group status, which would require abandoning the Yojoa name, we prefer to retain the Atima name in the Frey Pedro area. The reference section for the lower Atima Formation is section 24 (Figs. 6 and 8A), and for the upper Atima Formation it is section 7 (Figs. 6 and 12A).

Outcrop Distribution and General Stratigraphy

The lower Atima Formation is a gray, fossiliferous wackestone-packstone with uncommon lime mudstone intervals. In the Frey Pedro belt, the lower Atima Formation is exposed as a thick, massive limestone unit in fault contact with the basement schist along the San Francisco de la Paz–Gualaco road and in conformable contact with the underlying conglomerate of the Tepemechin Formation in streams north of Montaña Frey Pedro (Figs. 4 and 5). The lower Atima Formation forms the core of the northwest-plunging anticline of Montaña Agua Blanca (Fig. 4). Karst topography developed along the northwest continuation of the Montaña Frey Pedro anticline is interpreted as an extensive outcrop of the lower Atima Formation.

Lower Atima Formation on the Agua Blanca Block

Two-hundred and thirty meter thick limestone of the lower Atima Formation is in fault contact with basement schist north of the crest of the Montaña Frey Pedro (Fig. 8A, section 24). Bedding varies from massive to 10–20 cm thick (Figs. 8C and 8D). Bedding becomes more massive toward the base of the exposed section. Volcanic flows of the Manto Formation (discussed in a following section) are interbedded with strata of the Tayaco Formation and conformably overly the carbonate strata of the lower Atima Formation (Fig. 8B; Fig. 8A, sections 27 and 28).

Facies and Paleoenvironment of the Lower Atima Formation

Petrographic study of samples collected at regular intervals throughout the lower Atima Formation revealed the following carbonate facies: peloid-foram packstone; bioturbated, dolomitic mollusk-echinoid wackestone; mollusk mudstone with angular phosphate grains; requienid rudist-bearing peloid-foram packstone; and peloid-chondrodontid packstone with articulated bivalves and caprinid clasts (Table 1).

Figure 10. (A) Measured sections 36–47 from road outcrops along the San Francisco de la Paz–Gualaco road crossing the northern flank of the Montaña Frey Pedro (see Fig. 6 for locations of sections). Vertical scale is measured height above stratigraphic base (0 m base of section 24, Fig. 8A). D/U—fault with relative motion indicated; Kty—Tayaco Formation; Lsc—limestone; Qz—quartz; vc—volcanic; see Figure 8 for other lithologic abbreviation definitions. Sections 36–37 expose 250 m of middle to upper Albian-age upper Atima Formation (Kas) in fault contact with at least 275 m of lower Valle de Angeles Formation breccia (Kvai). A major unconformity occurs at the stratigraphic contact between the upper Atima Formation and the post-upper-Albian–age breccia of the lower Valle de Angeles Formation (sections 38–40). In section 37, 175 m of the upper Atima Formation overlies the Tayaco Formation. A 35-m-thick dacitic pyroclastic flow occurs near the base of the breccia of the lower Valle de Angeles Formation in section 36. (*Note:* The relative abundance of clasts in the sections is indicated by the order of the clast symbol, with the most abundant label first.) (B) Field photograph of matrix-supported gray limestone cobble (below notebook) of lower Valle de Angeles breccia. (C) Field photograph of weathered plagioclase-bearing andesite cobbles of lower Valle de Angeles breccia (to right of hammer).

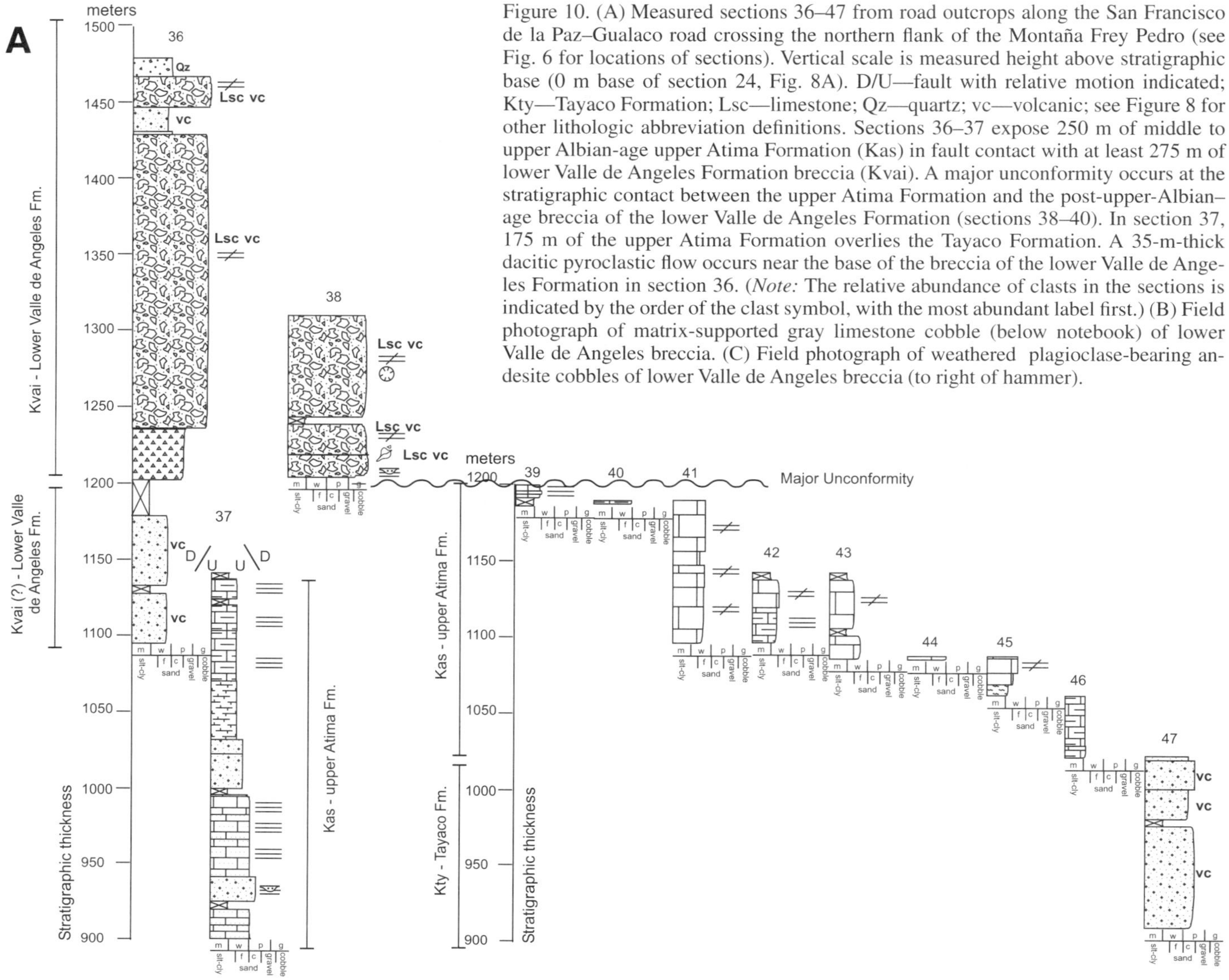

We interpreted the paleoenvironment of this assemblage of facies using a carbonate paleoenvironmental model based on the occurrence of key associations of calcareous algae as proposed by Conrad (1977). Key associations of algae from these samples include the following paleoenvironments: (1) shallow shelf lagoon: *Boueinia hochstetteri, Acicularia, Salpingoporella, Pycnoporidium lobatum*; (2) shelf margin buildups: *Rivularia piae, Lithocodium aggregatum, Parachaetetes texana, Pseudolithothamnium*; and (3) forereef–outer shelf: *Pithonella, Micritosphaera* (see Table 1 for occurrence in samples). The facies and the occurrence of diagnostic algae indicate an upward-shallowing environment from deeper marine shelf to shallow marine shelf to shallow marine shelf lagoon (Table 1).

Age of the Lower Atima Formation

While distinctive, short-ranging, age-diagnostic foraminifers are absent, the presence of *Cuneolina walteri*: middle–upper Albian; and *Praechrysalidina infracretacea*: Hauterivian–lower Albian (Scott and Gonzalez-Leon, 1991; Banner et al., 1991; Scott and Finch, 1999), indicate a mid-Cretaceous age for the lower Atima Formation (Table 1). Transgressive-regressive, deepening to shoaling represented by the lower Atima Formation of the Frey Pedro region is similar to the paleoenvironments reported for the lower Atima Formation in the Colon Mountains of eastern Honduras and from western Honduras (Scott and Finch, 1999).

Manto Formation

Separating the Atima Formation into upper and lower units are the Aptian- to Albian-age volcanic rocks of the Manto Formation and the mainly marine volcaniclastic strata of the Tayaco Formation. Both the Manto and Tayaco Formation names are introduced for the strata separating the upper and lower units of the Atima Formation in the Frey Pedro region.

Outcrop Distribution and General Description

We define the Manto Formation as a calc-alkalic volcanic complex of intermediate flows, dacitic pyroclastic flows and falls, and basaltic flows of Aptian-Albian age. The formation is named for the well-exposed volcanic sections near the town of Manto (Fig. 4). Predominate lithologies of the Manto Formation include plagioclase-bearing porphyritic andesite lava flows and shallow intrusive bodies; crystal-lithic dacite ignimbrite; aphanitic basaltic andesite lava flows; and autobrecciated andesite. Porphyritic andesite flows are the most common expression of the Manto Formation. The type section for the Manto Formation is in Section 24 (Figs. 6 and 8).

The extensive Manto Formation covers an area of ~1600 km^2 south of the Frey Pedro Mountains (Fig. 2). The Amacuapa Valley is composed entirely of the Manto Formation (Figs. 3 and 4). The hills south of the town of Manto expose a near-horizontal section of the Manto Formation (Fig. 13A). The valley and the base of the hills are porphyritic andesite flows overlain by aphanitic flows of basalt and minor andesite (Fig. 13B) and capped by 200 m of silicic tuffs with evidence of volcanic vents (Fig. 13C).

The volcanic rocks of the Manto Formation underlie the upper Atima Formation along the northern margin of the Amacuapa valley (Fig. 4). The Manto Formation extends southward from the Amacuapa valley to form the Montaña Casa de Tela (Figs. 2 and 3) where the Manto Formation is interbedded with massive limestone of the Atima Formation. Exposures of the Manto Formation extend westward past the village of Jano (Figs. 2 and 3). To the north, flows of the porphyritic andesite lava flows occur above the lower Atima Formation on the Agua Blanca block (Fig. 4) and in the Agalta valley (Figs. 3 and 4).

Manto Formation on the Agua Blanca Block

The measured sections of the volcanic rocks of the Manto Formation along the San Francisco de la Paz–Gualaco road reveal key contacts that confirm the stratigraphic position of the Manto Formation. Two flows of weathered andesite lava totaling 50 m in thickness occur at the base of the Tayaco Formation and overlie limestone of the lower Atima Formation (Fig. 8A, sections 24–28; Fig. 8B) on the Agua Blanca block. On the Frey Pedro block, interbedded lava flows of the Manto Formation and volcaniclastic strata of the Tayaco Formation occur (Fig. 12A).

Major and Trace Element Geochemistry of the Manto Formation

Geochemical analyses of volcanic rocks from the Manto Formation by L. Patino and T. Vogel indicate an arc affinity. Eight volcanic samples from the Manto Formation were analyzed for major, trace, and rare earth elements (sample location on Figs. 3 and 4) using XRF and ICP-MS techniques. Results from the analysis are displayed in Table 2. The analyzed samples vary from basaltic andesite to dacite in composition.

To ascertain the tectonic setting of the Manto Formation, multi-elemental plots of the trace element abundance from the samples were normalized to primitive mantle composition (Sun and McDonough, 1989) and compared to established geochemical relationships for volcanic rocks from various tectonic settings. The samples of the Manto Formation are enriched in large ion lithophiles (LIL) elements (Ba, K, Pb, Sr) while depleted in high field strength (HFS) elements (Nb, Zr, Ti) (Fig. 14A). LIL enrichment with HFS depletion is characteristic of evolved volcanic arc lavas worldwide (Perfit and Dickinson, 2000), indicating a volcanic arc setting for the Manto Formation.

Age Estimate and Correlation of the Manto Formation

The volcanic rocks of the Manto Formation occupy the same stratigraphic position as the Aptian- to Albian-age Tayaco Formation and are interbedded with the Aptian to Albian Tayaco Formation along the northern and western margins of the Amacuapa valley (Figs. 4 and 5). Additional field relations supporting an Aptian to Albian age for the Manto Formation include Manto volcanic flows interbedded with the Tayaco Formation and overlying the lower Atima Formation in measured sections 24–28

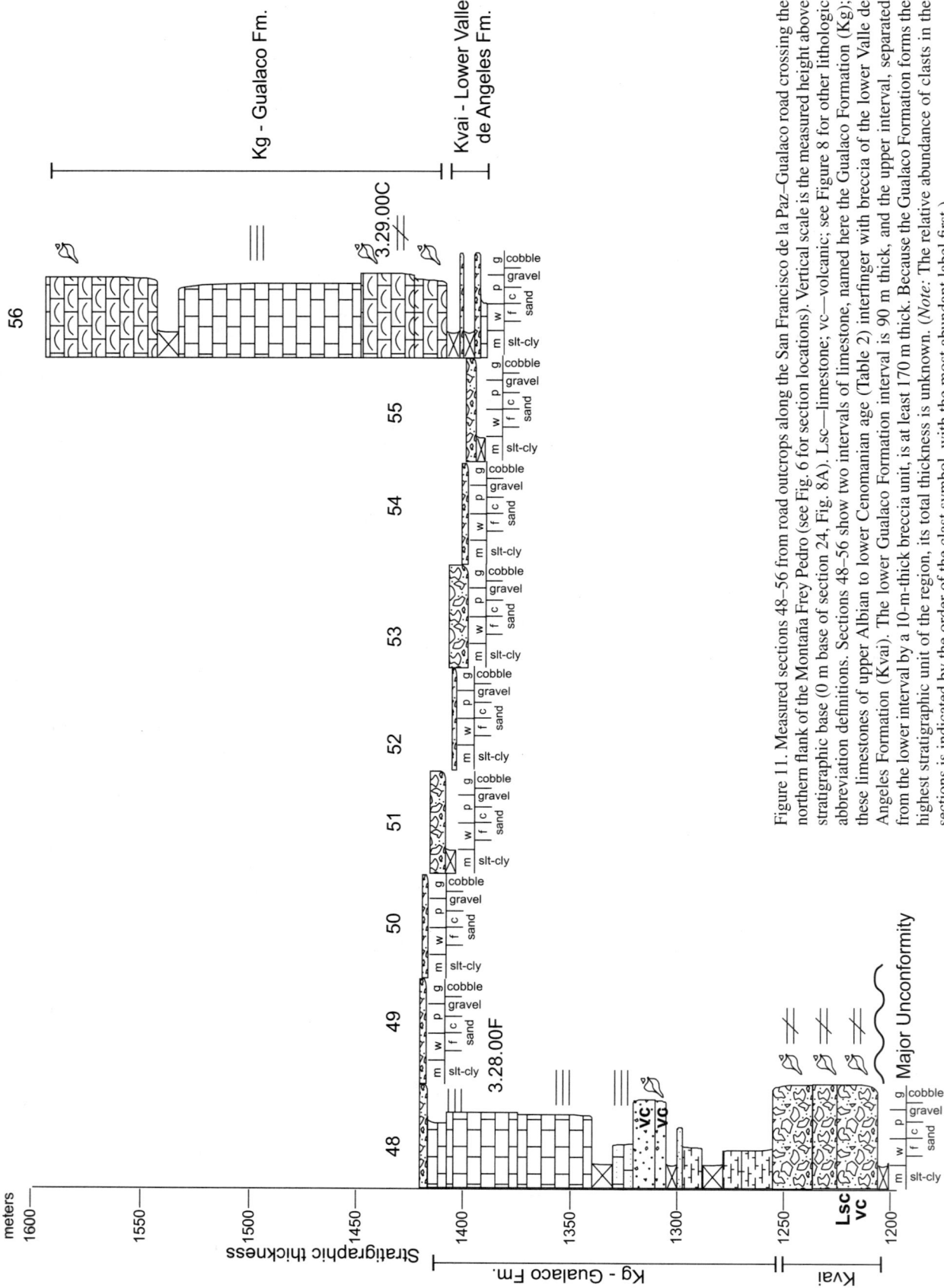

Figure 11. Measured sections 48–56 from road outcrops along the San Francisco de la Paz–Gualaco road crossing the northern flank of the Montaña Frey Pedro (see Fig. 6 for section locations). Vertical scale is the measured height above stratigraphic base (0 m base of section 24, Fig. 8A). Lsc—limestone; vc—volcanic; see Figure 8 for other lithologic abbreviation definitions. Sections 48–56 show two intervals of limestone, named here the Gualaco Formation (Kg); these limestones of upper Albian to lower Cenomanian age (Table 2) interfinger with breccia of the lower Valle de Angeles Formation (Kvai). The lower Gualaco Formation interval is 90 m thick, and the upper interval, separated from the lower interval by a 10-m-thick breccia unit, is at least 170 m thick. Because the Gualaco Formation forms the highest stratigraphic unit of the region, its total thickness is unknown. (*Note:* The relative abundance of clasts in the sections is indicated by the order of the clast symbol, with the most abundant label first.)

(Figs. 8A and 8B) and the upper Atima Formation overlying the Manto Formation north of the town of Manto (Fig. 4). Similar volcanic activity continued during the deposition of the Valle de Angles Formation as shown by the pyroclastic flow preserved within the breccia in measured section 36 (Fig. 10A).

The Manto Formation is correlated with andesite porphyry mapped by the MMAJ (1980) in the western part of the Frey Pedro study area (Fig. 3). Volcanic rocks of the Manto Formation also correlate with the Naranjal volcanics (Simonson, 1977), volcanics near the base of the Atima Formation (Gordon, 1990), and with an unnamed andesite mapped by Mohl (1969).

Other documented examples of Cretaceous intermediate volcanic rocks in Honduras include the Mal Paso complex (Atwood et al., 1976), the Hoya Grande andesite (Markey, 1995), the Wampu volcanics (Weiland et al., 1992; Rogers, 1994), the Campeche volcanics (Gordon, 1999, personal commun.), and Agua Dulce volcanics (Curran, 1981). The Turonian to Campanian Campeche and Wampu volcanics (Gordon, 1999, personal commun.; Weiland et al., 1992) would post-date the much older Manto Formation. Although the Hoya Grande andesite to the south of the Frey Pedro study area occupies a similar stratigraphic position as the Manto volcanics, correlation is not warranted by existing information. Similarly, presumed Cretaceous and Tertiary intermediate volcanics to the west and along-strike of the Manto volcanics cannot be confidently correlated with the Manto volcanics with existing age and geochemical information.

Tayaco Formation

The Tayaco Formation name is introduced for the mainly marine volcaniclastic strata of the Aptian to Albian age separating the Atima Formation into upper and lower units in the Frey Pedro region. The Tayaco Formation conformably overlies the lower Atima Formation on the Agua Blanca block (Fig. 4) where it reaches a thickness of 1100 m. On the southern flanks of the Jacaleapa and Frey Pedro blocks, the Tayaco Formation unconformably rests on metamorphic basement and is no more than 320 m thick. This unit is named the Tayaco Formation for its outcrop area in the Tayaco hills along the southern side of the Montaña Frey Pedro (Fig. 4). The type locality for this formation is designated as sections 30–35 (Figs. 6 and 9A).

Outcrop Distribution and General Stratigraphy

Overlying and in conformable contact with the lower Atima Formation is a mainly volcaniclastic unit that attains a maximum thickness of 1100 m (Fig. 5). Lithologies of the Tayaco Formation are a diverse mixture of rocks, including maroon to orange, muddy, coarse- to medium-grained sandstone; gravel conglomerate; and gray to dark brown mudstone and shale. The formation also includes thin interbeds of calcareous mudstone and limestone. On the northern side of Montaña Frey Pedro along the San Francisco de la Paz–Gualaco road, strata of the Tayaco Formation are in conformable contact with the lower Atima Formation. In contrast, on the south side of the range, the Tayaco Formation unconformably overlies the basement schist (Fig. 4). The Tayaco Formation overlying basement relationship continues westward along the southern side of Montaña Frey Pedro and the southern side of Montaña Jacaleapa (Fig. 4). The overlying carbonate strata of the upper Atima Formation are in conformable contact with the Tayaco Formation.

Volcanic rocks of the Manto Formation are interbedded with the Tayaco Formation near the basal contact with the lower Atima Formation along the north side of the Montaña Frey Pedro (Fig. 4). Along the northern margin of the Valle de Amacuapa, Tayaco Formation volcaniclastic strata and andesite flows are interbedded. This interbedded relation gives way to andesite flows and pyroclastic deposits in the valley to the south where only minor volcaniclastic intervals occur. Where intermediate volcanic flows and pyroclastic deposits are more abundant than clastic strata, the rocks are mapped as Manto Formation (Fig. 5).

Tayaco Formation on the Agua Blanca Block

A nearly continuous, 710-m-thick section of the Tayaco Formation on the northern flank of the Montaña Frey Pedro is conformable with underlying limestone of the lower Atima Formation (Figs. 8A and 9A, sections 24 through 35).

In addition to being interbedded with the Manto volcanic flows in the lower part of the sequence (sections 24, 27–30), the volcaniclastic nature of the deposit is shown by interbeds of volcanic ash with shale (at 425 and 990 m, section 30, Fig. 9A). Clasts of andesite are common in coarser horizons (Fig. 9D), and much of the sand-size detritus found throughout the measured section is volcanic in origin. Other lithic fragments include subrounded quartz and limestone pebbles.

Tayaco Formation on the Frey Pedro Block

In contrast to the continuity of the Tayaco Formation with the lower Atima Formation on the Agua Blanca block, the Tayaco Formation on the Frey Pedro block rests directly on basement schist (Fig. 12A, section 23). The thickness of the Tayaco Formation on the Frey Pedro block is 340 m.

As on the Agua Blanca block, volcanic clasts are common in the Tayaco Formation on the Frey Pedro block. Andesite pebbles are found in the coarser-grained horizons and volcanic sands are found in the finer beds that include calcareous-sandy mudstone (Fig. 12A, sections 15–23). Volcanic flows were not observed in the measured section on the Frey Pedro block (Fig. 4). A distinctive well-sorted, coarse-grained quartz arenite with scoured channels and trough crossbedding occurs in sections 15 through 17 and sections 20 through 23 (Fig. 12A). This tabular body of quartz arenite lies above calcareous siltstone and mudstone and below calcareous shale and sandy limestone. The arenite is interpreted as a shoreface deposit within the Tayaco Formation.

Facies and Paleoenvironment of the Tayaco Formation

On the Jacaleapa and Frey Pedro blocks, fluvial, shoreface, and shallow-marine facies record marine transgression during deposition of the Tayaco Formation. Fluvial facies of the Tayaco

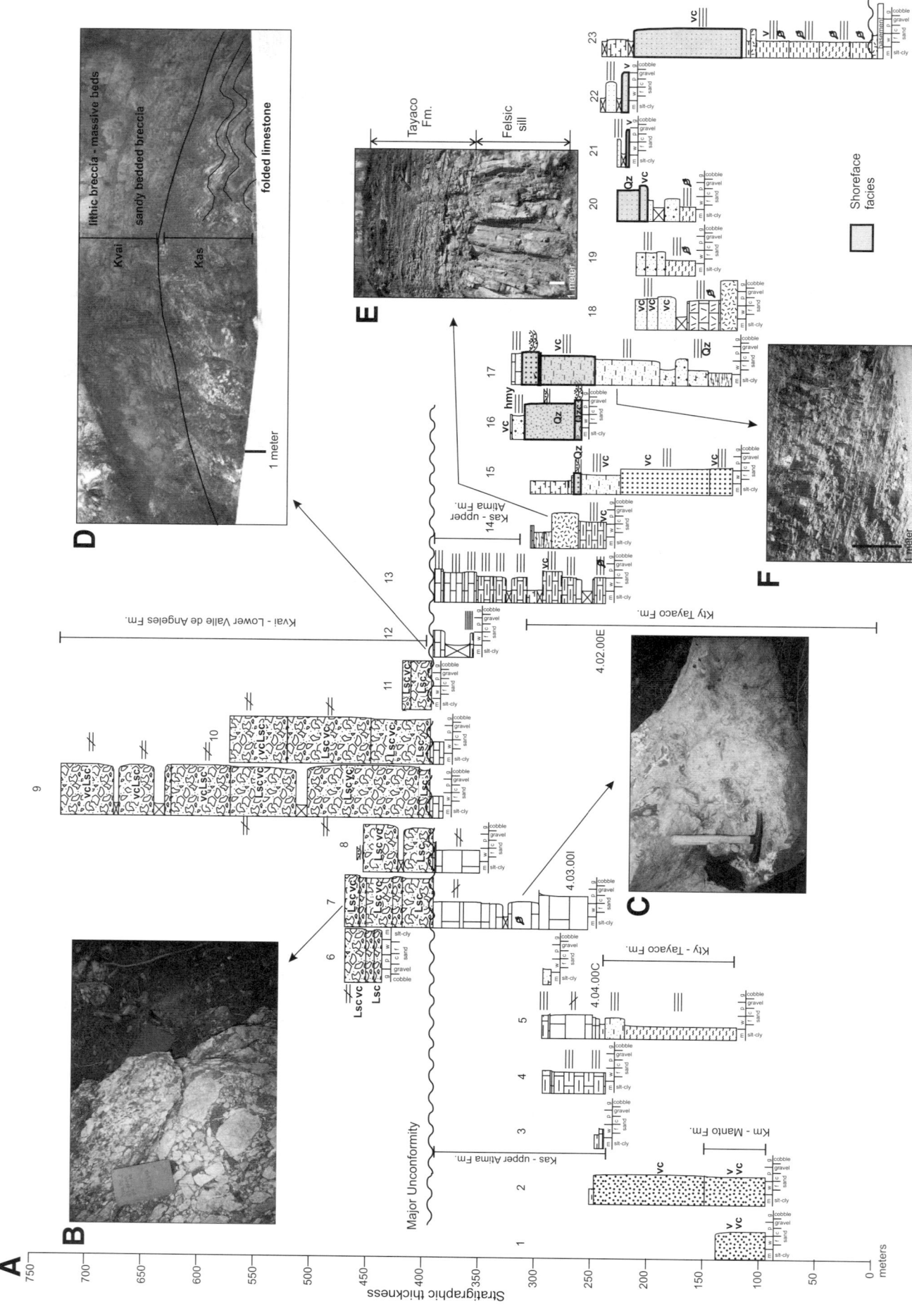

Figure 12. (A) Measured sections 1–23 from road outcrops along the San Francisco de la Paz–Gualaco road crossing the southern flank of the Montaña Frey Pedro (see Fig. 6 for locations of sections). Vertical scale is measured height above stratigraphic base (0 m base of section 23, Fig. 12A). Lsc—limestone; Qz—quartz; vc—volcanic; see Figure 8 for other lithologic abbreviation definitions. Three-hundred twenty-five meters of Aptian-Albian–age volcaniclastic Tayaco strata (Kty) unconformably overlie Paleozoic-age basement schist. Interpreted shoreface facies of the Tayaco Formation are highlighted by heavy boxes on sections 15–23. Middle to upper Albian–age limestone of the overlying upper Atima Formation (Kas) varies between 80 and 150 m in thickness. The unconformity between the upper Atima Formation and breccia of the lower Valle de Angeles Formation (Kvai) is both erosional and angular, as shown in sections 7–12. This erosion partially accounts for variability in the thickness of the upper Atima Formation. Breccia of the lower Valle de Angeles Formation is at least 330 m thick, with the proportion of volcanic lithic clasts increasing upsection (as seen on section 9). About 50 m of Manto volcanic mafic flows (Km) interfinger with Tayaco strata at sections 1 and 2. (*Note:* The relative abundance of clasts in the sections is indicated by the order of the clast symbol, with the most abundant label first.) (B) Field photograph of cobble-sized limestone clasts in breccia near the base of the Valle de Angeles Formation. (C) Gastropod-bearing horizon of middle to upper Albian age in limestone of upper Atima Formation. (D) Folded upper Atima Formation strata below folded unconformity with breccia of the Valle de Angeles Formation. (E) Felsic sill of Late Cretaceous–early Tertiary age at base of outcrop intruding shale of Tayaco Formation of Aptian to Albian age. (F) Characteristic outcrop of siltstone of the Tayaco Formation.

Formation occur along the southern flanks of Montaña Jacaleapa and in the Tayaco hills as small, upward-fining fluvial channel fills of trough-crossbedded medium-grained sands consisting of quartz and metamorphic lithic grains with a coarse, basal lag locally present (Fig. 4). The lag deposit contains limestone clasts along the south flank of Montaña Jacaleapa. Channels have scoured bases into overbank siltstone. White volcanic ash layers up to 5 cm thick occur within the overbank siltstone units.

In the Tayaco hills on the Frey Pedro block, a 3–5-m-thick, tan tabular body of well-sorted, coarse-grained, quartz and metamorphic clastic-bearing, lithic arenite occurs above the fluvial interval. This arenite contains symmetrical trough crossbeds and mollusks and other shell fragments, has little grading, and is interpreted as a shoreface deposit. Above the shoreface deposit, tan to orange calcareous sandstone, siltstone, and mudstone, containing bioturbated zones and non–age diagnostic marine snails, including *Turritella*, occur. A similar marine transgressive sequence was observed on the southern flank of the Jacaleapa block.

The shallow-marine paleoenvironment of the Tayaco Formation on the Frey Pedro and Jacaleapa blocks is represented by carbonate facies that include chondrodontid packstone, echinoderm-bearing sandy bivalve packstone, and bioclast grainstone. These facies represent paleoenvironments ranging from a marine shelf basin to a shallow marine shelf lagoon to a marginal marine, shallow marine shelf (Table 1). A quartz peloid grainstone with fragments of caprinid rudist (sample 2.28.00d) attests to the existence of a nearshore environment receiving mixed carbonate-clastic deposition. Overall, the Tayaco Formation represents relative sea-level regression followed by transgression prior to deposition of the upper Atima Formation.

The shoreface facies of the Tayaco Formation seen in the Tayaco hills appears equivalent to the shoreface facies of the measured section (Fig. 12A) that records marine regression followed by transgression.

On the Agua Blanca block, only marine and lower shoreface facies are evident in the Tayaco Formation. Marine facies include mainly bioturbated lime mudstone representing a deep marine shelf basin paleoenvironment (Table 1). A tabular body of moderately well-sorted, calcareous lithic sandstone and rounded lithic pebble conglomerate appears in the Tayaco Formation (sections 32–53, Fig. 9A) as an anomalously clean interval in the middle of an otherwise muddy formation. This sand body is interpreted as a lower shoreface deposit. The shoreface facies, bounded above and below by marine shelf facies environment of the Tayaco Formation on the Agua Blanca block, suggest marine regression followed by transgression during Tayaco Formation deposition.

Provenance of the Tayaco Formation

The clastic composition of coarse intervals within the Tayaco Formation indicates significant erosion of preexisting strata during deposition of the Tayaco Formation. Pebble-size to cobble-size clasts of biotite schist and gneiss dominate the basal, fluvial facies on the Jacaleapa and Frey Pedro blocks. As transport of schist clasts for long distances is unlikely; the source of the metamorphic clasts appears to be the emergent and proximal basement of Jacaleapa and Frey Pedro blocks. Limestone pebbles occur in the fluvial facies directly above the contact with the basement schist and gneiss on the Jacaleapa block. In contrast, limestone pebbles from the Agua Blanca block occur within the central part of the Tayaco Formation (Fig. 9B). The source of the limestone pebble is the lower Atima Formation, the only carbonate unit older than the Tayaco Formation.

Quartz sand, pebbles, and cobbles occur on all three blocks. While the basement metamorphic rocks appear to be the original source for these clasts, erosion and redeposition of the quartz pebble conglomerates of the Tepemechin Formation seems likely.

Intermediate (andesite and dacite) volcanic rock fragments are found on the Agua Blanca and Frey Pedro blocks, and volcanic sand forms occasional plagioclase-rich greywacke intervals within the Tayaco Formation on the Agua Blanca block. The most likely source of the volcanic clasts is the extensive exposure of the Manto Formation volcanic rocks in the Amacuapa valley (Fig. 4).

Age of the Tayaco Formation

An Aptian to Albian age of the Tayaco Formation is indicated by stratigraphic position between the upper Atima and lower Atima Formations. Samples from carbonate intervals within the

TABLE 1. FREY PEDRO AREA BIOSTRATIGRAPHIC DATA

Sample	Lithic Unit	Block	Age	Facies	Indet. Calcareous Algae	Indet. Benthic Forams	Indet. Planktic Forams	Indet. Radiolaria	Indet. Ostracodes	Indet. Bivalves	Indet. Chondrodontids	Indet. Caprinid Rudists	Indet. Radiolitid Rudists	Indet. Requienid Rudists	Indet. Gastropods	Indet. Echinoderms	Indet. Microcrinooids	Indet. Colonial Corals	Indet. Solitary Corals	Indet. Sponge Spicules	Indet. Serpulids	*Favreina* sp.	Miliolid Forams	*Cuneolina walteri* M-UAlb
03-29C	Kg	AB	M-U Albian	radiolitid packstone	x	x				x			x		x	x		x					x	x
03-28F	Kg	AB	?	bioclast-peloid wackestone-packstone					x															
02-28E	Kas	J	Apt-Alb	coral packstone-boundstone		x										x		x						
02-26A	Kas	AB	Apt-Alb?	planktic foram mudstone		x	x	x	x														x	
02-27C	Kas	J	Apt-Alb	bioclast packstone	x	x			x	x					x	x								x
02-28D	Kas	J	Apt-Alb	sdy qtz peloid grainstone		x			x	x					x									
04-03I	Kas	FP	Albian?	rudist packstone		x			?	x	?	x			x	x								
04-02E	Kas	FP		mollusk wackestone					x	x					x									
03-30D	Kas	AB	?	rudist packstone	x										x	x								
03-30C	Kas	AB	M-U Alb	rudist wackestone		x	?		x	x					x	x							x	x
03-05G	Kas	FP	?	biolime mudstone																				
03-05H	Kas	FP	Apt-Alb	chondrodont packstone					x	x	x													
04-09F	Kas	FP		peloid-bioclast packstone		x			x	x						x							x	
04-09D	Kas	FP	M-U Alb	peloid-req packstone-grainstone	x	x			x	x				x	x	x							x	x
04-07D	Kas	FP	?	peloid wackestone		?			x	x					x							x		
04-02E	Kas	FP		gastropod wackestone					x	x					x									
03-03J	Kas	FP		foram wackestone		x			x							x			x				x	x
03-03H	Kas	FP	Albian	foram-ostracode wackestone		x			x	x						x	x	x					x	
03-03F	Kas	FP	Albian	foram wackestone		x			x							x							x	
03-03E	Kas	FP	M-U Alb	mollusk wackestone	x	x			x	x					x									x
03-02L	Kas	FP	?	peloid-gastropod packstone		x									x	x					x	x		
03-02J	Kas	FP	?	sdy bioclast		x			x	x														
04-04C	Kas	FP	M-U Alb	foram-algal wackestone	x	x			x	x					x								x	x
04-06C	Kas	FP	U. Albian	rudist-bioclast packstone	x	x			x	x			x	x									x	
03-05A	Kty	FP	Apt-Alb	chondrodont packstone					x	x	x													
02-27D	Kty	J	Albian	bioclast grainstone						x		?				x								
02-28B	Kty	J	?	sdy qtz bivalve packstone						x						x								
03-09F	Kty	FP	?	dolo-pel grainstone, deformed																			x	
03-25B	Kty	AB	?	biolime mudstone					x	x					x									
03-24C	Kai	AB	Apt-Alb?	peloid-chondrodont packstone					x	x	x	x			x			x	x					
03-24B	Kai	AB	Apt-Alb	peloid-foram packstone		x			x	x				x	x	x							x	x
03-23B	Kai	AB	?	mollusk mudstone					x	x					x	x								
03-23-A	Kai	AB	?	dolo-mollusk-wackestone		x			x	x					x	x								
03-21A	Kai	AB	Apt-Alb	peloid-foram packstone		x			x	x													x	x

(Continued)

TABLE 1. FREY PEDRO AREA BIOSTRATIGRAPHIC DATA (Continued)

Sample	Lithic Unit	Block	Age	Facies	Hemicyclamina whitei Alb	Paracoskinolina coogani U Alb	Praechrysalidina infracretacea Hau-Alb	Pseudocyclammina hedbergi Apt-Alb	Pseudonummoloculina heimi Alb-Cen	Acicularia	Boueina hochsteteri Haut-Alb	Lithocodium aggregatum	Micritosphaera ovalis	Pithonella ovalis	Pithonella sphaerica	Parachaetetes texana	Pycnoporidium lobatum L Cret	Rivularia piae	Pseudolithothamnium alba	Salpingoporella muehlbergeri	Quartz	Feldspar	Rock Fragments
03-29C	Kg	AB	M-U Albian	radiolitid packstone						x						x							
03-28F	Kg	AB	?	bioclast-peloid wackestone-packstone																			
02-28E	Kas	J	Apt-Alb	coral packstone-boundstone								x											
02-26A	Kas	AB	Apt-Alb?	planktic foram mudstone										x	x								
02-27C	Kas	J	Apt-Alb	bioclast packstone				x												x			
02-28D	Kas	J	Apt-Alb	sdy qtz peloid grainstone					x			x								x			
04-03I	Kas	FP	Albian?	rudist packstone																			
04-02E	Kas	FP		mollusk wackestone																			
03-30D	Kas	AB	?	rudist packstone																			
03-30C	Kas	AB	M-U Albian	rudist wackestone				x															
03-05G	Kas	FP	?	biolime mudstone																			
03-05H	Kas	FP	Apt-Alb	chondrodont packstone																			
04-09F	Kas	FP		peloid-bioclast packstone																			
04-09D	Kas	FP	M-U Albian	peloid-req packstone-grainstone	?					x		x				x	x						
04-07D	Kas	FP	?	peloid wackestone																			
04-02E	Kas	FP		gastropod wackestone																			
03-03J	Kas	FP		foram wackestone	x			x				x											
03-03H	Kas	FP	Albian	foram-ostracode wackestone	x			x															
03-03F	Kas	FP	Albian	foram wackestone				x															
03-03E	Kas	FP	M-U Albian	mollusk wackestone																			
03-02L	Kas	FP	?	peloid-gastropod packstone																			
03-02J	Kas	FP	?	sdy bioclast																	x	x	x
04-04C	Kas	FP	M-U Alb	foram-algal wackestone									x		x					x			
04-06C	Kas	FP	U. Albian	rudist-bioclast packstone		x		x	x	x										x			
03-05A	Kty	FP	Apt-Alb	chondrodont packstone																			
02-27D	Kty	J	Albian	bioclast grainstone							x				x								
02-28B	Kty	J	?	sdy qtz bivalve packstone																x			
03-09F	Kty	FP	?	dolo-pel grainstone, deformed																			
03-25B	Kty	AB	?	biolime mudstone																			
03-24C	Kai	AB	Apt-Alb?	peloid-chondrodont packstone																	x		x
03-24B	Kai	AB	Apt-Alb	peloid-foram packstone				x				x											
03-23B	Kai	AB	?	mollusk mudstone																			
03-23-A	Kai	AB	?	dolo-mollusk-wackestone																			
03-21A	Kai	AB	Apt-Alb	peloid-foram packstone				x															

Note: Samples were collected along the San Francisco de la Paz-Gualaco road and adjacent region. Summary of age results is shown on the stratigraphic column in Figure 5. AB—Agua Blanca block (inverted rift); J—Jacapeapa block (northern rift shoulder); FP—Frey Pedro block (southern rift shoulder); Kai—"Lower Atima Formation"; Kty—Tayaco Formation; Kas—"Upper Atima Formation"; Apt—Aptian; Alb—Albian; M—middle; U—upper.

Figure 13. (A) Field photograph looking south across Valle de Amacuapa showing resistant hills south of the village on Manto formed by middle Cretaceous Manto volcanics. Photo taken at contact between andesite and limestone at the base of the upper Atima Formation. Note the lack of significant folding in Manto volcanics as a result of its position on the Frey Pedro basement block. (B) Field photograph showing thin lava flows of basaltic andesite sampled for geochemical analysis (sample 4.08.00E). (C) Field photograph showing columnar jointing defining small vent in upper parts of the Manto volcanic complex near the base of the silicic tuffs. Flow thickness indicates proximity to a middle Cretaceous eruptive center.

Tayaco Formation host *Boueina hochstetteri* and *Parachaetetes texana*, indicative of the Hauterivian?–upper Albian and Lower Cretaceous, respectively (Johnson, 1969; Scott and Gonzalez-Leon, 1991; Kuss and Senowbari-Daryan, 1992; Scott and Finch, 1999) (Table 1).

Upper Atima Formation

The middle to upper Albian–age upper Atima Formation conformably lies above the Tayaco and Manto Formations. This carbonate unit is estimated to be 620 m thick on the Agua Blanca block while 160 m thick on the southern flank of the Frey Pedro block.

Outcrop Distribution and General Stratigraphy

The upper Atima Formation is a gray, fossiliferous wackestone-packstone with local mudstone intervals. Bedding varies from massive to 10–20 cm thick. Locally, dark, calcareous shale and dark gray, thin-bedded argillaceous micrite appear near the upper 50 m of this unit. The upper Atima Formation occurs above and in conformable contact with the Tayaco Formation and the Manto Formation. Exposed in the hills north of the village of Manto, the upper Atima Formation directly overlies the porphyry andesite flows of the Manto Formation where the resistant limestone lithologies form a caprock above the volcanic flows (Fig. 4). Limestone layers 10–20 m thick occur within the volcanic flows below the caprock formed by the upper Atima Formation. The upper Atima Formation also forms a prominent hogback on the southern flank of Montaña Jacaleapa (Fig. 4).

Upper Atima Formation on the Agua Blanca Block

Two intervals of the upper Atima Formation are exposed. The first is a 240-m-thick, fault-bounded interval of sandy wackestone, volcaniclastic sandstone, and calcareous shale (Fig. 10A, section 37). The second interval is 180 m thick and consists mostly of massive wackestone that conformably overlies a coarse

TABLE 2. MAJOR* AND TRACE ELEMENT† CHEMISTRY OF MANTO VOLCANIC AND CHINDONA INTRUSIVE ROCKS

Sample	2.22a	3.31f	2.24c	2.24e	3.13i	4.08a	4.08c	4.08e	4.08h	4.09i
Unit	Chindona	Chindona	Manto	Manto	Manto	Manto	Manto	Manto	Manto	Manto
Rock type	granite	granite	basalt	rhyolite	dacite	dacite	rhyolite	andesite	andesite	andesite
Location§	15.186	15.325	14.972	14.925	14.804	14.900	14.897	14.897	14.835	14.939
	86.538	86.192	86.548	86.396	85.667	86.386	86.386	86.386	86.327	86.367
SiO_2	67.88	65.82	45.75	71.54	76.40	73.03	74.42	58.84	64.43	63.68
TiO_2	0.45	0.63	1.67	0.36	0.14	0.29	0.28	1.44	0.88	0.83
Al_2O_3	16.03	16.22	14.71	14.42	12.80	14.88	14.09	16.27	15.66	15.00
Fe_2O_3	3.32	4.05	9.03	1.99	1.44	1.92	1.78	7.44	4.63	4.59
MnO	0.07	0.06	0.20	0.03	0.01	0.04	0.02	0.05	0.05	0.06
MgO	1.21	1.98	6.43	0.38	0.16	0.51	0.60	2.94	2.17	2.39
CaO	3.11	4.45	9.66	2.17	0.09	1.97	1.99	4.82	3.51	4.74
Na_2O	3.81	3.86	3.94	3.32	3.76	3.68	3.53	3.81	4.64	3.99
K_2O	2.38	1.76	1.24	3.17	4.25	3.11	3.00	2.15	1.99	2.18
P_2O_5	0.14	0.16	0.55	0.18	0.04	0.10	0.10	0.55	0.24	0.24
Totals	98.40	98.99	93.18	97.56	99.09	99.53	99.81	98.31	98.20	97.70
Ni	bd	bd	152	bd	bd	bd	bd	25	20	17
Cu	bd	bd	26	bd	bd	15	bd	40	25	22
Zn	37	42	93	34	32	42	24	63	42	64
Rb	99	39	32	117	157	120	94	44	44	48
Zr	126	171	221	125	159	134	130	232	181	187
Y	20.8	12.84	24.18	13.59	26.9	5.58	6.15	36.7	14.06	15.55
Nb	14.48	13.65	20.97	20.23	14.68	6.94	6.85	29.74	17.39	15.58
Sr	208.98	428.58	552.8	235.3	67.39	408.74	505.66	1428.52	683.05	689.39
Ba	727.24	453.32	364.64	778.06	865.94	1473.32	1290.74	1180.34	816.72	853.16
La	17.86	24.48	31.98	21.48	19.27	23.95	23.03	42.56	27.57	26.31
Ce	42.5	52.99	62.06	47.31	31.17	49.15	47.62	85.58	56.11	56.72
Pr	4.72	5.58	8.49	5.39	5.79	4.83	4.73	11.3	6.02	6.27
Nd	17.55	18.91	34.81	18.37	20.36	15.81	15.08	46.08	20.89	22.07
Sm	4.2	3.67	6.68	3.67	4.65	2.67	2.58	9.29	3.89	4.19
Eu	1.07	1.15	2.02	1	0.91	0.94	0.9	2.7	1.34	1.35
Gd	3.95	3.3	6.36	3.22	4.37	2.32	2.26	8.64	3.65	3.98
Tb	0.62	0.48	0.89	0.47	0.73	0.29	0.29	1.26	0.5	0.58
Dy	3.45	2.21	4.46	2.3	4.28	1.06	1.08	6.47	2.4	2.75
Ho	0.73	0.45	0.85	0.45	0.97	0.21	0.22	1.29	0.49	0.55
Er	1.89	1.17	2.15	1.1	2.77	0.47	0.48	3.36	1.21	1.44
Yb	2.3	1.8	2.65	1.68	3.48	1.14	1.2	4.08	1.91	2.07
Lu	0.31	0.2	0.28	0.19	0.5	0.1	0.09	0.47	0.21	0.23
Hf	3.11	3.93	4.7	3.17	5.38	3.48	3.3	7.25	3.94	4.09
Ta	1.38	0.96	1.25	2.54	3.35	0.87	0.78	1.68	1.03	0.98
Pb	30.79	7.76	4.55	41.38	22.35	44.66	35.28	10.64	19.14	19.11
Th	6.82	5.66	4.75	5.75	13.5	8.04	7.66	5.85	5.35	7.16
U	4.76	2.21	0.82	9.99	5.46	4.87	5.28	1.76	2.71	3.82
V	51.68	105.68	101.67	43.06	10.01	44.68	39.14	204.11	125.92	123.46
Cr	24.62	38.12	324.75	24.16	13.99	22.37	20.6	88.33	76.01	100.36

Note: bd—below detection.
*XRF analysis by Thomas Vogel.
†ICP-MS analysis by Lina Patino, Michigan State University.
§latitude/longitude.

volcaniclastic interval of the Tayaco Formation (Fig. 10A, sections 39–47). The uppermost part of the upper Atima Formation is truncated by a major erosional unconformity with overlying sedimentary breccia of the Valle de Angeles Formation (Fig. 10A, section 38 and 39).

Upper Atima Formation on the Frey Pedro Block

The gradational contact between the Tayaco Formation and the upper Atima Formation is marked by a progressive decrease in clastic deposition and a progressive increase in biogenic deposition (Fig. 12A, section 13 and 5). As on the Agua Blanca block, the upper part of the upper Atima Formation is truncated by a major erosional unconformity with the overlying Valle de Angeles Formation (Fig. 12A, sections 7–12; Fig. 12D).

Facies and Paleoenvironment of the Upper Atima Formation

While the upper Atima Formation is mostly a shallow-marine carbonate with build-ups, locally restricted deeper-marine

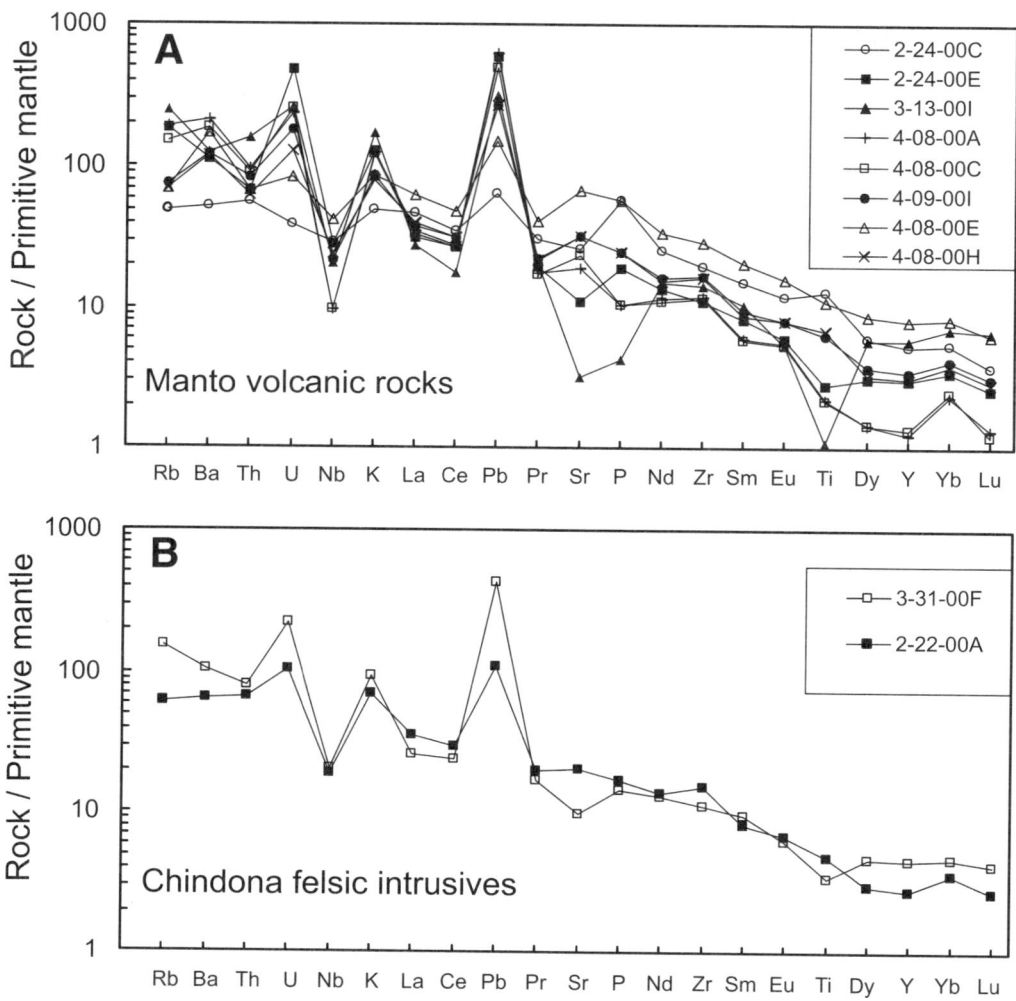

Figure 14. Multi-elemental geochemical pattern of igneous rocks from the Frey Pedro study area. Normalizing values for plots in A and B are for primitive mantle from Sun and McDonough (1989). Sample locations are shown in Figures 3 and 4 and are indicated on the stratigraphic column in Figure 5. (A) Results from eight samples from the Manto volcanic unit show enrichment of large ion lithophile elements (LILEs) (Ba, K, Pb, Sr) while depleted in high field strength elements (HSFEs) (Nb, Zr, Ti), which is indicative of a subduction-related origin. This unit occupies the Tayaco Formation stratigraphic position (Aptian to Albian age) and is spatially related to the formation of the Agua Blanca rift. (B) Results from two samples from the Chindona batholith show similar enrichment of LILEs and depletion of HSFEs, indicating association with a volcanic arc origin. This unit is correlated to felsic dikes cutting the folded Cretaceous strata of the Frey Pedro study area. These data are the first documentation of extensive volcanic arc rocks erupted on the Central Chortis terrane (see Fig. 1) in middle Cretaceous time and continuing during and after inversion of the Agua Blanca basin.

shelf facies occur on the Agua Blanca block (Table 1). Carbonate facies indicating shallow-marine shelf lagoon include rudist packstone, chondrodontid packstone, coral packstone, bioclast packstone, mollusk wackestone, and foram-algal wackestone (Table 1). Shallow marine shelf facies include quartz bioclast wackestone, foram wackestone, foram-ostracode wackestone, foram packstone, and peloid-bioclast packstone. The marine shelf environment is represented by bioturbated lime mudstone and deep marine shelf environments by planktic foram mudstone and *Favreina* (pellets)-bearing peloid grainstone-packstone. The deeper marine planktic foram mudstone facies occurs on the Agua Blanca block near the top of the section associated with dark, calcareous shale and dark gray, thin bedded argillaceous micrite, suggesting a period of local basin subsidence immediately prior to development of the erosional unconformity and deposition of the overlying Valle de Angeles Formation.

Age of the Upper Atima Formation

Foraminifers and calcareous algae identified in thin section from samples of the upper Atima Formation carbonates range from Albian to upper Albian (Table 1). Age-significant foraminifers include *Cuneolina walteri*: middle–upper Albian (Scott and Gonzalez-Leon, 1991; Scott and Finch, 1999); *Pseudocyclammina hedbergi*: Aptian–upper Albian (Banner and Highton, 1990; Scott and Gonzalez-Leon, 1991); *Hemicyclammina whitei*: Albian (Banner and Highton, 1990); *Paracoskinolina coogani*: upper Albian (Scott, 2002); and *Pseudonummoloculina heimi*: Albian-Cenomanian (Scott and Gonzalez-Leon, 1991).

Calcareous algae with diagnostic ranges found in the upper Atima Formation include *Parachaetetes texana*: Lower Cretaceous (Johnson, 1969: Scott and Finch, 1999); *Salpingoporella muehlbergeri*: Barremian-Aptian, Albian? (Conrad, 1977); and *Boueina hochstetteri*: Hauterivian?–upper Albian–Cenomanian? (Scott and Gonzalez-Leon, 1991; Kuss and Senowbari-Daryan, 1992).

Valle de Angeles Formation

The Upper Cretaceous Valle de Angeles Formation records a shift from carbonate to clastic deposition (Finch, 1981) (Fig. 5). The Valle de Angeles Formation contains a lower, generally coarser clastic interval, "lower redbeds," and an upper finer clastic interval, "upper redbeds," locally separated by shallow marine carbonate strata of the Cenomanian-age Jaitique and Esquias Formation (Horne et al., 1974; Finch, 1981; Scott and Finch, 1999). The Upper Cretaceous Valle de Angeles Formation is the most widely distributed Mesozoic unit on the Chortis block (Fig. 2).

Outcrop Distribution and General Stratigraphy

The clastic strata overlying the limestone of the upper Atima Formation in the Frey Pedro belt are identified as the Valle de Angeles Formation. The Valle de Angeles Formation in the Frey Pedro study area consists of a poorly consolidated, massive, cobble to boulder, polymictic breccia. The deposit attains a maximum thickness of 450 m. Bedding, where discernable, is massive, and single bedding units commonly encompass the entire exposure of the road cut. Tan to red, coarse-grained volcanic sand comprises the matrix that supports the limestone and volcanic lithic clasts (Fig. 10B and C). The basal contact of the Valle de Angeles Formation is a major erosional surface cut into the upper Atima Formation. Where the erosion surface intersects bedded intervals of the commonly massive upper Atima limestone, a slight angular unconformity is apparent (Fig. 12D).

The sedimentary breccia of the Valle de Angeles Formation is exposed along the north and south sides of Montaña Frey Pedro and is exposed as a fault-bounded sliver south of Montaña Agua Blanca (Fig. 4). Along with the upper Atima Formation, the Valle de Angeles Formation forms the prominent hogback of the southern flank of Montaña Jacaleapa (Fig. 4). Along the Río Telica and the southern side of the Montaña Casa de Tela, the base of the Valle de Angeles Formation rests unconformably on the Manto volcanic strata and at that location is composed of clasts of rounded volcanic and lithic clasts (Fig. 3).

Valle de Angeles Formation on the Agua Blanca Block

Observations from the measured sections on the north side of Montaña Frey Pedro indicate that deposition of the breccias of the Valle de Angeles Formation occurred in a marine setting and coincided with volcanic eruptions. A 275-m-thick section of the Valle de Angeles Formation breccia occurs above a 10-m-thick dacitic pyroclastic flow (Fig. 10A, section 36). Near the erosional base of the Valle de Angeles Formation, the deposit is dominated by limestone clasts that commonly include unconsolidated fragments of large, thick-shelled gastropods and corals (Fig. 10A, section 38). These shells appear to have been swept off the carbonate shelf of the upper Atima Formation and incorporated as clasts of the breccia. An additional indication of the marine setting of the Valle de Angeles Formation comes from the marine limestone of the Gualaco Formation that is interbedded with the clastic breccia near the top of the Valle de Angeles Formation (Fig. 11A, sections 48–56).

Valle de Angeles Formation on the Frey Pedro Block

Measured sections from the south side of Montaña Frey Pedro reveal a slightly angular erosional unconformity at the base of the Valle de Angeles Formation. Stratigraphic inversion is indicated by an upward variation in clast composition composed of progressively older stratigraphic units. The erosional base of the breccia is well exposed in sections 7–12 (Fig. 12A). At sections 11 and 12, the gently folded basal contact occurs above a series of folded beds of the underlying upper Atima Formation (Fig. 12D).

An exceptionally thick exposure of the Valle de Angeles Formation exhibits a decrease in the quantity of limestone clasts and an increase in volcanic rock fragment upsection (Fig. 12A, section 9). This progression indicates initial erosion of the limestone of the upper Atima Formation followed by erosion of the andesite of the Manto Formation.

Facies and Paleoenvironment of the Valle de Angeles Formation

The massive- to thick-bedded nature of the matrix-supported polymictic breccia of the Valle de Angeles Formation is consistent with a marine, debris flow origin on the Agua Blanca block. Other lines of evidence for a marine debris flow include the association of the deposit with marine limestone, erosion into underlying marine strata, and the presence of biogenic clasts near its base. On the Agua Blanca block, existing evidence supports shallow marine deposition in a subsiding basin.

On the Frey Pedro block, evidence for subaerial deposition of breccias of the Valle de Angeles Formation includes greater rounding of clasts and the appearance of fluvial channel fills. In the hills west of San Francisco de la Paz (Fig. 4), the breccia is mainly a limestone-clast, cobble conglomerate displaying sandy fluvial channels with trough crossbeds in the upper parts of otherwise massive beds deposited proximal to the eroding uplands underlain by upper Atima Formation limestone. These rocks are interpreted as alluvial fan facies representing the subaerial upland extent of the marine debris flows on the Valle de Angeles Formation on the Agua Blanca block.

Provenance of the Valle de Angeles Formation

Lithic clast provenance combined with facies variations of the Valle de Angeles Formation indicates uplift and erosion of the Manto Formation volcanic province centered in the Amacuapa valley. Volcanic clasts of the Valle de Angeles Formation include cobbles of plagioclase-bearing porphyritic andesite, which is the principal lithology of the Manto Formation. Limestone and marine fauna clasts of the Valle de Angeles Formation originate from the upper Atima Formation that underlies the breccia. On the Frey Pedro and Jacaleapa blocks, the occurrence of the Tayaco Formation in stratigraphic contact with the basement schist precludes the lower Atima Formation as the source of the limestone clasts.

Variation in the composition and facies of the Valle de Angeles Formation is attributed to its mode of deposition and the upland source rock exposed. From south to north, a transition from subaerial to marine facies occurs that coincides with an increasing contribution from erosion of the Manto Formation. Erosion of the underlying units on the Frey Pedro block occurred in a subaerial setting while marine deposition continued to the north on the Agua Fria block. This reflects the uplift of the southern paleogeography. Pyroclastic strata (Fig. 10A section 36) at the base of the Lower Valle de Angeles Formation suggest that uplift was associated with volcanic activity.

Age of the Valle de Angeles Formation

An early Late Cretaceous age is inferred for the Valle de Angeles Formation. This age is based on the upper Albian age of the underlying upper Atima Formation and the lower Cenomanian age of the overlying Gualaco Formation.

Gualaco Formation

On the Agua Blanca block, the breccia of the Lower Valle de Angeles Formation is conformably overlain by the limestone of the Gualaco Formation. The Gualaco Formation name is introduced for this carbonate interval of the Valle de Angeles Formation because it is geographically isolated from the Esquias and Jaitique Formations (Finch, 1981), the other intra-Valle de Angeles Formation carbonate units of Honduras.

Outcrop Distribution and General Stratigraphy

Directly south of the town of Gualaco on the Agua Blanca block, carbonate strata of the Gualaco Formation conformably overlie breccia of the Valle de Angeles Formation (Fig. 4). The tan to light gray limestone of the Gualaco Formation has 10- to 30-cm-thick beds.

Gualaco Formation on the Agua Blanca Block

The Gualaco Formation consists of a lower 90-m-thick limestone and an upper 190-m-thick limestone with an intervening lens of the polymicitic breccia (Fig. 11A, sections 48–56). The contact between the lower unit of the Gualaco Formation and the breccia is conformable, marked by an abrupt decrease in coarse clastic fragments.

Facies and Paleoenvironment of the Gualaco Formation

Carbonate facies of the Gualaco Formation are indicative of shallow-marine settings. A bioclast-peloid packstone (sample 3.28.00f) represents a restricted shelf environment, and a radiolitid packstone with miliolids and colonial corals (sample 3.29.00c) represents deposition on a shallow marine shelf (Table 1).

Age of the Gualaco Formation

The Gualaco Formation is one of several limestone intervals associated with the Valle de Angeles Formation. Others include the Esquias and Jaitique Formations, which both contain Cenomanian age fauna (Horne et al., 1974, Finch, 1981, Scott and Finch, 1999).

An age of upper Albian to Cenomanian for the Gualaco Formation is indicated by paleontologic identification and by regional stratigraphic correlation. Sample 3.29.00c contains the age-significant foraminifer *Cuneolina walteri*: middle–upper Albian (Scott and Gonzalez-Leon, 1991; Scott and Finch, 1999) and the calcareous algae *Parachaetetes texana*: Lower Cretaceous (Johnson, 1969: Scott and Finch, 1999).

STRUCTURE OF THE FREY PEDRO BELT

The intensity of shortening across the Frey Pedro belt is highly variable. Folding and faulting is less intense above metamorphic basement-cored blocks: the Jacaleapa to the north and the Frey Pedro to the south (Fig. 4). Deformation is most intense within the central Agua Blanca belt, where the sedimentary section is thickest and no basement is exposed. We infer that the central Agua Blanca block corresponds to an inverted Cretaceous rift (Agua Blanca rift), and the more rigid, northern and southern blocks are its former rift shoulders. During the inversion process, the thicker sedimentary rocks of the central rift block were structurally elevated relative to the former rift shoulders.

Structure Profiles of Frey Pedro Belt

The three structural profiles presented in Figure 15 were constructed across the Frey Pedro belt cross Montaña Frey Pedro, Montaña Agua Blanca, and the southern part of Montaña Jacaleapa; they illustrate the structure of the three blocks, with the central, rift block exhibiting reverse-faulted borders and more steeply dipping, intensively faulted and folded strata. The northern and southern basement-cored blocks display large anticlines defined by opposed dips in overlying Cretaceous strata. The intervening strata-filled Agua Blanca block is bounded to the north and south by high-angle reverse faults, with the sedimentary section containing steeply inclined strata, numerous folds, and more high-angle reverse faults.

Profile A–A′

This profile was drawn in the area north of the village of Manto (Fig. 4) and reveals a broad, open syncline cored by shallow-dipping strata of the upper Atima Formation. The fold is underlain to the south by volcanic rocks of the Manto Formation and to the north by the Manto Formation and the clastic strata of the Tayaco Formation (Fig. 15A). The blind thrust, producing a fault-bend fold, is inferred to explain observed bedding attitudes.

Profile B–B′

This section originates in the Valle de Amacuapa, south of the village of Guarizama, crosses Montaña Frey Pedro and Montaña Agua Blanca, and extends to the southern side of Montaña

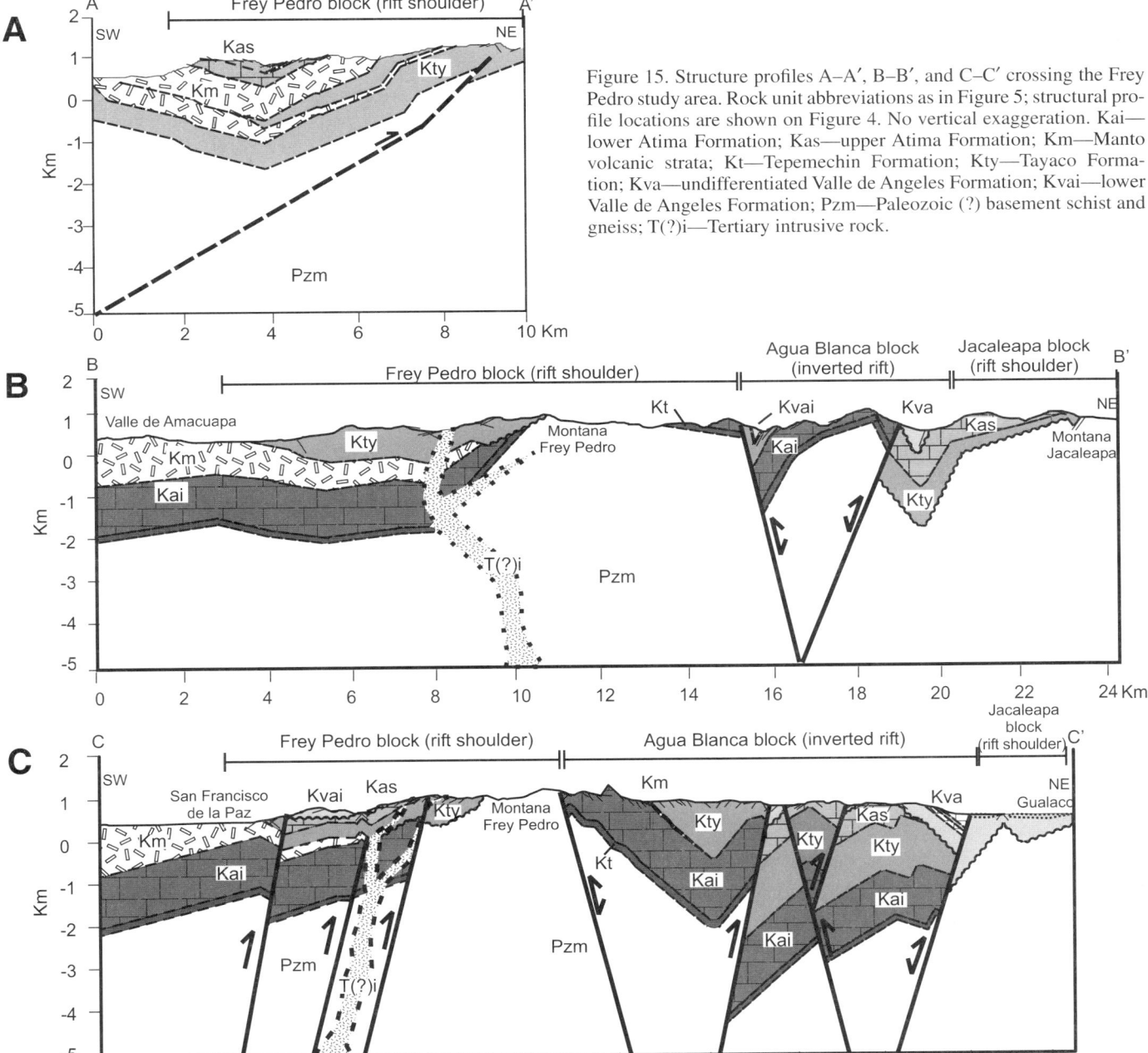

Figure 15. Structure profiles A–A′, B–B′, and C–C′ crossing the Frey Pedro study area. Rock unit abbreviations as in Figure 5; structural profile locations are shown on Figure 4. No vertical exaggeration. Kai—lower Atima Formation; Kas—upper Atima Formation; Km—Manto volcanic strata; Kt—Tepemechin Formation; Kty—Tayaco Formation; Kva—undifferentiated Valle de Angeles Formation; Kvai—lower Valle de Angeles Formation; Pzm—Paleozoic (?) basement schist and gneiss; T(?)i—Tertiary intrusive rock.

Jacaleapa (Fig. 4). Open anticlines and synclines, developed in the Tayaco Formation, occur on the southern flank of Montaña Frey Pedro (Fig. 15B). An undeformed, felsic intrusion cuts the folds and postdates the deformation. North of the basement-cored topographic crest of Montaña Frey Pedro, northeast-dipping strata of the Tepemechin Formation and lower Atima Formation are truncated by a steep reverse fault. The hanging wall of this reverse fault in the Agua Blanca hills contains a narrow syncline cored by the polymictic breccia of the Lower Valle de Angeles Formation. The breccia directly overlies the Tayaco Formation and the intervening upper Atima Formation is absent. A large asymmetric anticline cored by the lower Atima Formation is truncated by a steep reverse fault along the northern margin of Montaña Agua Blanca. A syncline folding the Valle de Angeles Formation occurs in the footwall of this reverse fault. Folds developed in the upper Atima Formation and the Tayaco Formation occur on the southern flank of Montaña Jacaleapa.

Profile C–C′

This profile parallels the stratigraphic section measured along the San Francisco de la Paz–Gualaco road and provides a nearly continuous structural transect across the Montaña Frey Pedro (Fig. 4). Open folds deform the Manto Formation, the Tayaco Formation, and the Lower Valle de Angeles Formation. Folds are also present on the southern flank of Montaña Frey Pedro, where several high angle reverse faults offset the stratigraphy. One undeformed felsic intrusion cuts the folds. North of the basement-cored crest of Montaña Frey Pedro, a high angle reverse fault places the lower Atima Formation in fault contact with basement schist. The hanging wall of this fault contains a thick sequence of lower Atima, Manto, and Tayaco Formations and forms a large syncline. A series of reverse faults repeats slices of folded strata of progressively higher stratigraphic levels north of the Agua Blanca hills. A large reverse fault at the foot of the Agua Blanca hills places hanging-wall rocks of the Gualaco Formation over rocks of the Tayaco Formation.

Structural Blocks of Frey Pedro Belt

The Frey Pedro block is defined as the crest and southern flank of Montaña Frey Pedro south of the large reverse fault and including the Valle de Amacuapa. Open symmetrical folds and moderately inclined strata characterize the structure of the Frey Pedro block (Fig. 15A). The stratigraphy below the Tayaco Formation is missing from the Frey Pedro block. Stratigraphic units are not repeated by reverse faults, indicating fault displacements of less than a few hundred meters, the thickness of the stratigraphic units. Observed faulting involving basement is limited to a reverse fault along the northern margin of the Frey Pedro Block.

Like the Frey Pedro block, the Jacaleapa block is a metamorphic basement-cored block with strata of the Tayaco and upper Atima Formation strata flanking the southern side of the block (Figs. 15B and 15C). The reverse faults to the south of the block mark its southern edge. The northern margin of the Jacaleapa block is defined by the southern edge of the Chindona batholith (Figs. 2 and 4).

The Agua Blanca block, bounded to the north and south by high-angle reverse faults, contains steeply inclined strata, numerous folds, and reverse faults (Figs. 15B and 15C). The Tayaco Formation thickens within the Agua Blanca block where older Cretaceous stratigraphic units have been exposed by the basin inversion process.

Structural Analysis of the Frey Pedro Belt

Structural observations are grouped by formation in Figure 16 in order to identify any older tectonic events that might have preceded the major Late Cretaceous shortening event recorded by the basin inversion seen in the structural profiles in Figure 15. Lower-hemisphere, equal-area stereoplots of poles to 590 measurements of bedding planes over all three blocks of the Frey Pedro belt indicate general dip directions to the northeast and southwest that correspond to the overall northwest trend of the Frey Pedro belt (Fig. 16A). The similarity in dip values of all stratigraphic units from the three blocks suggest that the final shortening event postdated the deposition of the youngest stratigraphic unit in the section, the Cenomanian Gualaco Formation (Fig. 5).

Fifty-six minor outcrop scale folds were identified (Fig. 16B). Minor fold profiles range from chevron to isoclinal. Thirty minor folds were measured from thin-bedded intervals of the upper Atima Formation, and 22 minor folds were measured from thin-bedded intervals of the lower Atima Formation (Fig. 5). Fold axes of minor folds from the lower Atima Formation scatter with the majority of the axes plunging to the northwest. Upper Atima Formation fold axes plunge to the west. The orientation of these minor fold axes is consistent with north-south shortening of the range and tightening of fold axes over in a southeastward direction.

Minor, outcrop-scale faults with normal displacements were identified in all stratigraphic horizons except the Gualaco Formation (Fig. 5). Sense of minor fault displacement was determined in the field using apparent offset in bedding planes and striated fault planes (Petit, 1987) (Fig. 16C). Calcite overgrowths on striated fault planes in the limestone of the upper and lower Atima Formation are common in the study area and account for most of the fault kinematic measurements. Fault orientations show considerable scatter. Right or left oblique-slip was observed on most normal faults. Down-to-the-southwest movement is suggested by the majority of the normal fault orientations from the upper Atima Formation.

Minor, outcrop-scale faults with reverse motion were identified in all stratigraphic horizons except the Gualaco Formation using both apparent offsets in bedding planes and fault striations (Fig. 16D). Right or left oblique-slip was observed for the majority of reverse faults. Up-to-the-northeast movement is suggested by reverse fault orientations in limestone of the upper Atima Formation. The youngest stratigraphic unit from which fault data was collected, the Lower Valle de Angeles Formation, displays

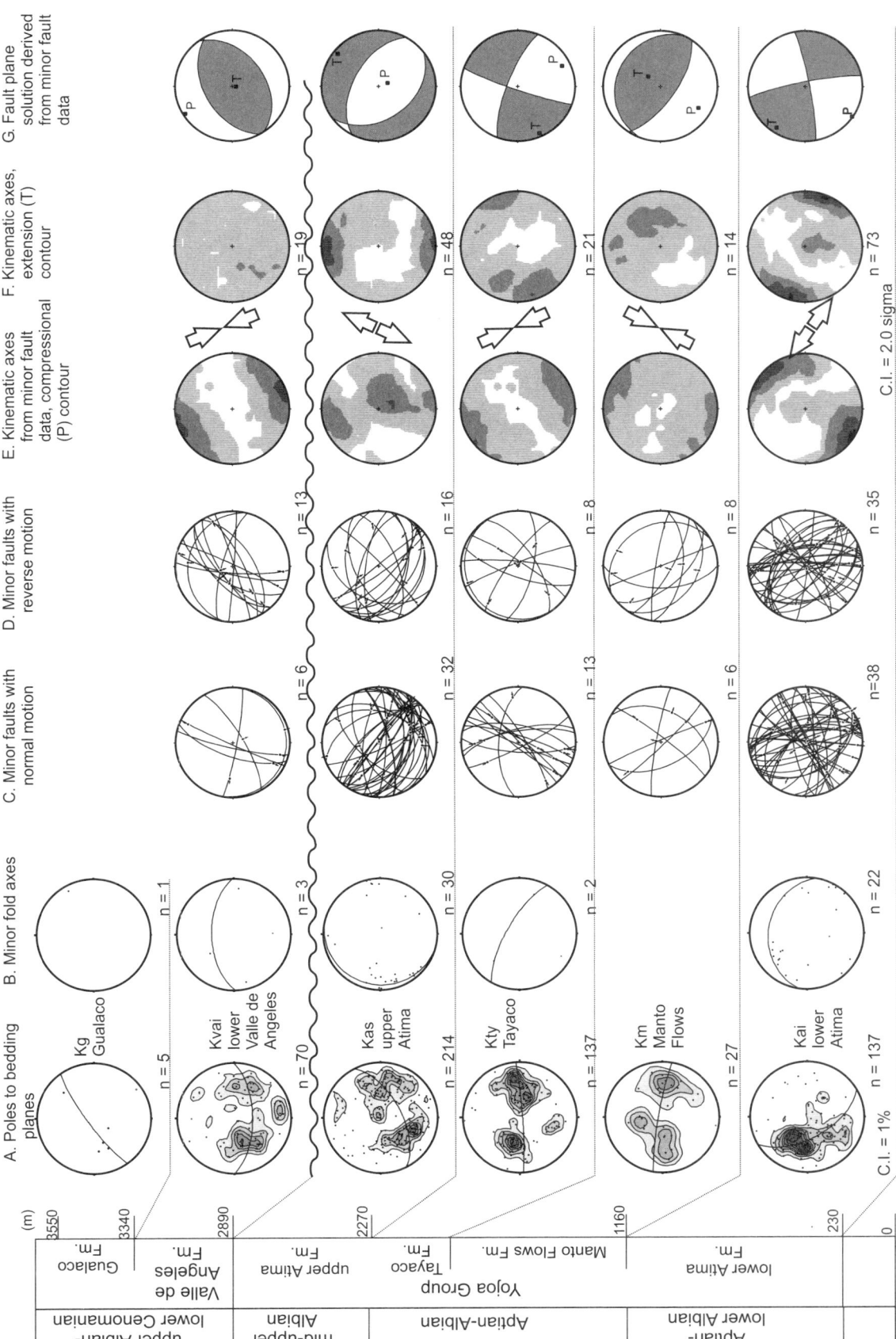

Figure 16. Summary of structural data for the Frey Pedro study area, grouped by stratigraphic unit in order to display variations in deformation through time. Plots are lower hemisphere Wulff projections with contour interval equal to 1% area for bedding poles and 2 sigma for kinematic axes. Contoured P and T kinematic axes and fault plane solution were calculated using the method of Marrett and Allmendinger (1990). For fault plane solutions, dark areas represent compressional quadrants. Units above and below a major mid-Cretaceous unconformity record similar deformational styles indicating a Late Cretaceous shortening event that inverted the Agua Blanca rift basin shown in the cross sections in Figure 15. See text for discussion.

reverse fault orientation consistent with north-south shortening of the Frey Pedro belt.

Using the method of Marrett and Allmendinger (1990), kinematic analysis of fault-slip data from the Frey Pedro belt is displayed in Figures 16E–16G. The kinematic axis of shortening, using the all fault-slip data from a stratigraphic interval is displayed as the P-contour (Fig. 16E), and the kinematic axis of extension as the T-contour (Fig. 16F). Arrows inserted between Figures 16E and 16F show the dominant extension or shortening direction derived from the kinematic analysis.

Shortening directions vary in trend from northeast to northwest (Fig. 16). One explanation for this variability is the oroclinal bend present in the study area with northwest-trending folds and faults along the northwest part of the range changing orientation to more east-west and northeast orientations in the eastern part of the study area. The San Francisco de la Paz–Gualaco road, along which most of these structural measurements were made, forms the approximate inflection point in the oroclinal bend of the range (Fig. 4). For this reason, the average shortening direction is taken as the north-south bisector of the northeast to northwest range of shortening directions.

The major shortening postdates the deposition of the youngest stratigraphic unit in the area, the upper Albian to lower Cenomanian Gualaco Formation (Fig. 5), and predates the intrusion of undated, felsic dikes on the Frey Pedro block (Fig. 4). Therefore, shortening is post-Cenomanian.

INTRUSIVE ROCKS IN THE FREY PEDRO STUDY AREA, EASTERN HONDURAS

Calc-alkalic volcanic rocks of the middle Cretaceous Manto Formation (see above) and the felsic intrusive rocks of the Late Cretaceous to early Tertiary Chindona batholith both indicate Cretaceous magmatic arc activity on the Central Chortis terrane. The Chindona batholith occupies an area of 1200 km² (with 197 km² of apparently related stocks, sills, and dikes) (Fig. 4).

Chindona Batholith

Outcrop Distribution and General Geology

Predominately a coarse-grained granodiorite, the Chindona batholith is the largest (1200 km²) of several major batholiths that trend eastward across north-central Honduras (Fig. 2). The Chindona batholith forms the youngest lithologic unit of the Frey Pedro study area because it intrudes the Mesozoic section, forming a roof pendant composed of altered Mesozoic strata, and andesite of the Manto Formation occurs north of the town of Chindona (Figs. 2 and 3).

Deeply weathered to grus, the Chindona batholith forms a distinctive badlands topography that is readily identifiable from LANDSAT imagery (e.g., area northwest of Agalta valley on Fig. 18B). Resistant bodies of the granodiorite occur locally in outcrops along the Río Manto, Río Grande, and Río Chindona and are found along the La Union–Olanchito road (Fig. 3). A second roof pendant composed of basement schist and gneiss is exposed south of the town of Esquipulas del Norte (Fig. 3).

Felsic Stocks, Dikes, and Sill

Within the Frey Pedro belt, dikes and at least one sill of the Chindona granodiorite intrude the Cretaceous sedimentary section. Emplacement of the sill produced contact metamorphism of the Tayaco Formation (Fig. 12A, section 14), resulting in a marble containing Cretaceous-age *Turritella* marine snails. Known felsic stocks, sills, and dikes total 197 km² of surface exposure in the Frey Pedro region. We interpret these intrusive rocks as coeval with the Chindona batholith based on similar composition and texture as well as common cross-cutting and contact metamorphism relationships.

Major and Trace Element Geochemistry of the Chindona Batholith

Geochemical analyses of the granodiorite from the Chindona batholith indicate a volcanic arc affinity. The two samples were analyzed for major and trace elements (sample location on Figs. 3 and 4) using XRF and ICP-MS. Results from the analysis are displayed in Table 2.

To ascertain the tectonic setting of the Chindona batholith, multi-elemental plots of the trace element abundance from the samples were normalized to primitive mantle composition, plotted (Sun and McDonough, 1989), and compared with established geochemical relations for volcanic rocks from various tectonic settings. Similar to the volcanic rocks of the Manto Formation, the samples of the Chindona batholith are enriched in LIL elements while depleted in HFS elements (Fig. 14B). The pattern of LIL entrenchment with HFS depletion is consistent with the association of the Chindona batholith with arc magmatism.

Age of the Chindona Batholith and Related Stocks, Dikes, and Sill

A Late Cretaceous to early Tertiary age is assigned to felsic intrusions associated with the Chindona batholith. This is based on contact metamorphism of Cretaceous strata by the felsic intrusion. Post-shortening emplacement is shown by dikes that cut the folded strata of the Frey Pedro belt.

TECTONIC AND STRATIGRAPHIC EVOLUTION OF THE FREY PEDRO BELT

Cretaceous Sedimentation in a Fault-Bounded Intra-Arc Basin

The distribution of Cretaceous strata across the Frey Pedro belt is consistent with the development of the Agua Blanca block as a rift bounded by the uplands of the Frey Pedro and Jacaleapa blocks during deposition of the Tayaco Formation. The thickening of the Tayaco Formation on the Agua Blanca block and the deposition of the Tayaco Formation on basement of the Jacaleapa and Frey Pedro blocks represents (1) subsidence and filling of the

Agua Blanca rift, and (2) uplift and erosion of the Jacaleapa and Frey Pedro blocks as rift shoulders.

Our interpretation of depositional units of the Frey Pedro belt is summarized in Figure 17. The major erosional unconformity at the base of the Valle de Angeles Formation serves as a horizontal datum to correlate the 56 measured sections. Pre-rift (i.e., pre–Tayaco Formation) stratigraphy is preserved in the Agua Blanca rift and eroded from the Frey Pedro and Jacaleapa blocks–rift shoulders. A key observation is that the Tayaco Formation thickens into the rift while thinning onto the rift shoulders (Fig. 17).

Rifting and basin development appears to begin with faulting, subsidence, and filling of the Agua Blanca rift with volcaniclastic sediment during Tayaco time (Aptian-Albian). The rift fills first with fluvial facies, alluvial fan facies, and lava flows marginal to the basin-bounding normal faults. Shelf facies dominate the central parts of the basin, while the appearance of shoreface facies reflects local sea-level fluctuations presumably driven by extensional basin subsidence. The lack of pre–Tayaco Formation strata on the Jacaleapa and Frey Pedro blocks precludes an estimate of dip-separation on the rift-bounding normal faults. Rift flank uplifts cannibalize pre-Tayaco strata and the metamorphic basements on the Frey Pedro and Jacaleapa blocks, producing the angular discordance between the lower Atima and Tayaco strata.

Subsidence, and presumably rifting, continued in the Agua Blanca rift through deposition of the shallow marine platform carbonates of the lower Cenomanian Gualaco Formation, although syn-rift deposition in this local rift appears to be overwhelmed by a much larger uplift and erosion event to the south in the Amapuca valley region (Fig. 4). This uplift of the Manto volcanic arc by Lower Valle de Angeles time is preserved in the development of a prominent angular unconformity across the Frey Pedro block with deposition of the alluvial fan facies proximal to the subaerially exposed volcanic highlands to the south (Figs. 4, 15C, and 17). Uplift of the Manto Arc is recorded on the Agua Blanca block by the upward transition to the submarine debris flows (including rudist and gastropod clasts stripped from the platform deposits of the upper Atima Formation) deposited on the shelf environs of the Agua Blanca rift.

Development of the Agua Blanca rift is accompanied by eruption of calc-alkaline volcanic rocks of the Manto Formation immediately to the south in the Amacuapa valley (Fig. 4). This is evident by the interbedded relationship between the lava flows of the Manto Formation and the volcaniclastic strata of the Tayaco Formation, indicating that the Agua Blanca rift formed as an intra-arc basin.

Late Cretaceous Inversion of the Agua Blanca Rift

The intensity of deformation on the Agua Blanca block relative to the Frey Pedro and Jacaleapa blocks (Figs. 4 and 15) is consistent with regional shortening and inversion of a sediment-filled rift between two basement-cored blocks. The large reverse faults bounding the Agua Blanca block coincide with thickening wedges of syn-rift Tayaco Formation strata. These reverse faults are in a plausible position to have originated as normal faults bounding the Agua Blanca rift and may have served as conduits for at least some Manto lava flows near the southern margin of the Agua Blanca rift.

The entire Mesozoic section of central Honduras was shortened in a regional, late Cretaceous, Laramide-age event discussed at length previously by many authors (cf. Mills et al., 1967; Wilson, 1974; Horne et al., 1974; Burkart et al., 1987; Donnelly et al., 1990). The Frey Pedro belt is the easternmost of four northwest-trending structural belts in Honduras (Fig. 2). Of the remaining three belts, only part of the Comayagua belt in the Agalteca Range of central Honduras has been mapped (Emmet, 1983). Emmet (1983) documented a Laramide-age "flower structure" with deformation similar to that described here for the inverted Agua Blanca rift. We suggest that the Comayagua belt may also be an inverted rift structure, which are commonly confused with strike-slip flower structures in many areas of the world (Harding, 1985; Lawton, 2000). Pending future mapping, a rift inversion origin for the Montaña de la Flor and Minas de Oro belts is speculative, but consistent with the known geology of the other two parallel belts of deformed Cretaceous rocks (Fig. 2).

We propose that the flat-lying limestone capping Yoro Mountain of the Montaña de la Flor belt is the only pre-Neogene stratigraphic unit in Honduras that post-dates Laramide-age deformation (Fig. 18A). Field investigation of Yoro Mountain reveals that the nearly horizontal-bedded limestone lies above steeply dipping strata correlated with the Valle de Angles Formation. Offshore hydrocarbon exploration on the Nicaragua Rise has revealed that the Eocene limestone of the Punta Gorda Formation lies above deformed Late Cretaceous Valle de Angeles Formation (Alivia et al., 1984; Rogers et al., this volume, Chapter 6). If the Yoro Mountain limestone correlates with the Punta Gorda Formation, it indicates a pre-Eocene age of deformation for the northwest-trending structural belts of Honduras assuming deformation was synchronous.

Post-Inversion Modification of the Frey Pedro Belt

The easternmost part of the Frey Pedro belt has been affected by post-shortening deformation related to displacement along the Guayape fault system (Finch and Ritchie; Gordon and Muehlberger, 1994). Just west of the San Francisco de la Paz–Gualaco road, the northwest-trending Montaña Frey Pedro bends sharply to the east, merging with the northeast-trending Sierra de Agalta (Fig. 18B). This trend is repeated by Montaña Jacaleapa to the north and to a lesser extent by Montaña Casa de Tela to the south (Fig. 3). The physiographic expression of these ranges is consistent with oroclinal bending of the eastern part of the Frey Pedro belt by left-lateral strike-slip motion on the Guayape fault, as first suggested by Finch and Ritchie (1991). East of the Guayape fault is a region of northwest-directed, latest-Cretaceous shortening that reaches full expression in the Colon fold belt of eastern Honduras (Figure 2 in Rogers et al., this volume, Chapter 6). Strain partitioning of the northwest-directed shortening of the Colon belt along the left-slip Guayape fault accounts for (1) the oroclinal

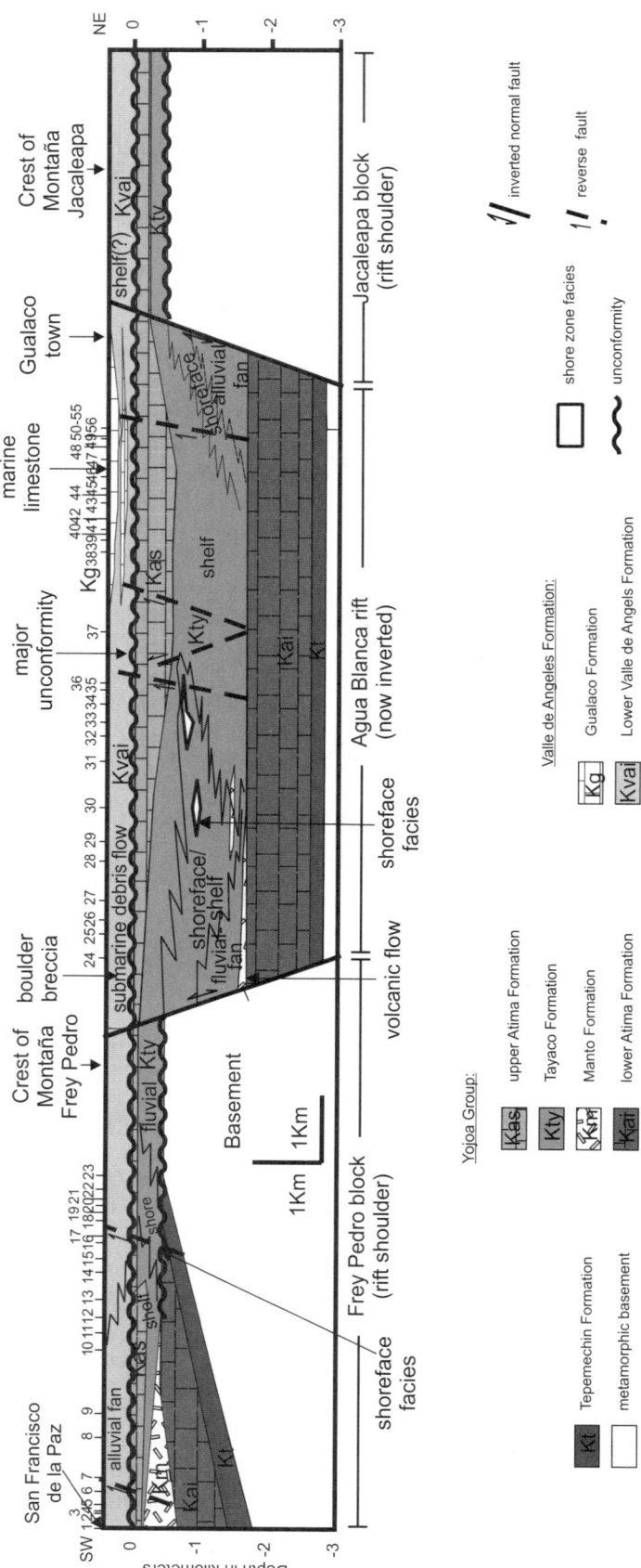

Figure 17. Restored appearance (no vertical exaggeration) of the Agua Blanca rift and northern Jacaleapa basement block and southern Frey Pedro basement block based on correlation of measured sections 1–56 shown in Figures 8–12 (see Fig. 6 for locations of individual sections along the San Francisco de la Paz–Gualaco road). The vertical position of the sections was determined by using the unconformity at the base of the upper Albian to lower Cenomanian Valle de Angeles Formation as a reference point. Note that the Tayaco Formation (Kty) thickens between the Frey Pedro and Jacaleapa basement blocks, forming a "steer's head" rift profile. An unconformity between the Tayaco Formation and the lower Atima (Kai) and Tepemechin (Kt) Formations is inferred to account for limestone clasts of the lower Atima Formation present within the Tayaco Formation (cf. Fig. 9B).

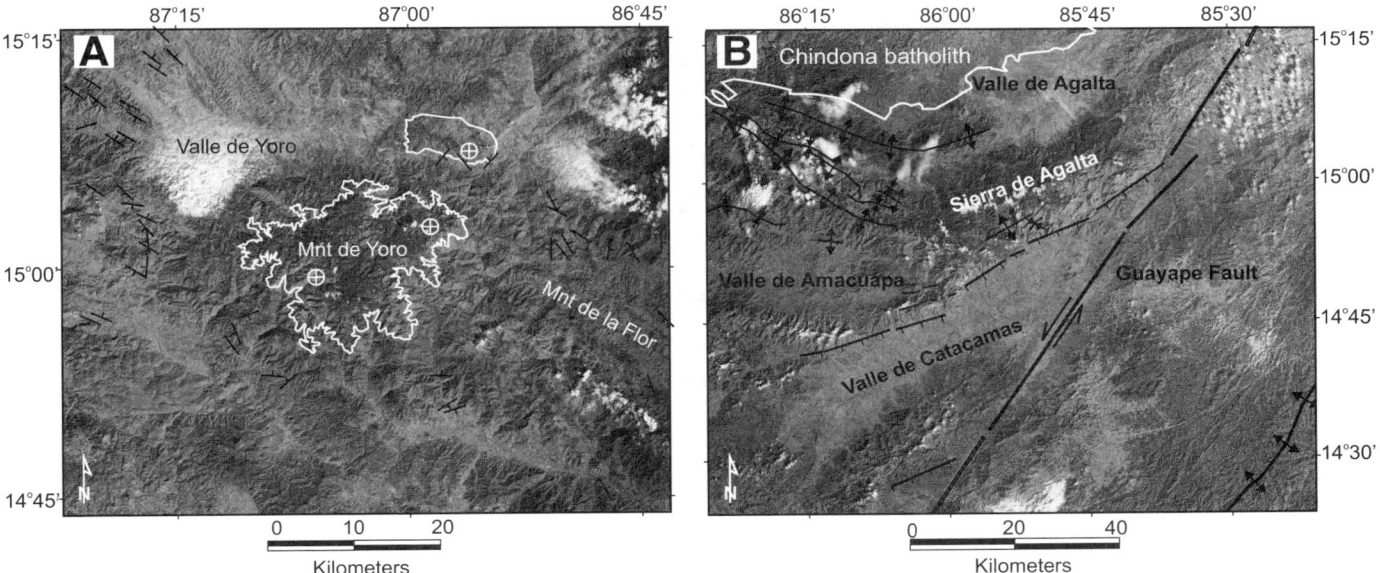

Figure 18. (A) LANDSAT image of Montaña de Yoro in northern Olancho province (see Fig. 2 for location of image), where subhorizontal, massive limestone caps the folded and tilted Cretaceous strata of the Montaña de la Flor belt. Bedding attitudes with dips ranging from 30° to 80° are common to the northwest and southeast of Montaña de Yoro (Mohl, 1969; MMAJ, 1980; R. Finch, 1999, personal commun.) and are related to northwest-trending folds of Cretaceous strata. Based on seismic data from the northern Nicaragua Rise, we infer that the upper carbonate unit is Eocene in age and overlies an angular unconformity on folded rocks as young as the Late Cretaceous. (B) LANDSAT image of Sierra de Agalta showing oroclinal bending of topographic range and of trends of folds in the Frey Pedro belt. Oroclinal bending is consistent with interpretation of an early phase of left-lateral motion on the Guayape fault system (Finch and Ritchie, 1991; Gordon, 1993a; Gordon and Muehlberger, 1994). These authors proposed that a later phase of right-lateral strike-slip motion produced the large normal fault system shown bounding the northern edge of the Valle de Catacamas. The normal fault system truncates the Agalta orocline. Montaña Casa de Tela, south of the Amacuapa Valley, may also reflect a truncated orocline. Mnt—Montaña.

bending of the Sierra de Agalta, and (2) the near-perpendicular trends of shortening on either sides of the Guayape fault (Fig. 2).

Neogene Fault Truncation of the Frey Pedro Belt

Neogene right-lateral strike-slip motion on the Guayape fault resulted in down-to-the-southeast normal faults that formed the Catacamas valley (Gordon and Muehlberger, 1994) (Fig. 18B). These faults abruptly truncate the southern edge of the Sierra de Agalta orocline.

PIERCING LINES BETWEEN SOUTHWESTERN MEXICO AND THE CHORTIS BLOCK

Many workers have previously speculated that the Chortis block originated along the southwestern margin of Mexico prior to Cenozoic translation from the North America plate to the Caribbean plate (Dengo, 1985; Ross and Scotese, 1988; Pindell and Barrett, 1990; Riller et al., 1992; Herrman et al., 1994; Pindell, 1994; Schaaf et al., 1995; Dickinson and Lawton, 2001; Cerca et al., 2004). Others have taken an alternative view that southern Mexico and the Chortis were never joined (Keppie and Moran-Zenteno, 2005). Results from this study, combined with the regional geology of the Chortis block and southern Mexico, provide sufficient detail to compare the features of the Chortis block to southern Mexico.

Five features common to southern Mexico and the Chortis block are (1) Precambrian basement, (2) similar Mesozoic cover, (3) mid-Cretaceous arc geochemistry, (4) alignment of late Cretaceous structural trends, and (5) a common magnetic signature. Our best fit alignment of the five features places the Chortis block along the southern margin of Mexico, as displayed in Figure 19A. Figure 19B displays a more interpretive view of the tectonic configuration of the southwestern margin of North America in the latest Cretaceous prior to translation of the Chortis block to the east.

Precambrian Basement

The Precambrian basement of Mexico is documented in the Oaxaca region and is believed to represent the southward continuation of North American Gondwana elements into Mexico (Ortega-Gutierrez et al., 1995; Dickinson and Lawton, 2001). The distribution of Precambrian basement exposure in Mexico is shown in Figure 19A and correlates with the Oaxaca terrane in Figure 19B. On the Chortis block, Manton (1996) reports Precambrian basement of 1.0 Ga in the Yoro region of Honduras (Fig. 19A, near the circled 3, and Figure 19B, label pC).

The inexact alignment of the Precambrian basement of Mexico with that of Honduras is attributed to a period of Late Jurassic to Early Cretaceous rifting of the Chortis block from Mexico during opening of the Americas (Pindell and Kennan, 2001), prior to suturing of the Chortis block back against southern Mexico (Harlow et al., 2004).

Mesozoic Stratigraphy

A similar sequence of Late Cretaceous clastic, marine sandstone and shale over Early Cretaceous shallow-water platform carbonate rocks occurs both on the Chortis block and in southern Mexico (see generalized Cretaceous strata distribution in Fig. 19A). In Honduras, the Late Cretaceous flysch is represented by the clastic strata of the Valle de Angeles Formation (Figs. 20A–20C) (Mills et al., 1967; Wilson, 1974; Horne et al., 1974; Finch, 1981; Rogers et al., this volume Chapter 6; this study) and in Mexico by the Mexicala Formation (Fig. 20D) (Johnson et al., 1991; Lang et al., 1996; Cabral-Cano et al., 2000; Cerca et al., 2004). Mid-Cretaceous carbonate deposition resulted in deposition of the Morelos Formation in Mexico and the Atima Formation on the Chortis block.

Geochemical Trends between Early Cretaceous Arc Rocks

The Early Cretaceous age and arc affinity of the Manto Formation on the Chortis block is similar to rocks of Early Cretaceous arc activity in southern Mexico (Fig. 19A). In southwestern Mexico, three separate Early Cretaceous arcs have been recognized: the Arteaga, Arcelia, and Teloloapan (Mendoza and Suastegui, 2000).

We compared the multi-elemental geochemical patterns of the Manto Formation volcanic rocks of Honduras with the reported values for the three Early Cretaceous volcanic arcs in Mexico in order to evaluate the possibility that the Manto Formation represents the southern continuation of one or more of the three Mexican arcs (Fig. 21). In this comparison, all data is normalized to the primitive mantle composition of Sun and McDonough (1989).

The easternmost Teloloapan arc sequence shows a multi-elemental geochemical pattern similar to that of the Manto Formation (Fig. 21A). The Arcelia arc sequence displays deviation from the Manto arc values for Rb, Ba, Th, Nd, K, La, Ce, and Zr (Fig. 21B). The pattern of the Zihuatanejo arc sequence deviates from the Manto arc values for practically all elements (Fig. 21C). In summary, the geochemical data most support a direct correlation between the Teloloapan arc of Mexico and the Manto Formation of Honduras.

Late Cretaceous (Laramide) Shortening Structures

Late Cretaceous shortening, attributed to Laramide-age deformation in Mexico, is observed both in southern Mexico (Burkart et al., 1987; Campa, 1985; Lang et al., 1996) and on the Chortis block (Horne et al., 1974; Donnelly et al., 1990). Our best-fit positioning of the Chortis block along the southern margin of Mexico aligns the Late Cretaceous north-trending fold thrust belts of Mexico with the four Late Cretaceous structural belts in Honduras (Fig. 19A).

Inverted rift basins such as the Agua Blanca block of the Frey Pedro belt (circled 4 on Fig. 19A) have not been described in southern Mexico. However, an inverted rift can be inferred for the subsequent Late Cretaceous shortening of an extensional basin described by Monod et al. (2000) for the carbonate to clastic transition (Morelos to Mexicala Formations) in the Teloloapan region (Fig. 19A).

Correlation of Magnetic Province Boundary

Prominent magnetic boundaries exist on both the Chortis block and in western Mexico (position of main boundaries shown in Figure 19A; Rogers, 2003; Rogers et al., this volume, Chapter 4). The sources of magnetic data are the North American Magnetic Anomaly Group (2002) and the Dirección General de Minas e Hidrocarburos, Honduras.

In Mexico, the prominent magnetic boundary separates the accreted Guerrero terrane from the autochthonous terranes

Figure 19. (A) Restoration of the Chortis block to a position adjacent to southwest Mexico in order to realign similar, pre-Tertiary structures and geologic units on the Chortis block and in southwestern Mexico. Key features in Honduras and southern Mexico: (I) Mesozoic volcanic arc (see comparison of geochemistry in Fig. 21); (II) Late Cretaceous and Jurassic stratigraphy (see comparison of stratigraphic columns in Fig. 20); (III) Precambrian basement exposure; (IV) aligned structural belts formed in Late Cretaceous times; and (V) a distinct magnetic boundary between the accreted Guerrero terrane and nuclear Mexico and between the Southern and Central Chortis terranes (Rogers et al., Chapter 4, this volume). Structural belts on the Chortis block: 1—Comayagua belt; 2—Minas de Oro belt; 3—Montaña de la Flor belt; 4—Frey Pedro belt. GFS—Guayape Fault; SPS—San Pedro Sula. (B) Latest Cretaceous tectonic setting of the southwest corner of North America prior to the eastward displacement of the Chortis block and Caribbean plate based on the geologic features shown in A. Following Dickinson and Lawton (2001), Moores (1998), and Tardy et al. (1994), the Guerrero arc and Great Arc of the Caribbean are assumed to form a continuous, west- and north-facing arc system wrapping around the continental promontory of southwestern North America. Late Cretaceous accretion of this arc system appears coeval with Laramide-age shortening in southern Mexico (Lang et al., 1996; Cabral-Cano et al., 2000) that is proposed to have inverted the Agua Blanca rift basin in the Frey Pedro study area of Honduras. This inversion event terminated Cretaceous-Manto extensional arc activity on the Chortis block. The accretion of the Guerrero Arc and Great Arc of the Caribbean to southwestern North America added the Guerrero terrane, the forearc basins of the Middle America Trench, and the Siuna terrane. Labeled Mexican terranes: J—Juarez; M—Mixteca; O—Oaxaca; T—Teloloapan; Xo—Xolopa. SPS—San Pedro Sula; Late K—Late Cretaceous; pC—Precambrian.

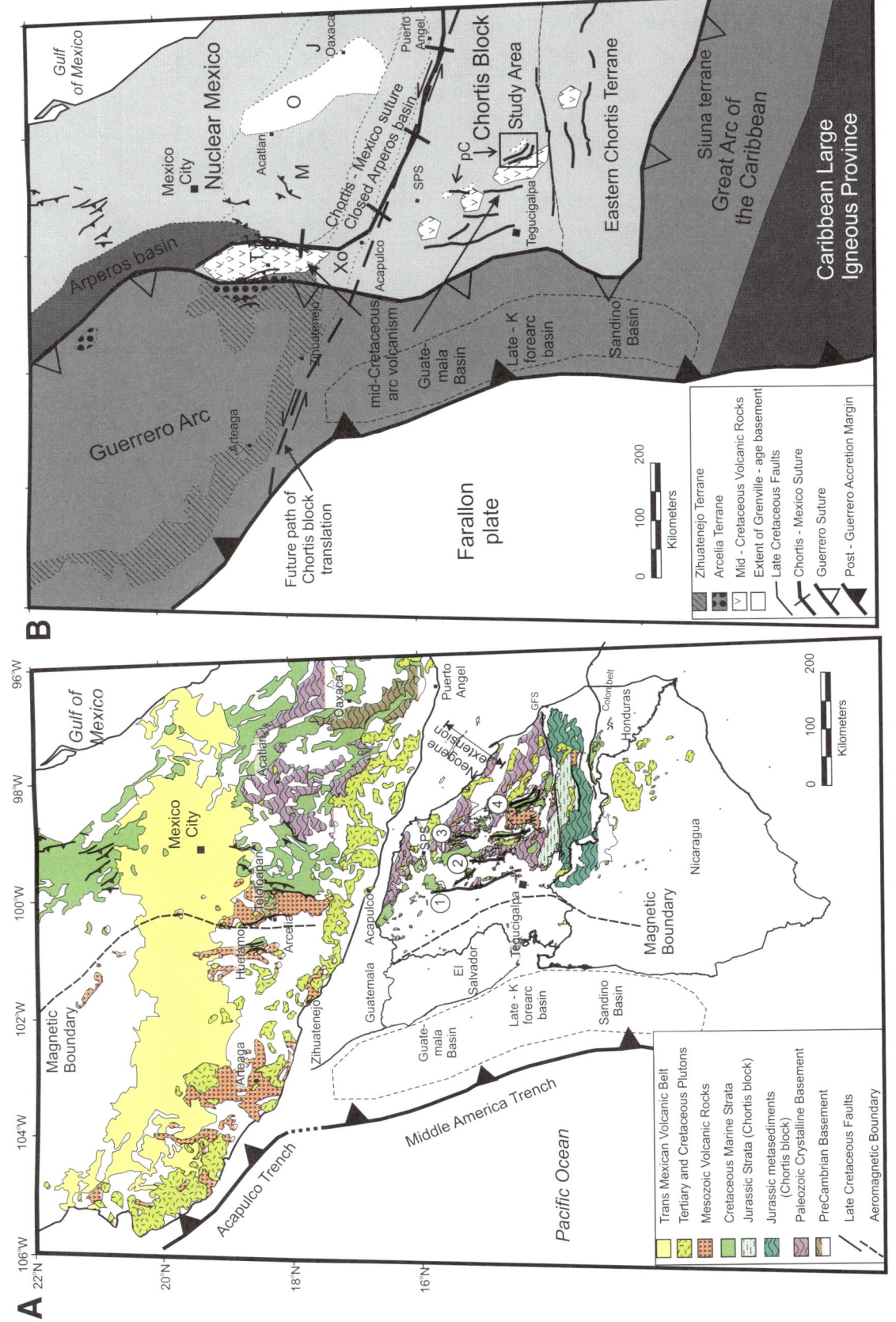

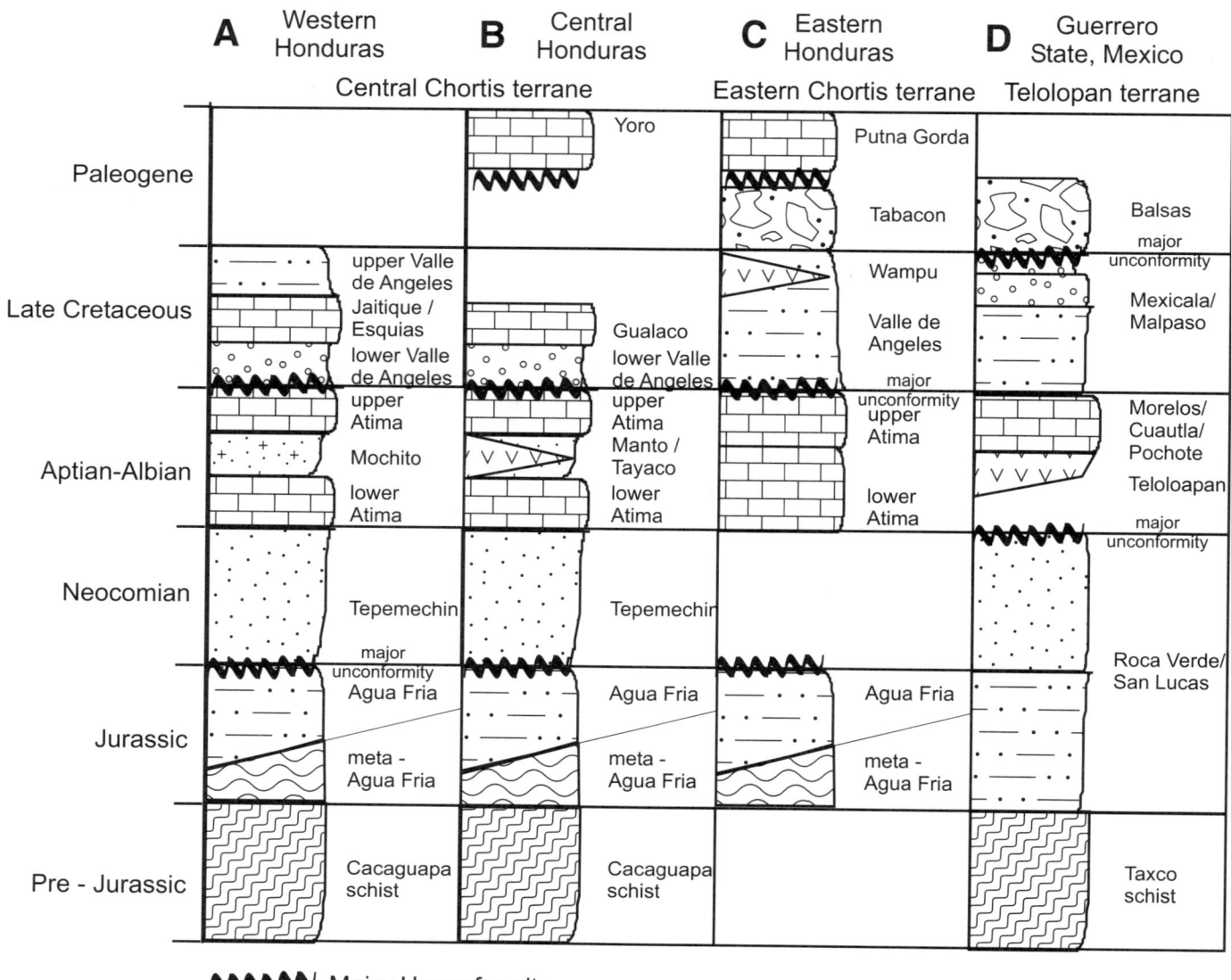

Figure 20. Correlation of late Mesozoic and Paleogene stratigraphy from Honduras and from the Guerrero state of southern Mexico. Data sources: western Honduras—Scott and Finch (1999) and Kozuch (1991); central Honduras—this study; eastern Honduras—Rogers et al. (Chapter 6, this volume); Guerrero State—Lang et al. (1996), Johnson et al. (1991), and Cabral-Cano (2000). See text for discussion.

of nuclear Mexico (North American Magnetic Anomaly Group, 2002) (Fig. 19B). A similar relationship is inferred on the Chortis block, where the boundary separates the Southern and Central Chortis terranes (Rogers, 2003). Although the magnetic boundary is buried beneath Miocene volcanic strata on the Chortis block, geochemical data from the modern Central American volcanic arc on the southern Chortis terrane shows that the basement on which the arc is built is post-Paleozoic and likely to be an older volcanic arc (M. Carr, 2003, personal commun.).

Positioning the Chortis block along the southern margin of Mexico based on the best-fit alignment of features common to both regions results in the following configuration of the southwestern corner of North America prior to Cenozoic translation of the Chortis block:

1. Continuity of the Chortis block and the autochthonous terranes of nuclear Mexico, providing a common basement and platform for the deposition of late Mesozoic strata and its subsequent Late Cretaceous shortening (Fig. 19B);
2. Continuity of the geochemically-similar Cretaceous Teloloapan and Manto arcs (Fig. 19B); and
3. An open western and southern margin along the southwest corner of North America to which the Guerrero arc and the Caribbean arc accretes. Components of Caribbean

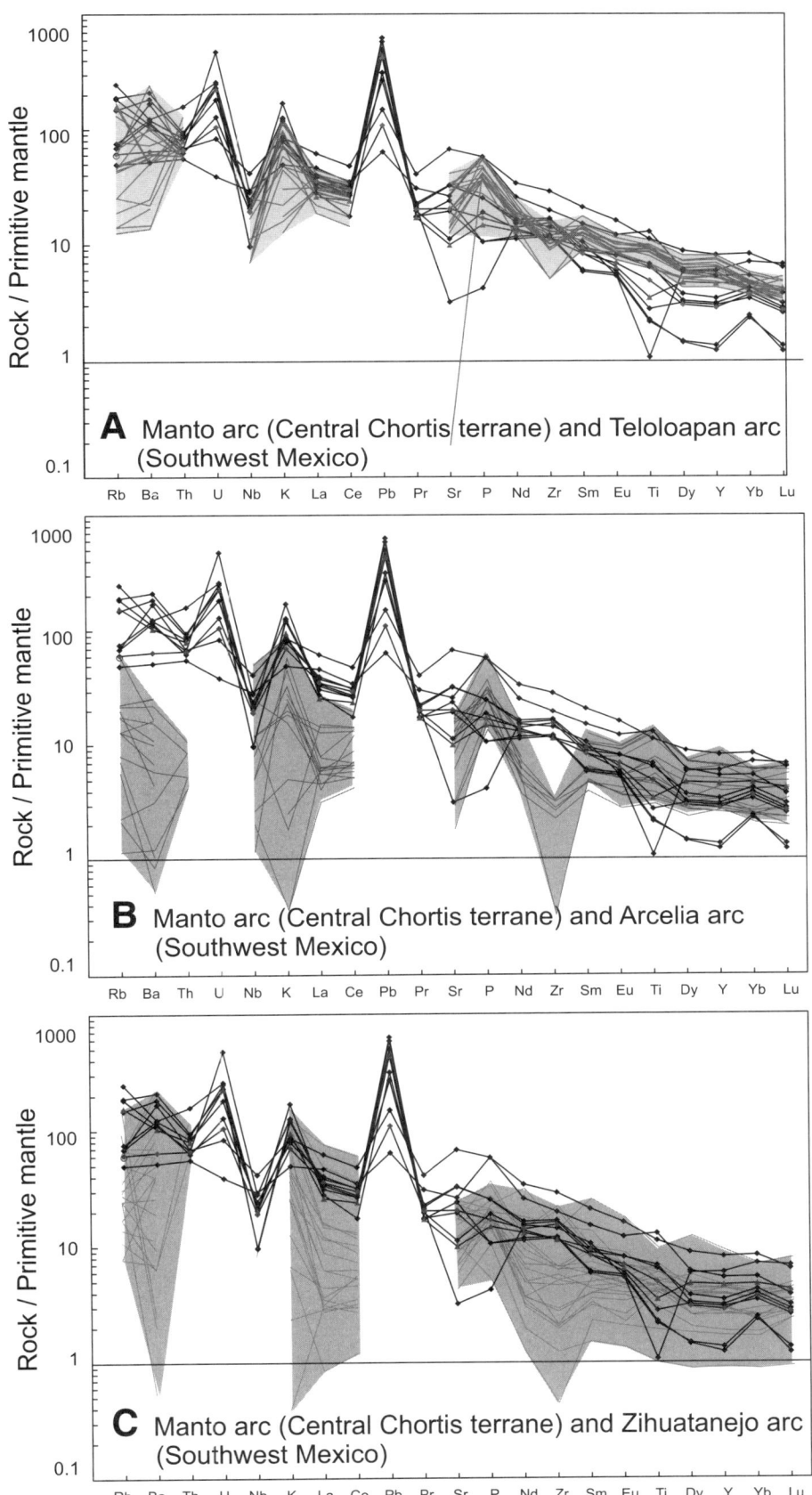

Figure 21. Multi-elemental geochemical pattern of igneous rocks from the Manto arc in Honduras compared to Early Cretaceous arc sequences from southern Mexico (reported by Mendoza and Suastegui, 2000) is consistent with the interpretation that the Manto arc was the southern continuation of the Teloloapan arc. Location of Mexican data shown in Figure 19A, corresponds to the Guerrero subterranes labeled in Figure 19B. Normalizing values for plots are for primitive mantle from Sun and McDonough (1989). (A) Comparison of geochemistry of the Manto arc rocks of Honduras with the Teloloapan arc rocks of Mexico (gray shaded region) shows comparable pattern and range of elements. The similarity of multi-elemental patterns is consistent with continuity of the Manto and Teloloapan arcs. (B) Comparison of geochemistry of the Manto arc rocks of Honduras with the Arcelia arc rocks of Mexico (gray shaded region) shows deviation in pattern and range of elements, notably lower values of Rb, Ba, Th, Nd, K, La, Ce, and Zr. The dissimilarity of multi-elemental patterns is inconsistent with continuity of the Manto and Arcelia arcs. (C) Comparison of geochemistry of Manto arc rocks of Honduras with the Zihuatanejo arc rocks of Mexico (gray shaded region) shows deviation in pattern and range of elements, notably lower values of Rb, Ba, Nd, K, La, Ce, and Zr. The dissimilarity of multi-elemental patterns is inconsistent with continuity of the Manto and Zihuatanejo arcs.

arc include the Guerrero terrane, the southern Chortis terrane, and the Siuna terrane (Fig. 19B).

CONCLUSIONS

The main conclusions of this study are as follows:
1. **A Middle Cretaceous volcanic arc existed on the Chortis block.** The Aptian-Albian stratigraphic position of the calc-alkalic volcanic rocks of the Manto Formation, their extensive distribution, and their volcanic arc geochemical affinity confirms the presence of a middle Cretaceous volcanic arc on the Chortis block.
2. **Middle Cretaceous carbonate and clastic sedimentation occurred in a fault bounded intra-arc basin.** Thickening of the Tayaco Formation strata into the inferred Agua Blanca rift (now inverted) is concurrent with erosion of its rift shoulders and coeval eruption of the calc-alkalic volcanic rocks of the Manto Formation. These relations suggest an intra-arc setting for the central Chortis block during the middle Cretaceous.
3. **Intra-arc basins were inverted by regional shortening in the Late Cretaceous.** In the Frey Pedro belt, Late Cretaceous inversion of the Agua Blanca rift between the basement-cored Frey Pedro and Jacaleapa blocks produced intense deformation of the rift, while the rift shoulders experienced less intense deformation. Similar deformation is assumed to have produced the Comayagua, Minas de Oro, and Montaña de la Flor belts that parallel the Frey Pedro belt to the east (Fig. 2).
4. **Geologic history of Chortis block suggests that Chortis and southern Mexico terranes were once aligned.** Alignment of features common to the Chortis block and to southern Mexico constrains the pre-Tertiary configuration of the southwestern corner of North America. Common features shared by the Chortis and terranes in southern Mexico include (1) Precambrian basement, (2) Mesozoic stratigraphy, (3) late Mesozoic volcanic arc geochemistry, (4) Late Cretaceous structural trends, and (5) a common magnetic signature.

ACKNOWLEDGMENTS

Funding for R. Rogers and P. Mann was provided by the Petroleum Research Fund of the American Chemical Society (grant 33935-AC2 to P. Mann). T. Vogel provided X-ray fluorescence spectrometry analysis. We thank B. Simonson and C. Busby for reviews and the Instituto Geographic Nacional of Honduras for field support for R. Rogers. Thanks also to the following people for reviews of earlier drafts of this paper: W. Dickinson, R. Finch, M. Gordon, T. Lawton, and W. Muehlberger. The authors acknowledge the financial support for publication costs provided by the University of Texas at Austin's Geology Foundation and the Jackson School of Geosciences–University of Texas Institute for Geophysics contribution 1707. Special thanks to Lisa Bingham for her help with text and figures.

REFERENCES CITED

Alivia, F., Tappmeyer, D., Aves, H., Gillett, M., and Klenk, C., 1984, Recent studies of basins are encouraging for future exploration in Honduras: Oil and Gas Journal, v. 17, p. 139–149.

Atwood, M., Cullen, C., Smith, C., and Simonson B., 1976, Mapa Geológica de Honduras, Minas de Oro: Tegucigalpa, Honduras, Instituto Geográfico Nacional, escala 1:50,000, 1 sheet.

Azéma, J., Biju-Duval, B., Bizon, J.J., Carfantan, J.C., Mascle, A., and Tardy, M., 1985, Le Honduras (Amerique Centrale Nucleaire) et le bloc d'Oaxaca (Sud du Mexique): Deux ensembles comparables du continent Nord-Americain separés par le jeu decrochant senestre des failles du système Polochic-Motagua, in Mascle, A., ed., Géodynamique des Caraibes: Paris, Editions Technip, p. 427–438.

Banner, F., and Highton, J., 1990, On Everticyclammina (foraminfera), especially *E. Kelleri* (Henson): Journal of Micropaleontology, v. 9, p. 1–14.

Banner, F., Simmons, M., and Whittaker, J., 1991, The Mesozoic *Chrysalidinidae* (foraminifera, *Textulariacea*) of the Middle East; the Redmond (Aramco) taxa and their relatives: Bulletin of the British Museum, Natural History, Geology Series, v. 47, p. 101–152.

Barckhausen, U., Ranaero, C., von Huene, R., Cande, S., and Roeser, H., 2001, Revised tectonic boundaries in the Cocos plate off Costa Rica: Implications for the segmentation of the convergent margin and for plate tectonic models: Journal of Geophysical Research, v. 106, p. 19,207–19,222, doi: 10.1029/2001JB000238.

Burkart, B., Denton, B., Dengo, C., and Moreno, G., 1987, Tectonic wedges and offset Laramide structures along the Polochic fault of Guatemala and Chiapas, Mexico: Reaffirmation of large Neogene displacement: Tectonics, v. 9, p. 411–422.

Cabral-Cano, E., Lang, H., and Harrison, C., 2000, Stratigraphic assessment of the Arcelia-Teloloapan area, southern Mexico: Implications for southern Mexico's post-Neocomian tectonic evolution: Journal of South American Earth Sciences, v. 13, p. 443–457, doi: 10.1016/S0895-9811(00)00035-3.

Campa, M., 1985, The Mexican thrust belt, in Howell, D.G., ed., Tectonostratigraphic terranes of the Circum-Pacific Region: Houston, Texas, Circum-Pacific Council on Energy and Mineral Resources, Earth Science Series, no. 1, p. 299–313.

Campa, M., and Coney, P., 1983, Tectono-stratigraphic terranes and mineral resource distributions in Mexico: Canadian Journal of Earth Sciences, v. 20, p. 1040–1051.

Case, J., MacDonald, W., and Fox, P., 1990, Caribbean crustal provinces; seismic and gravity evidence, in Dengo, G., and Case, J.E., eds., The Caribbean Region: The Geology of North America: Boulder, Colorado, Geological Society of America, v. H, p. 15–36.

Cerca, M., Ferrari, L., Bonini, M., Corti, G., and Manetti, P., 2004, The role of crustal heterogeneity in controlling vertical coupling during Laramide shortening and the development of the Caribbean—North American transform boundary in southern Mexico: Insights from analogue models: Geological Society [London] Special Publication 227, p. 117–140.

Conrad, H., 1977, The Lower Cretaceous calcareaous algae in the area surrounding Geneva (Switzerland); biostratigraphy and depositional environments, in Fluegel, E., ed., Fossil Algae; Recent Results and Developments: Springer-Verlag, Berlin, p. 295–360.

Curran, D., 1981, Mapa Geológica de Honduras, Taulabe: Tegucigalpa, Honduras, Instituto Geográfico Nacional, escala 1:50,000, 1 sheet.

DeMets, C., 2001, A new estimate for present-day Cocos-Caribbean plate motion: Implications for slip along the Central American volcanic arc: Geophysical Research Letters, v. 28, p. 4043–4046, doi: 10.1029/2001GL013518.

DeMets, C., Jansma, P., Mattioli, G., Dixon, T., Farina, F., Bilham, R., Calais, E., and Mann, P., 2000, GPS geodetic constraints on Caribbean-North American plate motion: Geophysical Research Letters, v. 27, p. 437–440, doi: 10.1029/1999GL005436.

Dengo, G., 1985, Mid America: Tectonic setting for the Pacific margin from southern Mexico to northwestern Columbia, in Nairn, A., and Stehli, F., eds., The Pacific Ocean: The ocean basins and margins: New York, Plenum Press, v. 7, p. 123–180.

Dickinson, W., and Lawton, T., 2001, Carbonaceous to Cretaceous assembly and fragmentation of Mexico: Geological Society of America Bulletin, v. 113, p. 1142–1160, doi: 10.1130/0016-7606(2001)113<1142: CTCAAF>2.0.CO;2.

Donnelly, T., Horne, G., Finch, R., and López-Ramos, E., 1990, Northern Central America: The Maya and Chortis blocks, in Dengo, G., and Case J., eds., The Caribbean Region: Boulder, Colorado, The Geological Society of America, The Geology of North America, v. H. p. 37–76.

Emmet, P., 1983, Geology of the Agalteca Quadrangle, Honduras, Central America [M.S. thesis]: Texas, The University of Texas at Austin, 201 p.

Fakundiny, R.H., 1970, Geology of the El Rosario Quadrangle, Honduras, Central America [Ph.D. dissertation]: Texas, The University of Texas at Austin, 234 p.

Finch, R., 1981, Mesozoic stratigraphy of central Honduras: AAPG Bulletin, v. 65, p. 1320–1333.

Finch, R., 1985, Mapa Geológica de Honduras, Santa Barbara: Tegucigalpa, Honduras, Instituto Geográfico Nacional, escala 1:50,000, 1 sheet.

Finch, R.C., and Ritchie, A.W., 1991, The Guayape fault system, Honduras, Central America: Journal of South American Earth Sciences, v. 4, p. 43–60, doi: 10.1016/0895-9811(91)90017-F.

Gordon, M.B., 1990, Strike-slip faulting and basin formation at the Guayape Fault–Valle de Catacamas Intersection, Honduras, Central America [Ph.D. thesis]: Austin, Texas, University of Texas at Austin, 259 p.

Gordon, M.B., 1993a, Revised Jurassic and Early Cretaceous (Pre-Yojoa Group) stratigraphy of the Chortis block: Paleogeographic and tectonic implications, in Pindell, J.L., and Perkins, R.F., eds., Mesozoic and Early Cenozoic development of the Gulf of Mexico and Caribbean Region: A context for hydrocarbon exploration: Austin, Texas, Gulf Coast Section, Society of Economic Paleontologists and Mineralogists Foundation, p. 143–154.

Gordon, M.B., 1993b, Mapa Geológica de Honduras, Hoja de Santa Maria del Real: Tegucigalpa, Honduras, Instituto Geográfico Nacional, escala 1:50,000, 1 sheet.

Gordon, M., and Muehlberger, W., 1994, Rotation of the Chortís block causes dextral slip on the Guayape fault: Tectonics, v. 13, p. 858–872, doi: 10.1029/94TC00923.

Hannah, R.S., Vogel, T.A., Patino, L.C., and Alvarado, G.E., 2002, The Origin of the chemically variable 0.33 Ma Valle Central ash-flow tuff, Costa Rica: Bulletin of Volcanology, v. 64, p. 117–133.

Harding, T.P., 1985, Seismic characteristics and identification of negative flower structures, positive flower structures, and positive structural inversion: AAPG Bulletin, v. 69, p. 582–600.

Harlow, G.E., Hemming, S.R., Avé Lallemant, H.G., Sisson, V.B., and Sorensen, S.S., 2004, Two HP-LT serpentinite-matrix mélange belts, Motagua fault zone, Guatemala: A record of Aptian and Maastrichtian collisions: Geology, v. 32, p. 17–20, doi: 10.1130/G19990.1.

Herrmann, U., Nelson, B., and Rathschbacher, L., 1994, The origin of a terrane: U/Pb zircon geochronology and tectonic evolution of the Xolapa complex (southern Mexico): Tectonics, v. 13, p. 455–474, doi: 10.1029/93TC02465.

Horne, G., Atwood, M., and King, A., 1974, Stratigraphy, sedimentology, and paleoenvironment of Esquias Formation of Honduras: AAPG Bulletin, v. 58, p. 176–188.

Horne, G., Clark, G., and Pushkar, P., 1976, Pre-Cretaceous rocks of northwestern Honduras: Basement terrane in Sierra de Omoa: AAPG Bulletin, v. 60, p. 566–583.

INETER (Instituto Nicaragüense de Estudios Territoriales), 1995, Mapa Geológica Minero de la Republica de Nicaragua: Managua, Nicaragua, escala 1:500,000, 4 sheets.

Johnson, C., Lang, H., Cabal-Cano, E., Draper, G., Harrison, C., and Barros, J., 1991, Preliminary assessment of stratigraphy and structure, San Lucas region, Michoacan and Guerrero states, SW Mexico: Mountain Geologist, v. 28, p. 121–135.

Johnson, J., 1969, A review of the Lower Cretaceous algae: Professional Contributions of the Colorado School of Mines: Golden, Colorado, Colorado School of Mines, 180 p.

Keppie, J., and Moran-Zenteno, D., 2005, Tectonic implications of alternative Cenozoic reconstructions for southern Mexico and the Chortis block: International Geology Review, v. 47, p. 473–491.

Klitgord, K., and Mammerickx, J., 1982, Northern East Pacific Rise: magnetic anomaly and bathymetric framework: Journal of Geophysical Research, v. 87, p. 6725–6750.

Kozuch, M., 1990, Mapa Geológica de Honduras, Hoja de San Francisco de Becerra: Tegucigalpa, Honduras, Instituto Geográfico Nacional, escala 1:50,000, 1 sheet.

Kozuch, M., 1991, Mapa Geológica de Honduras: Tegucigalpa, Honduras, Instituto Geográfico Nacional, escala 1:500,000, 3 sheets.

Kuss, J., and Senowbari-Daryan, B., 1992, Anomuran coprolites from Cretaceous shallow water limestones of NE Africa: Cretaceous Research, v. 13, p. 147–156, doi: 10.1016/0195-6671(92)90032-L.

Lang, H., Barros, J., Cabal-Cano, E., Draper, G., Harrison, C., Jansma, P., and Johnson, C., 1996, Terrane deletion in northern Guerrero State: Geofisica Internacional, v. 35, p. 349–359.

Lawton, T.F., 2000, Inversion of Late Jurassic–Early Cretaceous extensional faults of the Bisbee basin, southeastern Arizona and southwestern New Mexico: New Mexico Geological Society Guidebook, v. 51, p. 95–102.

Leroy, S., Mauffret, A., Patriat, P., and Mercier de Lepinay, B., 2000, An alternative interpretation of the Cayman Trough evolution from a reidentification of magnetic anomalies: Geophysical Journal International, v. 141, p. 539–557, doi: 10.1046/j.1365-246x.2000.00059.x.

Manton, W., 1987, Tectonic interpretation of the morphology of Honduras: Tectonics, v. 6, p. 633–651.

Manton, W., 1996, The Grenville of Honduras: Geological Society of America, Annual Meeting Abstracts with Programs, v. 28, no. 7, p. A-493.

Markey, R., 1995, Mapa Geológica de Honduras, Hoja de Morocelli: Tegucigalpa, Honduras, Instituto Geográfico Nacional, escala 1:50,000, 1 sheet.

Marrett, R., and Allmendinger, R., 1990, Kinematic analysis of fault slip data: Journal of Structural Geology, v. 12, p. 973–986, doi: 10.1016/0191-8141(90)90093-E.

Marshall, J.S., Fisher, D.M., and Gardner, T.W., 2000, Central Costa Rica deformed belt: Kinematics of diffuse faulting across the western Panama block: Tectonics, v. 19, p. 468–492, doi: 10.1029/1999TC001136.

Mendoza, T.O., and Suastegui, M., 2000, Geochemistry and isotopic composition of the Guerrero Terrane (western Mexico): Implications for the tectono-magmatic evolution of southwestern North America during the Late Mesozoic: Journal of South American Earth Science, v. 13, p. 297–324.

Mills, R., Hugh, K., Feray, D., and Swolfs, H., 1967, Mesozoic stratigraphy of Honduras: AAPG Bulletin, v. 51, p. 1711–1786.

Metal Mining Agency of Japan (MMAJ), 1980, Report on geology survey of the western area (Olancho): Government of Japan, Japan International Cooperation Agency, vol. 5, 138 p.

Mohl, J., 1969, Geologic maps of: Locomapa, La Habana, and Barrosa (Honduras): Tegucigalpa, Honduras, unpublished ASARCO Inc. report to Direccion General de Minas, 1:50,000 scale.

Monod, O., Busnardo, R., and Guerrero-Suastegui, M., 2000, Late Albian ammonites from the carbonate cover of the Teloloapan arc volcanic rocks (Guerrero State, Mexico): Journal of South American Earth Sciences, v. 13, p. 377–388, doi: 10.1016/S0895-9811(00)00030-4.

Moores, E., 1998, Ophiolites, the Sierra Nevada, "Cordillera," and orogeny along the Pacific and Caribbean margins of North and South America: International Geology Review, v. 40, p. 40–54.

North American Magnetic Anomaly Group, 2002, Magnetic anomaly map of North America: U.S. Geological Survey Special Map, 1: 10,000,000.

Ortega-Gutierrez, F., Ruiz, J., and Centeno-Garcia, E., 1995, Oaxaquia, a Proterozoic microcontinent accreted to North America during the late Paleozoic: Geology, v. 23, p. 1127–1130, doi: 10.1130/0091-7613(1995)023<1127: OAPMAT>2.3.CO;2.

Perfit, M., and Dickinson, J., 2000, Plate tectonic and volcanism, in Sigurdsson, H., ed., Encyclopedia of Volcanology: San Diego, California, Academy Press, p. 89–114.

Petit, J., 1987, Criteria for the sense of movement on fault surfaces in brittle rocks: Journal of Structural Geology, v. 9, p. 597–608, doi: 10.1016/0191-8141(87)90145-3.

Pindell, J., 1994, Evolution of the Gulf of Mexico and the Caribbean, in Donovan, S., and Jackson, T., eds., Caribbean geology: An introduction: Jamaica, University of the West Indies Publisher's Association, p. 13–39.

Pindell, J., and Barrett, S., 1990, Geological evolution of the Caribbean region; a plate-tectonic perspective, in Dengo, G., and Case, J.E., eds., The Caribbean Region: Boulder, Colorado, Geological Society of America, The Geology of North America, v. H, p. 405–432.

Pindell, J., and Kennan, L., 2001, Kinematic Evolution of the Gulf of Mexico and the Caribbean, in Fillon, R., et al., eds., Petroleum Systems of deep-water basins: Global and Gulf of Mexico Experience, 21st Annual

Research Conference: Austin, Texas, Gulf Coast Section, Society of Economic Paleontologists and Mineralogists Foundation, CD-ROM.

Riller, U., Ratschbacher, L., and Frisch, W., 1992, Left-lateral transtension along the Tierra Colorada deformation zone, northern margin of the Xolapa magmatic arc of southern Mexico: Journal of South American Earth Sciences, v. 5, p. 237–249, doi: 10.1016/0895-9811(92)90023-R.

Rogers, R.D., 1994, Preliminary stratigraphy and structure along the Rió Patuca and Rió Wampú, La Mosquitia, Honduras: Geological Society of America Abstracts with Programs, v. 26, no. 7, p. 247.

Rogers, R.D., 1995. Mapa Geológica de Honduras, Hoja de Valle de Jamastran: Tegucigalpa, Honduras, Instituto Geográfico Nacional, escala 1:50,000, 1 sheet.

Rogers, R.D., 2003, Jurassic–Recent tectonic and stratigraphic history of the Chortis block of Honduras and Nicaragua (northern Central America) [Ph.D. dissertation]: Austin, Texas, University of Texas at Austin, 289 p.

Rogers, R.D., and Mann, P., 2007, this volume, Transtensional deformation of the western Caribbean-North America plate boundary zone, in Mann, P., ed., Geologic and tectonic development of the Caribbean plate in northern Central America: Geological Society of America Special Paper 428, doi: 10.1130/2007.2428(03).

Rogers, R.D., Kárason, H., and van der Hilst, R., 2002, Epeirogenic uplift above a detached slab in northern Central America: Geology, v. 30, p. 1031–1034, doi: 10.1130/0091-7613(2002)030<1031:EUAADS>2.0.CO;2.

Rogers, R.D., Patino, L., and Scott, R., 2003, The Cretaceous margins of the extreme southwest corner of the North American plate: Geological Society of America Cordillera section Abstracts with Programs, v. 37, no. 4, p. 75.

Rogers, R.D., Mann, P., Emmet, P.A., 2007, and Venable, M.E., 2007, this volume, Colon fold belt of Honduras: Evidence for Late Cretaceous collision between the continental Chortis block and intraoceanic Caribbean arc, in Mann, P., ed., Geologic and tectonic development of the Caribbean plate in northern Central America: Geological Society of America Special Paper 428, doi: 10.1130/2007.2428(06).

Rogers, R.D., Mann, P., Emmet, P.A., 2007, this volume, Tectonic terranes of the Chortis block based on integration of regional aeromagnetic and geologic data, in Mann, P., ed., Geologic and tectonic development of the Caribbean plate boundary in northern Central America: Geological Society of America Special Paper 428, doi: 10.1130/2007.2428(04).

Rosencrantz, E., 1994, Opening of the Cayman Trough and the evolution of the northern Caribbean Plate boundary: Geological Society of America Abstracts with Programs, v. 27, no. 7, p. 153.

Rosencrantz, E., Ross, M., and Sclater, J., 1988, Age and spreading history of the Cayman Trough as determined from depth, heat flow, and magnetic anomalies: Journal of Geophysical Research, v. 93, p. 2141–2157.

Ross, M., and Scotese, C., 1988, A hierarchical tectonic model of the Gulf of Mexico and Caribbean region: Tectonophysics, v. 155, p. 139–168, doi: 10.1016/0040-1951(88)90263-6.

Schaaf, P., Morán-Zenteno, D., Hernández-Bernal, M., Solís-Pichardo, G., Tolson, G., and Kohler, H., 1995, Paleogene continental margin truncation in southwestern Mexico: Geochronological evidence: Tectonics, v. 14, p. 1339–1350, doi: 10.1029/95TC01928.

Scott, R., and Finch, R., 1999, Cretaceous carbonate biostratigraphy and environments in Honduras, in Mann, P., ed., Caribbean Basins: Sedimentary Basins of the World series: Amsterdam, Elsevier, v. 4, p. 151–166.

Scott, R., and Gonzalez-Leon, C., 1991, Paleontology and biostratigraphy of Cretaceous rocks, Lampazos area, Sonora, Mexico, in Pérez-Segura, E., and César, J.A., eds., Studies of Sonoran Geology: Geological Society of America Special Paper 254, p. 51–67.

Scott, R.W., 2002, Upper Albian Benthic foraminifers new in West Texas: The Journal of Foraminiferal Research, v. 32, p. 43–50, doi: 10.2113/0320043.

Sedlock, R., Oretega-Gutierez, F., and Speed, R., 1993, Tectonostratigraphic Terranes and Tectonic Evolution of Mexico: Geological Society of America Special Paper 278, 153 p.

Simonson, B., 1977, Mapa Geológica de Honduras, El Porvenir: Tegucigalpa, Honduras, Instituto Geográfico Nacional, escala 1:50,000, 1 sheet.

Southernwood, R., 1986, Late Cretaceous limestone clast conglomerates of Honduras [M.S. thesis]: Dallas, University of Texas at Dallas, 300 p.

Sun, S., and McDonough, W., 1989, Chemical and isotopic systematics of oceanic basalts; implications for mantle composition and processes, in Saunders, S., ed., Magmatism in the ocean basins: Geological Society [London] Special Publication 42, p. 313–345.

Tardy, M., Lapierre, H., Freydier, C., Coulon, C., Gill, J., Mercier de Lepinay, B., and Beck, C., Martinez-R., Talavera-M., Ortiz-H., E., Stein, G., Bourdier, J., and Yta, M., 1994, The Guerrero suspect terrane (western Mexico) and coeval arc terranes (the Greater Antilles and the Western Cordillera of Colombia): A late Mesozoic intra-oceanic arc accreted to cratonal America during the Cretaceous: Tectonophysics, v. 230, p. 49–73.

Weiland, T., Suayah, I., and Finch, R., 1992, Petrologic and tectonic significance of Mesozoic volcanic rocks in the Río Wampú area, eastern Honduras: Journal of South American Earth Sciences, v. 6, p. 309–325, doi: 10.1016/0895-9811(92)90049-5.

Williams, H., and McBirney, A., 1969, Volcanic history of Honduras: University of California Publications in Geological Sciences, no. 85, 101 p.

Wilson, D., 1996, Fastest known spreading on the Miocene Cocos-Pacific plate boundary: Geophysical Research Letters, v. 23, p. 3003–3006, doi: 10.1029/96GL02893.

Wilson, H., 1974, Cretaceous sedimentation and orogeny in nuclear Central America: AAPG Bulletin: v. 58, p. 1348–1396.

Manuscript Accepted by the Society 22 December 2006

Colon fold belt of Honduras: Evidence for Late Cretaceous collision between the continental Chortis block and intra-oceanic Caribbean arc

Robert D. Rogers*
Paul Mann
Institute for Geophysics, Jackson School of Geosciences, University of Texas at Austin, J.J. Pickle Research Campus, Bldg. 196 (ROC), 10100 Burnet Road (R2200), Austin, Texas 78758-4445, USA

Peter A. Emmet
Cy-Fair College, Fairbanks Center, 14955 Northwest Freeway, Houston, Texas 77040, USA

Margaret E. Venable[†]
Consultant, Exploration Geology, 3000 Brady Hoffman Road, Lincolnton, North Carolina 28092-8220, USA

ABSTRACT

We document a previously unrecognized, thin-skinned arc-continental collisional zone, termed here the Colon fold-thrust belt, which trends northeastward for 350 km near the Honduras-Nicaragua border region. The Colon belt occurs in three collinear segments: (1) a 200-km-long belt of remote but well-exposed Jurassic–Late Cretaceous rock outcrops described from original geologic mapping presented in this study; (2) a 75-km-long subsurface belt of Jurassic–Late Cretaceous rocks known from onland seismic reflection studies and exploration drilling for oil; and (3) an offshore 75-km-long subsurface belt of Late Cretaceous to Eocene rocks known from exploration studies. These three segments share a continuity of the deformation front and associated folds, as well as a similar timing of fold-thrust deformation (segment one: post-Campanian; segment two: post–Late Cretaceous; segment three: post-Cretaceous and possible to Eocene); and all segments display southeastward-dipping thrusts and related northeastward-verging folds that structurally elevate Cretaceous rocks.

The structural position of the Siuna belt of oceanic island arc affinity to the south of the Colon fold-thrust belt, its association with calc-alkaline volcanic rocks of the Caribbean arc, and its Campanian (75 Ma) emplacement age, suggest that the Siuna belt was overthrust to the north and northwest onto the hanging wall of the Colon fold-thrust belt. The northwestward-transported Colon fold-thrust belt and adjacent Siuna belt document a Late Cretaceous collisional event between a south-facing continental

---

*Present address: Department of Geology, California State University Stanislaus, 801 West Monte Vista Avenue, Turlock, California 95382, USA; e-mail: rrogers@geology.csustan.edu.
[†]mevenable@fastmail.fm

Rogers, R.D., Mann, P., Emmet, P.A., and Venable, M.E., 2007, Colon fold belt of Honduras: Evidence for Late Cretaceous collision between the continental Chortis block and intra-oceanic Caribbean arc, *in* Mann, P., ed., Geologic and tectonic development of the Caribbean plate boundary in northern Central America: Geological Society of America Special Paper 428, p. 129–149, doi: 10.1130/2007.2428(06). For permission to copy, contact editing@geosociety.org. ©2007 The Geological Society of America. All rights reserved.

margin of the Chortis block of northern Central America and an eastward and northeastward-moving, Early to Late Cretaceous Caribbean arc system.

Keywords: collision, fold-thrust belt, Chortis, Caribbean arc.

INTRODUCTION

Late Mesozoic subduction arc systems of the western margin of the Americas are coeval with the opening of the North Atlantic Ocean and the South Atlantic Ocean basins. Triassic-Jurassic opening in the north Atlantic produced the separation between the Americas that formed the Gulf of Mexico and the Caribbean region (e.g., Bullard et al., 1965; Pindell and Dewey, 1982; Pindell and Barret, 1990). Early Cretaceous entry of the Caribbean arc (including the modern Lesser Antilles arc) between the Americas is recorded by arc-continent collisional deformation and emplacement of ophiolites in Guatemala (Rosenfeld, 1981), as well as in Colombia and western Venezuela (Kerr et al., 1998; Mann, 1999). The history of this diachronous collision along the southern North America margin is poorly constrained because of the lack of detailed study of many regions of Central America, particularly in remote areas. This study provides new constraints on the geologic evolution of understudied regions of eastern Honduras and northern Nicaragua and the record of collision deformation along the southern margin of North America in the Late Cretaceous (Fig. 1).

Tectonic and Geological Setting

Northern Central America straddles the North America and Caribbean plates and is divided into the Maya block and Chortis block (Dengo, 1973) (Fig. 1). The Chortis block is a Precambrian-Paleozoic continental block that presently occupies the northwestern corner of the Caribbean plate (Gordon, 1990). The Chortis block is bounded to the north by left-lateral strike-slip faults of the present-day North America–Caribbean plate boundary (Burkart and Self, 1985; Rogers and Mann, this volume) and to the southwest by the Middle America trench and volcanic arc of the present-day Cocos-Caribbean plate boundary. The southern and eastern edge of the Chortis continental block is not well defined because (1) it is located in remote areas of eastern Honduras and northern Nicaragua; (2) it is covered by lowlands of eastern Central America; or (3) it is blanketed by waters of the Caribbean Sea and the Cenozoic carbonate platform of the Nicaragua Rise (Fig. 1). South of the Chortis block is the Chorotega block of Costa Rica and Panama with an oceanic island-arc affinity (Dengo, 1973). The Chortega block developed on the western margin of the Caribbean large igneous province (Case et al., 1990; Sinton et al., 1997).

Most previous work on the Chortis block has focused on the northern edge of the block where it is juxtaposed with the Maya block of southern Guatemala across the Motagua and Polochic left-lateral strike-slip faults of the North America–Caribbean plate boundary (Fig. 1). Donnelly et al. (1990) and Burkart (1994) have documented a Late Cretaceous collisional orogeny that emplaced ophiolites along the northern edge of the Motagua fault valley and produced north-south shortening of pre-Cenozoic strata in eastern Guatemala and Belize. Reconstructing the two sides of the Late Cretaceous collision across the Motagua suture zone is complex because as much as 1100 km of late Eocene to Recent left-lateral motion accompanying the opening of the Cayman trough has been superimposed onto the suture zone (Rosencrantz et al., 1988; Leroy et al., 2000; Harlow et al., 2004) (Fig. 1). This large-scale eastward migration of the Chortis block along these strike-slip faults is supported by detailed fault kinematic and radiometric dating of igneous rocks in the zone of southernmost Mexico affected by the lateral shear (Riller et al., 1992; Schaaf et al., 1995). Pre-Tertiary reconstructions of the Chortis block place it ~1100 km farther to the west along the southern margin of Mexico (e.g., Azéma et al., 1985; Dengo, 1985; Pindell and Barrett, 1990; Tardy et al., 1994). Cretaceous-age piercing lines between the Chortis block and southwestern Mexico are proposed by Rogers et al. (Chapter 5, this volume) and include the northwest-trending foldbelts of central Honduras with north trending foldbelts of Guerrero State of Mexico, the Honduras Olancho arc with the Teloloapan arc, and a prominent magnetic boundary along the Pacific-facing margin of continental crust in both Honduras and Mexico.

There have been few previous efforts to constrain in detail the location of the southern and eastern margins of the continental Chortis block and its tectonic relationships with oceanic and island arc terranes of southern Central America. Dengo (1985) placed the southern boundary of the Chortis block near the Honduras-Nicaragua political boundary and its offshore boundary along the Hess escarpment. Pindell and Barrett (1990) and Tardy et al. (1994) have shown reconstructions for a Chortis block suturing against arc and oceanic plateau terranes in southern Central America, but these reconstructions have remained largely conjectural since there are few published field observations from this area. Case et al. (1990) compiled crustal refraction lines that allowed the inference of a boundary between continental rocks of the Chortis terrane and arc rocks of the eastern Nicaragua Rise and southern Central America. Venable (1994) was the only field-based effort in northern Nicaragua to define a suture zone between the Chortis block and an oceanic terrane to the south, which she named the Siuna terrane (Fig. 1).

The specific objectives of this study include the following:
1. Presentation of new geologic field observations from the northeast-trending Colon Mountains of the eastern

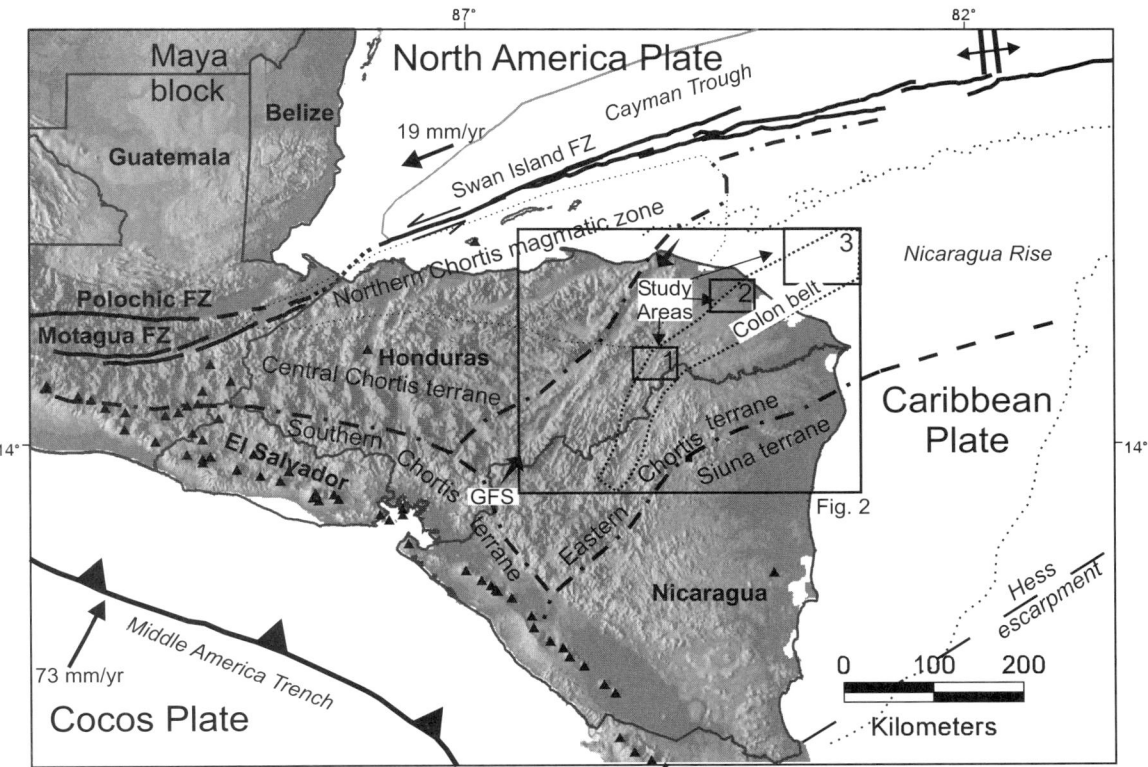

Figure 1. Topographic and tectonic setting of the Late Cretaceous Colon fold-thrust belt. Individual study areas are (1) the Colon Mountains, (2) the Mosquito plains, and (3) the offshore region of the Nicaragua Rise. Box displays geologic map location of the area shown in Figure 2. The Maya block on the North America plate and the Chortis block on the Caribbean plate are the two main continental blocks of northern Central America. The two blocks are separated by the Motagua, Polochic, and Swan Islands left-lateral strike-slip faults. Black triangles are active volcanoes of the Middle America arc. Note the orthogonal relationship between northeast-trending topography of the East Chortis terrane and Colon fold-thrust belt with the active Middle America arc. Plate velocities are relative to a fixed Caribbean Plate (DeMets et al., 2000). GFS—Guayape fault system; FZ—fault zone.

Chortis block. We have also incorporated the paleontological results of Scott and Finch (1999), which are based on Cretaceous sedimentary rock samples collected in the Colon study area. The only previous work in eastern Honduras was reconnaissance in nature and was restricted to outcrops along major rivers (Mills and Hugh, 1974).

2. Integration of unpublished geologic field mapping, isotopic, and radiometric results from Venable (1994) from the Siuna terrane of Nicaragua into a regional tectonic framework. These results are the only modern geologic data and isotopic dating results from northeastern Nicaragua. Previous studies relied only on reconnaissance traverses along major rivers (Zoppis Bracci and del Giudice, 1958; Paz-Rivera, 1963).

3. Integration of published onland seismic reflection and well data from the along-strike continuation of the Colon Mountains beneath the Mosquitia coastal plain of easternmost Honduras. We have correlated the stratigraphy encountered in these wells and tied to seismic reflection lines with the stratigraphy mapped in the Colon Mountains 70 km to the southwest.

4. Integration of subsurface seismic reflection and well data from an unpublished industry report by Rockwell (1985) from the along-strike continuation of the Colon Mountains beneath the submarine area of the eastern Nicaragua Rise. These geophysical data are inferred to correlate both to the Mosquitia plains subsurface study of Mills and Barton (1996) and to the geologic study of the Colon Mountains.

Together, these objectives address the significance of the Colon belt as a useful constraint for late Mesozoic reconstructions of the region and its usefulness in determining when the Siuna terrane to the south of the Colon Belt accreted to the Chortis block.

Geology of Eastern Honduras and Northern Nicaragua

A geologic map summarizing the geologic setting of the Colon Mountains study area is shown in Figure 2. This map, showing all pre-Tertiary geologic units of eastern Honduras and

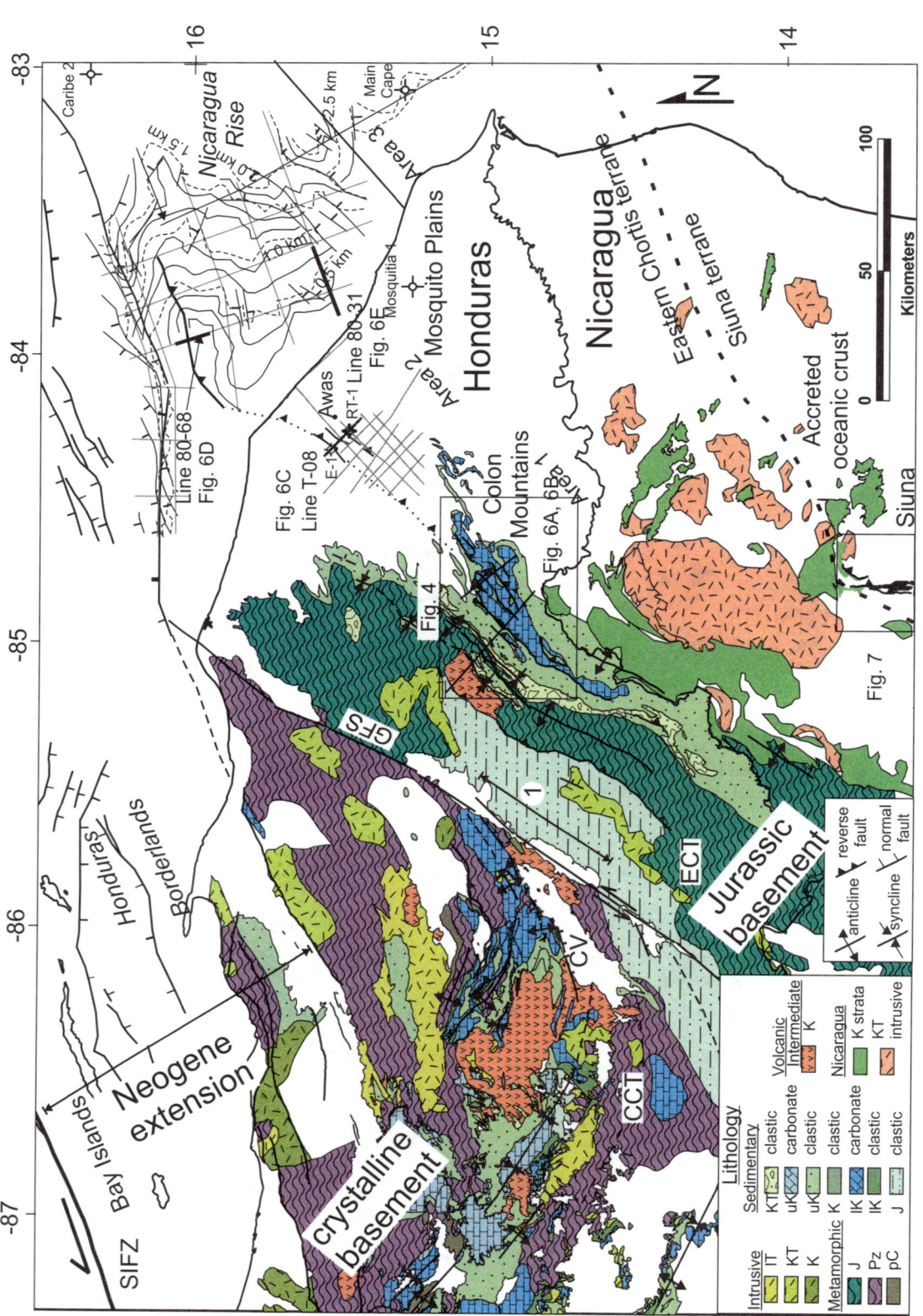

Figure 2. Pre-Tertiary geology of the continental Chortis block (Eastern Chortis terrane and Siuna terrane) of eastern Honduras and northeastern Nicaragua compiled from Kozuch (1991), Rockwell (1985), Mills and Barton (1996), INETER (1995), and this study. The Colon fold-thrust belt is described in this study in three along-strike segments: Area 1—the Colon Mountains of eastern Honduras; Area 2—the Mosquitia Plains of eastern Honduras; and Area 3—the Nicaragua Rise (Caribbean Sea). The Siuna terrane of northeastern Nicaragua is inferred to represent the leading edge of the collided Caribbean arc system. The Guayape fault system (GFS) forms a major terrane boundary between the Eastern Chortis terrane (ECT—basement of Jurassic metasedimentary rocks) and the Central Chortis terrane (CCT—basement of Precambrian and Paleozoic crystalline rocks). Contour lines (dashed at 0.5 km interval) on the Nicaragua Rise show depth in kilometers to top of the Cretaceous (Rockwell, 1985), indicating a structurally elevated block along the trend of the Colon belt. RT-1—Raite Tara 1 well; E-1—Embarcadero 1 well. CV—Catacamas valley; line 1 (noted in the blank circle) denotes the amount of left-lateral displacement on the GFS. The white parts of map are post-Cretaceous in age. SIFZ—Swan Island fault zone.

northeastern Nicaragua, was compiled from Kozuch (1991) and INETER (1995), supplemented with surface and subsurface information from Rockwell (1985), Mills and Barton (1996), and this study. Eastern Honduras is an ideal area to examine pre-Tertiary tectonic history of the Chortis block because the area has remained tectonically stable in the Cenozoic and therefore has not been overprinted by tectonic events affecting either the North American–Caribbean strike-slip boundary in the Honduran borderlands (Rogers and Mann, this volume) or by tectonic events associated with the subduction of the oceanic Cocos plate beneath the Caribbean plate (Rogers et al., 2002; Jordan et al., this volume). One disadvantage of geologic studies in eastern Honduras is that there are few units of Tertiary age. For this reason, correlations must be made into the subsurface of the Mosquitia plain or of the Nicaragua Rise in order to establish the age of the Cretaceous deformation seen in the Colon Mountains.

The Colon belt of fold-thrust deformation is described in this study in three along-strike segments shown on Figure 2: the Colon Mountains of eastern Honduras; the Mosquitia Plains of eastern Honduras near Awas; and offshore along the Nicaragua Rise (Caribbean Sea). The Siuna belt of northeastern Nicaragua is also described using data from Venable (1994) and is inferred to represent the leading edge of the collided Caribbean arc system.

Geologic mapping and radiometric studies have shown that the Northern Chortis Magmatic Zone and Central Chortis terrane (areas north of the Guayape fault system in Fig. 2) have a Grenville- to Paleozoic-age basement composed of gneiss and schist (Case et al., 1990; Donnelly et al., 1990; Manton 1996; Manton and Manton, 1999; Nelson, et al., 1997; Rogers, 2003; Rogers et al., this volume, Chapters 4 and 5) (Fig. 1). Seismic refraction studies compiled by Case et al. (1990) show that these regions are underlain by a continental crust ~45 km in thickness.

Thinner, 30–35-km-thick continental to transitional crust consisting of Jurassic sedimentary and metasedimentary rocks make up the basement of the Eastern Chortis terrane southeast of the Guayape fault system (Case et al., 1990; Gordon, 1993a, 1993b; Rogers, 1995; Viland et al., 1996). The continental to transitional affinity of the crust is based on refraction data and gravity models summarized in Case et al. (1990). Mapping in the Valle de Jamastran at the SW termination of the Guayape fault, Rogers (1995) observed a gradational vertical contact between the Bathonian-age sandstone and shale of the Agua Fria Formation (Gordon and Young, 1993) and greenschist-facies phyllite and quartzite. A major nonconformity between the Agua Fria Formation and underlying Paleozoic metamorphic basement is inferred from the much lower degree of deformation in the Agua Fria than observed in surrounding basement outcrops. These metasedimentary rocks, with metamorphic grade increasing to the east, were followed east of the Guayape fault along the Río Patuca into the metamorphic rocks previously assumed by Kozuch (1991) to be the Paleozoic Cacaguapa Group (Fig. 2).

The Siuna terrane is an oceanic terrane defined and named for the open-pit mining Siuna mining district of northeastern Nicaragua by Venable (1994) (Figs. 1 and 2). Case et al. (1990) compiled seismic refraction data from the area of the Siuna terrane and proposed a 20–25-km island-arc crust built on oceanic crust. A fundamental observation from the map shown in Figure 1 is that the overall structural strike and related topography of the Eastern Chortis continental terrane and the Siuna oceanic terrane is at right angles to the trend of the Middle America trench and volcanic arc. Moreover, trends of the Eastern Chortis and Siuna terranes are at a high angle to the Central Chortis terrane (Rogers et al., Chapter 4, this volume).

The linear and topographically prominent Guayape fault system (Finch and Ritchie, 1991; Gordon and Muehlberger, 1994) forms a major terrane boundary between the northeasterly striking rocks in the Eastern Chortis terrane (basement composed of Jurassic metasedimentary rocks) and more eastward-striking rocks in the Central Chortis terrane underlain by basement composed of Precambrian and Paleozoic crystalline rocks (Rogers, 2003; Rogers et al., this volume, Chapter 4). Gordon and Muehlberger (1994) documented several kilometers of right-lateral strike-slip motion along the fault during Neogene time. Finch and Ritchie (1991) proposed ~50 km of left-lateral motion based in part on the apparent lateral offset of the Agua Fria Formation on either side of the fault. The map compilation shown in Figure 2 indicates that the apparent left-lateral offset is closer to 60 km (length of line indicated by "l" in Fig. 2). The sense of oroclinal bending of the parallel ranges northwest of the fault in the Central Honduras terrane also supports the left-lateral sense of slip.

Mesozoic Stratigraphy of the Central and Eastern Chortis Terranes

Despite the lack of crystalline basement east of the Guayape fault and evidence of an apparent 60 km of left-lateral offset, a very similar Mesozoic stratigraphy occurs on both the Central and Eastern Chortis terranes on both sides of the fault. Figure 3 compares the names, thicknesses, and ages of the main Mesozoic formations found on both sides of the fault; these are shown in map view on Figure 2.

Agua Fria Formation

The middle Jurassic Agua Fria Formation forms the basal clastic unit on both terranes to the northwest and southeast of the Guayape fault system. The formation is at least 1700 m thick and consists of coastal plain fluvial deposits, minor shallow-marine carbonate rocks, and rhythmically-bedded siliciclastic sedimentary rocks that have been interpreted as marine turbidites by Ritchie and Finch (1985), Gordon (1990), and Rogers (1995). The basal contact of this formation on older rocks has never been observed. Gordon (1993b) inferred an underlying metamorphic basement in the Catacamas Valley of the Central Chortis terrane based on the presence of recycled, metamorphic clasts in conglomerate of the Agua Fria Formation. Viland et al. (1996) document regional deformation in the Late Jurassic that metamorphosed parts of the Agua Fria Formation prior to deposition of Cretaceous strata.

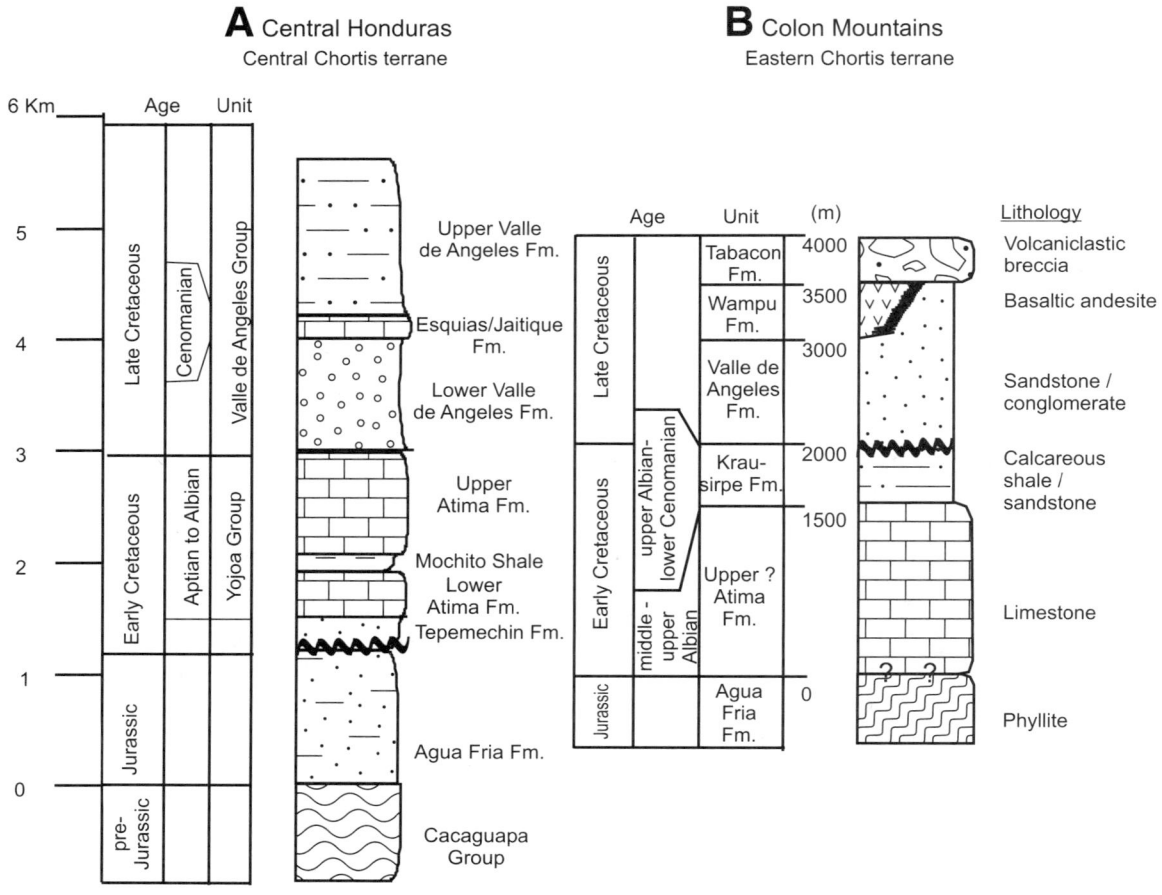

Figure 3. Comparison of Mesozoic stratigraphic units, nomenclature, and thicknesses in meters northwest and southeast of the terrane boundary (Guayape fault system) separating Central and Eastern terranes (A) compiled from previous workers (Finch, 1981; Donnelly et al., 1990; Gordon, 1993a; Scott and Finch, 1999) and Rogers (2003) and (B) the Eastern Chortis terrane exposed in the Colon Mountains. Ages in the Colon Mountains are based on paleontological dating by Scott and Finch (1999) and radiometric dating of the Wampu volcanics by Weiland et al. (1992).

Tepemechin Formation

The Tepemechin Formation is a thin conglomerate that unconformably overlies the Agua Fria Formation and forms the base of the overlying Cretaceous carbonate stratigraphy of the Yojoa Group on the Central Chortis terrane (Gordon, 1993a). This unit includes the unnamed siliciclastics strata of Simonson (1977) and the Todos Santos Formation of Mills et al. (1967).

Yojoa Group

Early Cretaceous-Cenomanian shallow-marine limestone of the Yojoa Group overlies the basal clastic rocks of the Agua Fria Formation on both sides of the Guayape fault (Mills et al., 1967; Scott and Finch, 1999; Rogers et al., this volume, Chapter 5) (Fig. 3). The Yojoa Group is divided into the upper and lower Atima Formations by the intervening Mochito shale and other units. The shallow water–shelf limestone of the Atima Formation is locally up to 1400 m thick. Volcanic rocks, including andesite and dacite flows and pyroclastic rocks, occur within the Atima Formation on the Central and Eastern Chortis terranes (Carpenter, 1954; Simonson, 1977; Gordon, 1990; Rogers, 2003; Rogers et al., this volume, Chapter 5).

Valle de Angeles Formation

Overlying the Yojoa Group is a thick sequence of coarse-grained, terrigenous redbeds of the Valle de Angeles Formation (Mills et al., 1967; Finch, 1981) (Fig. 3). Clastic rocks of the formation include clay-rich, lithic conglomerate (both matrix- and clast-supported) and sandstone and shale deposited as submarine and subaerial debris flows (Rogers and O'Conner, 1993). The conglomerate contains clasts of all underlying lithologies including the Yojoa Group and redbed clasts derived from syndepositional reworking (Mills et al., 1967; Finch, 1981; Rogers, 1996). Thickness of the lower redbeds of the Valle de Angeles Formation can range from several hundred up to 1000 m. Where exposed, the contact between the Valle de Angeles redbeds and the limestone of the underlying Yojoa Group is unconformable, with paleo-epikarst developed at the contact in one location in the Colon Mountains.

Discontinuous shallow marine carbonate strata of Campanian age (Esquias-Jaitique Formation) is recognized in central Honduras and was used as a datum to subdivide the Valle de Angeles Formation into a lower and upper sequence (Finch, 1981; Scott and Finch, 1999) (Fig. 3). Limestone associated with minor gypsum deposits occurs locally in this unit and suggests an isolated marine depositional basin (Horne et al., 1974; Finch, 1981). Limestone of this unit varies in thickness from a few tens of meters to several hundred meters.

The upper redbeds of the Valle de Angeles Formation are generally finer-grained than the lower redbeds. Upper redbeds are interbedded with minor mafic volcanic rocks that thicken to >300 m in eastern Honduras (Weiland et al., 1992; Rogers, 1996). Scott and Finch (1999) proposed that the upper Valle de Angeles redbeds rapidly blanketed the carbonate rocks of the Valle de Angeles Formation without producing an erosional unconformity. The grain size transition from lower, coarse-grained to upper, fine-grained redbeds of the Valle de Angeles Formation is gradational, and diminishing grain size and bedding thickness are noted over several hundred meters of section (Rogers and O'Conner, 1993). Like the lower redbeds, the upper redbeds have been interpreted as debris flows in the Tegucigalpa region (Rogers and O'Conner, 1993).

Intrusive Rocks

Felsic plutons ranging in age from Cretaceous to early Tertiary intrude the stratigraphic units described above and are shown on the compilation map in Figure 2 (Southernwood, 1986; Kozuch, 1991; INETER, 1995). Ages of the plutons are known mainly from field relations and from radiometric dating (cf. Rogers et al., Chapter 4, this volume, Table 1 therein) for a compilation of radiometric ages from Honduras).

THE COLON FOLD-THRUST BELT IN EASTERN HONDURAS

The Colon fold-thrust belt is expressed as folded, massive limestone beds of the Cretaceous Atima Formation (Fig. 3B) in the Colon Mountains between the Río Coco and Río Patuca (Fig. 4). Maximum elevation of the range is 880 m above sea level (masl); average elevation of the area of the surrounding range the varies from 100 m in the lowlands along the Patuca and Coco Rivers to an average of 300–400 masl in the Patuca Mountains northwest of the Colon Mountains. A broad zone of open folds affecting all Cretaceous units shown on the column in Figure 3B parallels the Colon belt and extends at least 20 km to the northwest of the belt into the Patuca Mountains (Fig. 4).

As seen on the LANDSAT image in Figure 4, the Río Patuca closely follows the structural grain of the Colon fold-thrust belt defined by a reverse fault bounding the northeast side of the Colon Mountains (Fig. 5). Both the Río Patuca and Río Coco south of the Colon Mountains are entrenched bedrock rivers showing no signs of deflection or offset by Quaternary faults. Moreover, the regional geomorphology of eastern Honduras supports the interpretation that the area is a tectonically stable part of the Caribbean plate (Rogers et al., 2002; Rogers and Mann, this volume). The northeast flow directions of the Río Patuca and Río Coco reflect their origins as northeast-flowing alluvial rivers prior to the late Neogene epeirogenic uplift that incised their channels into bedrock canyons (Rogers et al., 2002).

The Colon Mountains were originally described as block faulted by Mills and Hugh (1974). However, parallelism between the northeast-trending Guayape strike-slip fault system and the northeast-trending Colon Mountains (Fig. 2) has previously led workers to interpret the deformation of the Colon Mountains as a large, topographically-elevated, "flower structure" produced by lateral shearing and transpression on vertical strike-slip fault planes (Gordon and Muehlberger, 1994; Mills and Barton, 1996). This interpretation was mainly supported by interpretation of satellite imagery (Fig. 4) and aerial photographs and not by detailed structural observations in the field.

Riverbank and stream exposures in the area of the confluence of the Río Wampu and Río Patuca provide excellent cross sectional exposures of the structural and stratigraphic units of the Colon Mountains (Fig. 5). Mapping and paleontological dates from sedimentary units sampled in the Colon Mountains (Scott and Finch, 1999) reveal a pre-Cenomanian, south-facing, continental margin setting. Shallow-water Cretaceous carbonate rocks and clastic strata totaling 4 km in thickness were deformed by thin-skinned, northwestward-directed thrusting following Campanian time.

Stratigraphy of the Colon Fold-Thrust Belts

Basement and Overlying Yojoa Group

Low-grade metasedimentary basement of the area is composed of phyllite and quartzite of presumed Jurassic age (metamorphosed Agua Fria Formation) (Fig. 3B). This metasedimentary unit crops out in the north-central part of the study area (Fig. 5) and extends to the northeast to the Guayape fault (Gordon, 1993a) (Fig. 2). Agua Fria Formation metasedimentary rocks are moderately resistant and, coupled with their diverse bedding plane foliations, produce a rugged upland topography with a dendritic drainage (Fig. 4).

An estimated 1500 m of massive shallow water limestone of the Atima Formation of the Yojoa Group overlies the Agua Fria Formation, although this contact has not been observed in the field. The resistant limestone of the Atima Formation is a prominent ridge-former, and extensive karst topography is developed on landscapes underlain by the Atima Formation. The limestone contains upper Albian foraminifera as well as Albian to lower Cenomanian foraminifera, indicating a stratigraphic position equivalent to the Upper Atima Formation (Scott and Finch, 1999).

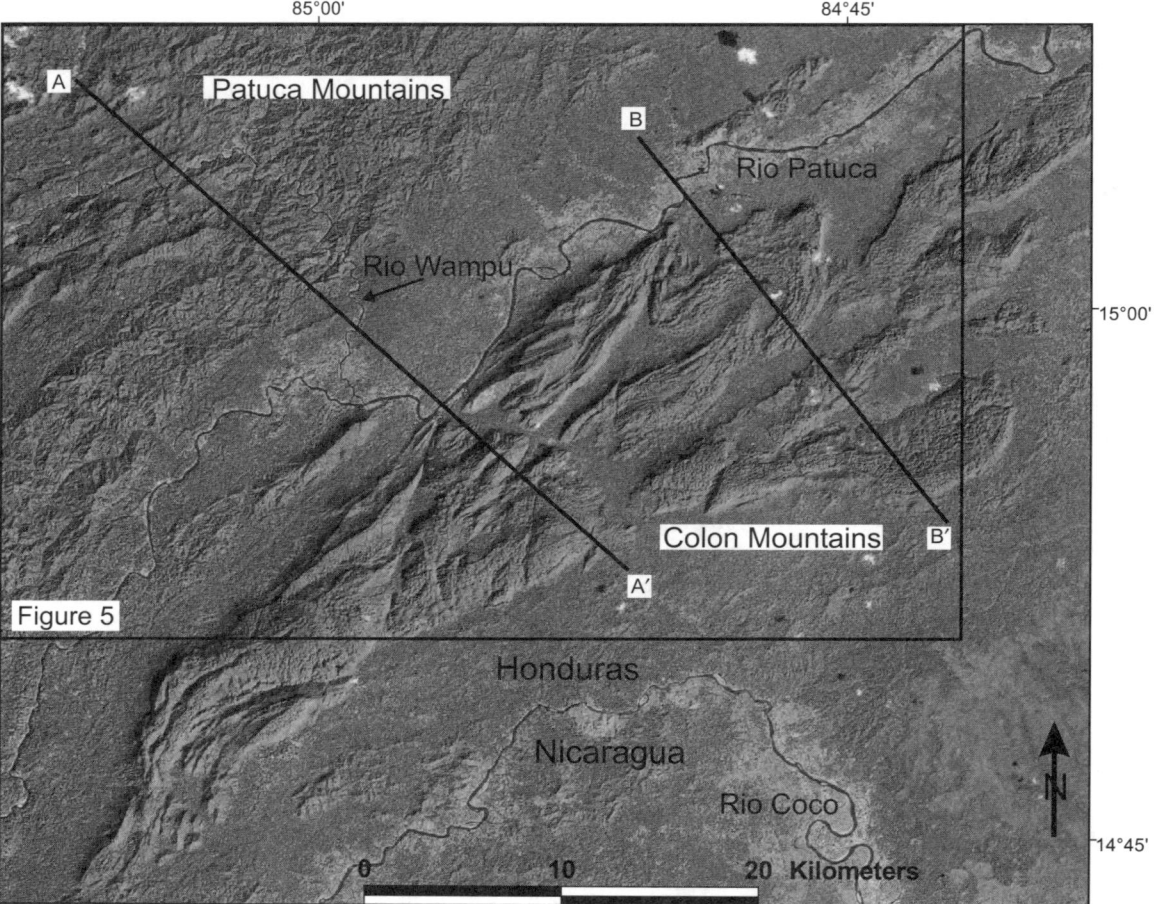

Figure 4. Landsat image of the Colon Mountains of eastern Honduras showing large-scale, northwest-verging folds in Mesozoic sedimentary units (shown on Figure 3B), major rivers, and the location of the geologic map in Figure 5. The inner box shows the extent of field-based mapping. Lines A–A′ and B–B′ show the locations of cross sections in Figure 6. The Río Coco forms the political boundary between Nicaragua and Honduras.

Based on carbonate petrography, Scott and Finch (1999) infer an upward-shoaling carbonate succession. Biofacies indicate that this succession occurred as the result of marine transgression from middle to inner shelf paleodepths that gave rise to a "keep-up" carbonate platform that aggraded during a relative sea-level rise. Atima Formation strata are entirely missing north of the Río Patuca (Figs. 4 and 5).

Krausirpe Formation

Calcareous marine shale containing carbonaceous detritus of the Krausirpe Formation conformably overlie the carbonate rocks of the Upper Atima Formation (Rogers, 1996) (Fig. 3B). Rocks of the Krausirpe Formation are not resistant and form the strike valleys of the Colon Mountains (Fig. 4). The Krausirpe Formation is estimated to be 500 m thick, and its floral and faunal components include late Albian to early Cenomanian palynomorphs as well as late Albian to early Cenomanian foraminifera (Scott and Finch, 1999). Biofacies indicate that these beds represent a transition from carbonate shelf (Atima lithology) to terrigenous shelf (Krausirpe lithology) with paleowater depths of up to 50 m as indicated by the presence of planktic foraminifers and dinoflagellates.

Late Cretaceous Valle de Angeles Formation

The Valle de Angeles Formation overlies marine strata of the Yojoa Group and Krausirpe Formation and contains medium to coarse-grained immature sandstone and conglomerate with subrounded clasts of metamorphic, volcanic, carbonate, and redbed lithologies. Rocks of the Valle de Angeles Formation are not resistant and form the non-alluvial lowlands of the region. Total thickness of the Valle de Angeles redbeds is estimated to be 1500 m in the eastern part of the study area and decreases to several hundred meters in the western part of the study area below the volcanic flows of the Wampu Formation (Fig. 5). Carbonate clasts were derived from limestone of the Atima Formation and contain late Aptian to late Albian foraminifera Cenomanian calcareous Dasyclad alga and the late Albian rudist *Mexicaprina* sp. (Scott and Finch, 1999).

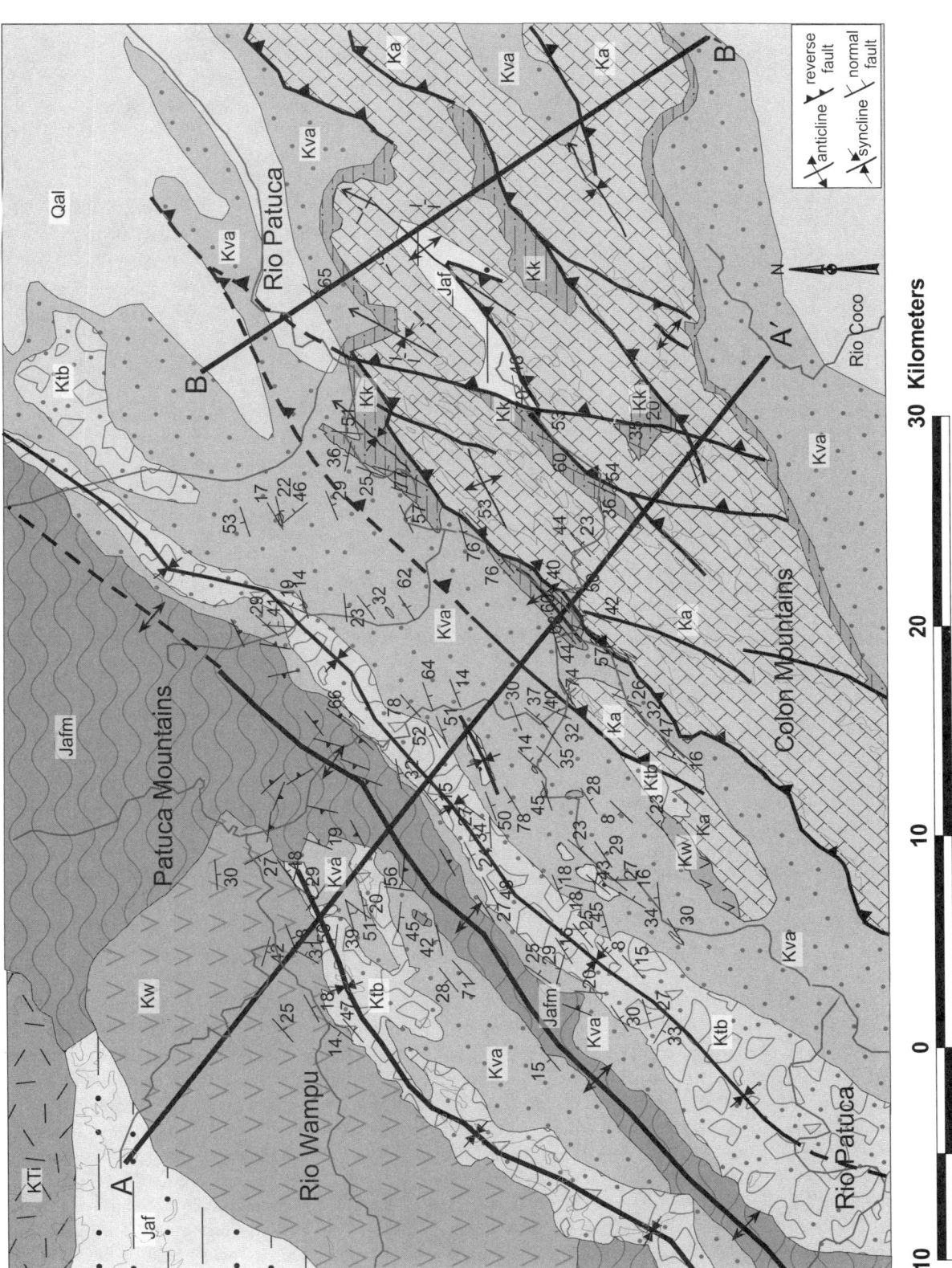

Figure 5. Geologic map of Colon Mountains study area near the confluence of the Río Wampu and Río Patuca (see Figure 4 for map location area relative to the entire Colon fold-thrust belt). Southeastward-dipping thrust faults transport early Cretaceous massive limestone of Atima Formation to the northwest over Late Cretaceous Valle de Angeles redbeds. A–A' and B–B' show the locations of the geologic cross sections in Figure 6. Jaf—Jurassic Agua Fria Formation; Jafm—Jurassic metamorphosed Agua Fria Formation; Ka—Cretaceous Atima Formation; Kk—Cretaceous Krausirpe Formation; Ktb—Cretaceous Tabacon Formation; KTi—Cretaceous-Tertiary intrusive rocks; Kva—Cretaceous Valle de Angeles Formation; Kw—Cretaceous Wampu volcanics.

Limestone-clast conglomerate in the lower part of the Valle de Angeles Formation is abundant proximal to the limestone massif of the Colon Mountains (Fig. 3B). In areas to the northwest of the Colon Mountains, metamorphic clasts predominate and limestone clasts are absent to rare. The Atima Formation (source of the limestone clasts) is not present northwest of the Río Patuca (Figs. 4 and 5). Subangular pebble clasts of red sandstone appear throughout the section becoming more prevalent toward the top of the redbeds.

Wampu Volcanic Unit

Flows of basaltic andesite within clastic strata of the Valle de Angeles Formation are exposed along the Río Patuca upstream of its confluence with the Río Wampu (Fig. 5). Wampu volcanic flows are somewhat less resistant than the Agua Fria metasedimentary rocks and develop a rectangular drainage (Fig. 4). These volcanic exposures define a northeast-trending outcrop belt that is an outlier of the basaltic andesite flows of the main Wampu volcanic field to the northwest that was studied by Weiland et al. (1992) (Fig. 5). Weiland et al. (1992) reported K/Ar ages of 70.4 (±3.4) Ma and 80.7 (±4.3) Ma from volcanic flows in this northern area. In the upper 500 m of the Valle de Angeles Formation, Wampu volcanic flows are interbedded with the clastic strata (Fig. 3B).

Tabacon Formation

The Tabacon Formation is a 500 m-thick, massive bedded, subangular cobble breccia, composed of volcanic clasts derived from the flows of the Wampu Formation and reworked clasts from the redbed strata (Rogers, 1996) (Fig. 2 and 5). Its age is taken as Late Cretaceous (<70 Ma) based on its stratigraphic position above the radiometrically dated Wampu volcanic rocks (70–80 Ma) (Weiland et al., 1992). It is the youngest stratigraphic unit present in the study area and is involved in the folding and thrusting. The volcanic breccia of the Tabacon Formation is a prominent ridge former that is distinguishable from the ridge-forming Atima Formation by the lack of karst development (Fig. 4).

Mills and Hugh (1974) mapped the Late Jurassic Todos Santos Formation and Atima Formations along the Río Wampu and west of the Río Patuca in an area where Rogers (1996) mapped Late Cretaceous breccia of the Tabacon Formation (Fig. 5). We suggest that the resistant beds of the Tabacon breccia were incorrectly identified by Mills and Hugh (1974) because the Tabacon unit forms strike ridges remarkably similar in appearance to the ridge-forming Atima limestone when viewed from a distance or on aerial photographs. Because Tabacon breccia overlies the metasedimentary Agua Fria Formation along the Río Wampu, Mills and Hugh (1974) may have simply assumed that the first clastic unit would be the Todos Santos Formation.

An angular unconformity with 10–15 degrees of discordance separates the Krausirpe Formation from the overlying clastic strata of the Valle de Angles Formation. A zone of paleo-epikarst is developed where the Valle de Angeles Formation unconformably blankets the massive limestone of the Atima Formation.

Paleo-epikarst at the contact and the complete erosion of the underlying Krausirpe strata record a period of subaerial exposure of the limestone prior to the beginning of the Late Cretaceous Valle de Angeles deposition. North of the Río Patuca, limestone of the Atima Formation is absent, and the Valle de Angeles redbeds were observed in direct unconformable contact with the metasedimentary strata of the Agua Fria Formation. Patchy deposition of Atima Formation limestone has been reported across the Chortis block (cf. Finch, 1981; Donnelly et al., 1990; Rogers, 2003), and its complete absence north of the Río Patuca (Fig. 5) could signify its nondeposition rather than its deposition and subsequent erosion. However, the angular unconformity between the Krausirpe and Valle de Angeles Formation, the Atima clasts within the Valle de Angeles redbeds, and the karstified erosion surface at the Valle de Angeles–Atima contact are evidence for a peripheral bulge developed in front of the advancing Colon thrust belt. The antiformal basement exposure northwest of the Río Patuca marks the trace of the bulge (Figs. 2 and 5).

Because no obvious angular unconformity is observed in the Late Cretaceous section, we infer that the main shortening event that formed the Colon fold-thrust belt occurred after the deposition of the Tabacon Formation ca. 70 Ma (Fig. 3B). Because the contact between the Valle de Angeles redbeds and the Tabacon breccia is gradational with upward coarsening, progressive increase in clast angularity, and increasing volcanic clast content, it is possible that the Tabacon Formation is an early syntectonic deposit related to the early phase of the shortening event. To the northeast and southwest, away from the Wampu volcanic field, the breccia clasts in the Tabacon Formation contain a greater abundance of metamorphic rock fragments, suggesting a regional uplift and source from the northwest in latest Campanian to Maastrichtian time. Because the Tabacon breccia blanketed the latest Cretaceous landscape and rests unconformably on the Agua Fria phyllite and schist, we interpret the Tabacon Formation as a synorogenic clastic wedge shed southeastward from subaerially exposed highlands formed during Late Cretaceous–early Tertiary shortening as the peripheral bulge formed in front of the advancing Colon thrust sheet.

Structural Geology of the Colon Fold-Thrust Belt

The Colon belt is a fold-thrust belt of imbricate, southeast-dipping reverse faults that place Cretaceous carbonate strata of the Atima Formation over post-Cenomanian Valle de Angeles Formation (Fig. 6A and 6B). Broad, open folds are found 20 km northwest of the thrust front, indicating a broad foreland zone of convergent deformation (Figs. 4 and 5). The northeast-striking frontal thrust parallels the Río Patuca on the west flank of the Colon Mountains (Fig. 4). Reverse faults and northeastward-plunging folds of the Colon Mountains coincide with the outcrop area of shale of the Krausirpe Formation, which forms an incompetent horizon on which northwest directed shortening is facilitated (Figs. 4 and 5). West of the Río Patuca where the Krausirpe Formation is absent, large folds with northeastward-trending axial surfaces deform Jurassic metasedimentary rocks and overlying Cretaceous strata.

At least four thrust sheets comprise the Colon Mountains as displayed in the structure profile (Figs. 6A and 6B). The location of the faults is revealed by the repetition of the distinctive Atima-Krausirpe stratigraphy visible in the strike valleys of the Colon Mountains (Fig. 4). In the Patuca Mountains west of the Río Patuca, the ridges composed of volcanic breccia of the Tabacon Formation form two open synclines separated by a metamorphic-cored anticline flanked by Valle de Angeles strata dipping off the anticline (Fig. 6A). This represent the peripheral bulge in front of the main thrust sheets (Fig. 5).

THE SIUNA TERRANE OF NICARAGUA

Seventy kilometers southeast of the Colon fold-thrust belt of Honduras is the Siuna mining district (Figs. 2 and 7) (Venable, 1994). Here serpentinite bodies containing high nickel and chromium, ultramafic cumulates, and podiform chromite are thrust to the north and west and are imbricated with Cretaceous sedimentary strata and calc-alkaline volcanic rocks. Wehrlite is associated with the serpentinite. X-ray diffraction analysis shows the serpentinite bodies to be composed of lizardite and chrysotile with minor magnetite and chromite (Venable, 1994). Undeformed diorite and granodiorite plutons intrude strata and volcanic rocks and postdate the shortening deformation. Venable (1994) reports an $^{40}Ar/^{39}Ar$ age of 59.89 (±0.47) Ma for a biotite separate from a diorite pluton. Whole rock analysis of the diorite yielded a $^{147}Sm/^{144}Nd$ ratio of 0.135624 and a present day $^{143}Nd/^{144}Nd$ of 0.512985, with a present day epsilon Nd value of +6.8, corresponding to an initial epsilon Nd value of +7.2 at 60 Ma, indicating lack of contamination from continental crust. Small, undeformed hornblende andesite dikes intrude the faulted contacts between the serpentinites and sedimentary strata. $^{40}Ar/^{39}Ar$ dating of a hornblende separate from the andesite dikes yielded an age of 75.62 (±1.33) Ma.

Venable (1994) defined the Siuna terrane as an oceanic island arc developed on an oceanic basement that was active from the Early to middle Late Cretaceous and subsequently deformed in the Late Cretaceous as the Siuna arc accreted to the southern margin of the continental Chortis block. We propose that the Siuna terrane formed the leading edge of the Guerrero-Caribbean island arc that accreted to the margin of southern Mexico and to the Chortis block in Late Cretaceous times. This arc system formed the leading edge of the Caribbean oceanic plateau province, which formed in Late Cretaceous time (95–88 Ma) (Tardy et al., 1994; Hoernle et al., 2002; Mann, this volume).

THE COLON FOLD-THRUST BELT BENEATH THE MOSQUITIA COASTAL PLAIN OF EASTERN HONDURAS

The location of the subsurface seismic reflection grid and well study of Mills and Barton (1996) is shown on Figure 2 and covers a large area of the Mosquitia alluvial plain of Honduras centered on the village of Awas. The area is underlain by a large Quaternary alluvial plain related to fluvial deposition from the combined outflow of the Río Patuca and Cocos (Fig. 2). The overall elevation of the plain is a few tens of meters above sea level with the highest points confined to bedrock hills (Mills and Hugh, 1974). A small part of the Mosquitia alluvial plain is apparent on the southeast corner of the LANDSAT image shown in Figure 4.

Seismic interpretations tied to the two wells drilled along multichannel seismic line T-08 of Mills and Barton (1996) (Fig. 6C) identified several of the same lithologic formations described herein from outcrops in the Colon Mountains. These relationships are summarized on the stratigraphic column in Figure 3B.

Our interpretation of the units encountered in the two wells differs significantly from the stratigraphic interpretation by Mills and Barton (1996). Their approach was to adopt new formation names for units that we consider to be correlatable to the stratigraphic section of the Colon Mountains (Rogers, 1996) (Fig. 3B). Part of the correlation problem with both wells was related to the fact that no paleontological age determinations were made, so none of the units described by Mills and Barton (1996) were dated. Our correlations rely solely on lithostratigraphic correlation. The first author visited the well sites in February 1992 and examined cuttings from the wells. The stratigraphic units described in the two wells included the units summarized in Table 1, which are shown in detail on Figure 13 *in* Mills and Barton (1996) and schematically on Figure 6C herein.

Adopting these stratigraphic correlations, the Embarcadero well and multichannel seismic line T-108 document duplex thrusting of Jurassic Agua Fria Formation over a fault sliver of Late Cretaceous Valle de Angeles Formation, which in turn is thrust over Jurassic Agua Fria Formation (Fig. 6C). The northern part of the seismic line shows repetitions in the unit we interpret as the Late Cretaceous (post-80 Ma) Tabacon Formation. These observations constrain the age of thrusting in eastern Honduras as post-dating the deposition of the Tabacon Formation, or Late Cretaceous–Tertiary (post-80 Ma).

The structural style displayed on the seismic line consists of duplexed, south-dipping reverse faults. The hanging wall of these faults produce a structural high centered near the Embarcadero well seen in Figure 6C. This elevated hanging wall in the subsurface is directly along-strike from the topographically-elevated hanging wall of the Colon Mountains (Fig. 2).

Aeromagnetic data provided to us by the Dirección General de Energia (Honduras) was used to trace the frontal thrust along-strike in the subsurface from the Colon Mountains, across the Mosquitia Plains and the coast of Honduras, to the Nicaraguan Rise offshore (Fig. 8). The magnetic character of the autochthonous hanging wall of the Colon fold-thrust belt is dominated by low amplitude (<100 gamma) short-to-medium wavelength (5–10 km peak to trough distance) anomalies that range in color from orange to dark green in the area to the south and southeast of the

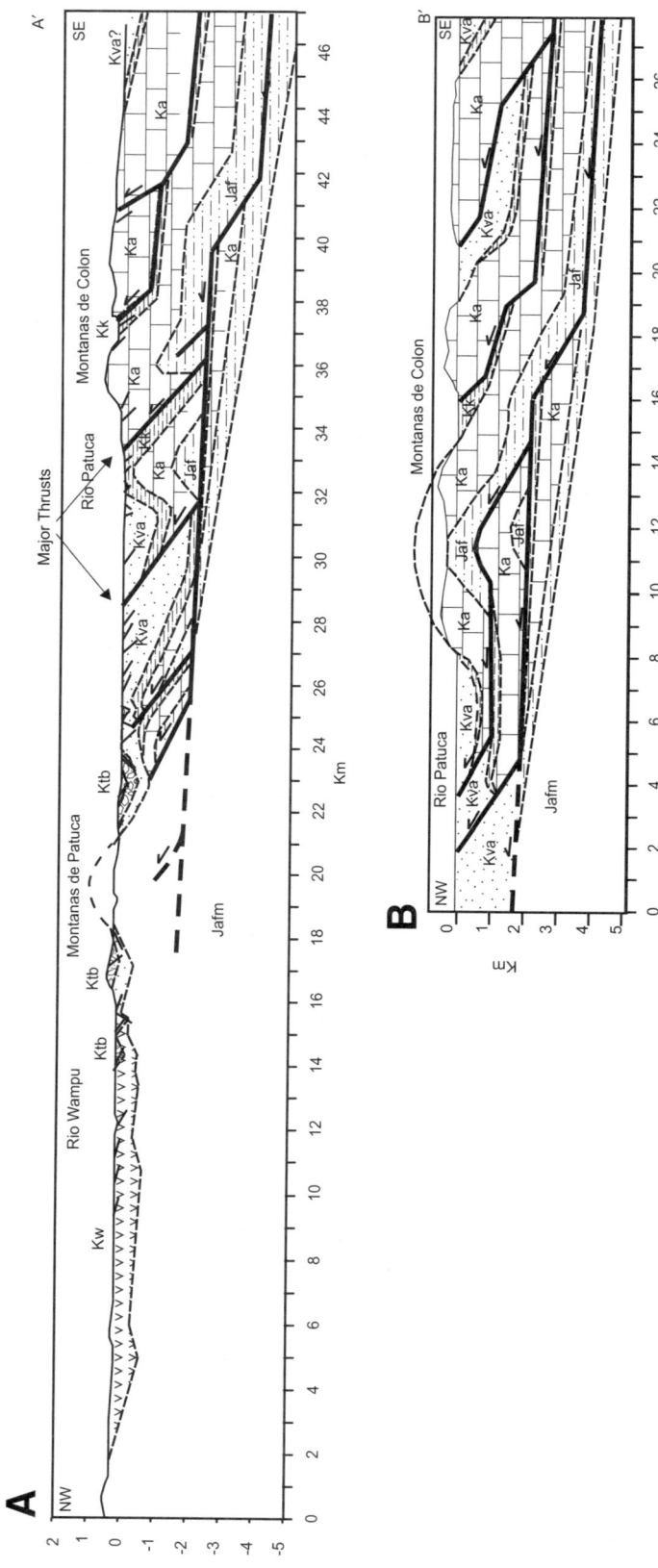

Figure 6 (*on this and following page*). Structural cross sections and multi-channel seismic profiles across all three segments of the Colon fold-thrust displayed at the same horizontal scale (except for C, which is at 2×) to facilitate structural comparisons between segments. (A and B) Geologic cross sections through the Colon Mountains segment, eastern Honduras, based on 1:50,000 scale geologic mapping shown in Figure 5. Scale is in kilometers, and there is no vertical exaggeration. Jaf—Jurassic Agua Fria Formation; Jafm—Jurassic metamorphosed Agua Fria Formation; Ka—Cretaceous Atima Formation; Kk—Cretaceous Krausirpe Formation; Ktb—Cretaceous Tabacon Formation; KTi—Cretaceous-Tertiary intrusive rocks; Kva—Cretaceous Valle de Angeles Formation; Kw—Cretaceous Wampu volcanics. Measured dips are plotted along ground surface. (C) Onland multi-channel seismic reflection section across the Mosquitia Plains near the village of Awas modified from seismic line T-08 in Figure 7 *in* Mills and Barton (1996) (see Fig. 2 for location of line, position of frontal thrust, and associated grid of seismic data). Vertical scale is in two-way travel time. Age control and position of some of the higher level, southeast-dipping thrusts are constrained by well logs from Raiti-Tara-1 and Embarcadero-1. Main thrust in Raiti-Tara-1 well places lithologic equivalent of Tabacon Formation over Cretaceous Valle de Angeles Formation; upper thrust in Embarcadero-1 places Jurassic Agua Fria Formation over Cretaceous Valle de Angeles Formation. (D) Offshore multi-channel seismic reflection section across the leading edge of the Colon fold-thrust belt beneath the Nicaragua Rise from Rockwell (1985) (see Fig. 2 for location of line, position of frontal thrust, and associated grid of seismic data). (E) Offshore multi-channel seismic reflection section displays Cretaceous strata (G-horizon) elevated to near seafloor beneath the Nicaragua Rise and the unconformity at the base of the Tertiary from Rockwell (1985) (see Fig. 2 for location of line and associated grid of seismic data). P—top of Eocene; B—top of Atima Formation; dashed line is a Miocene marker. Inset map shows location of seismic lines and wells.

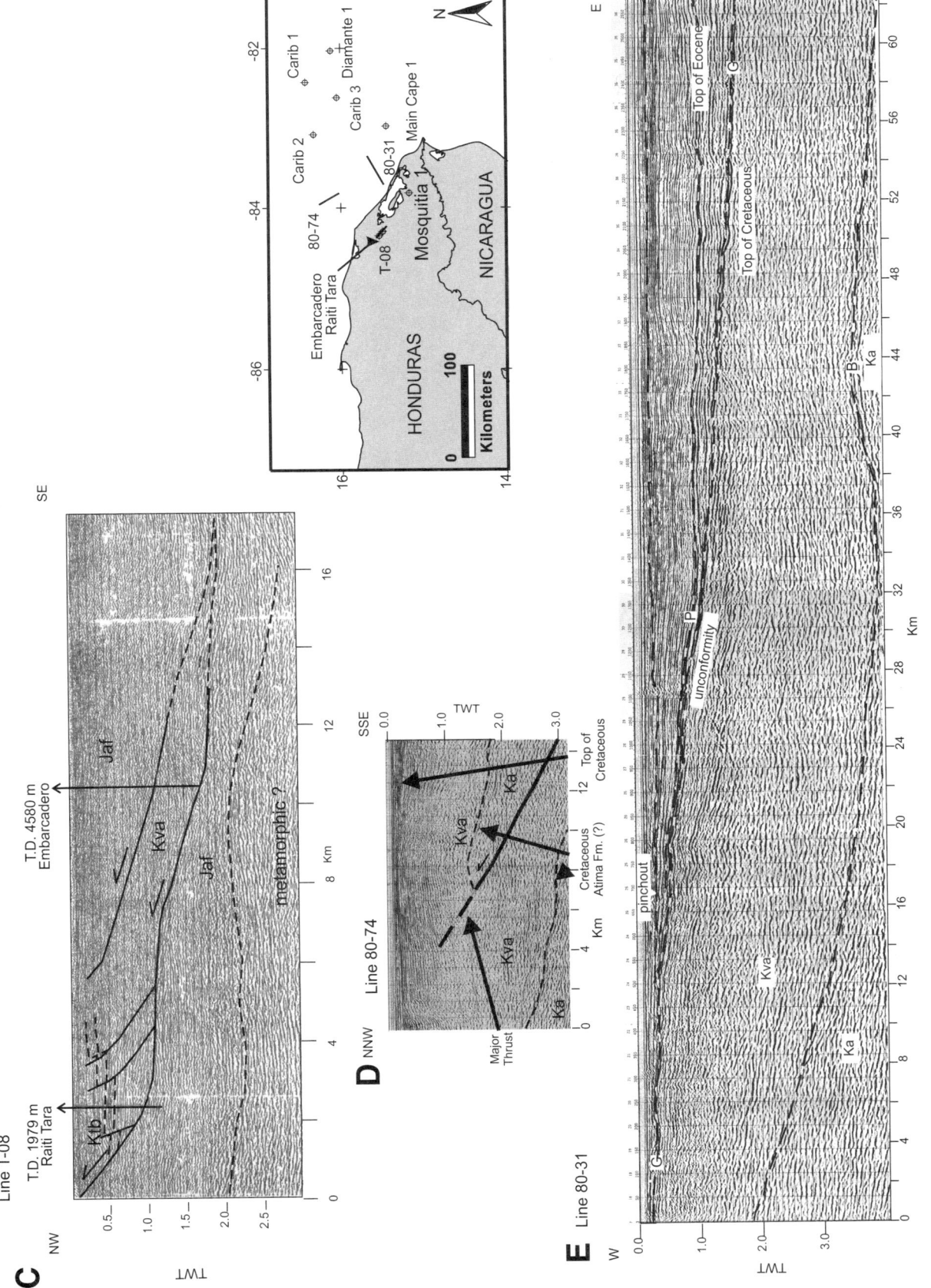

Figure 6 (*continued*).

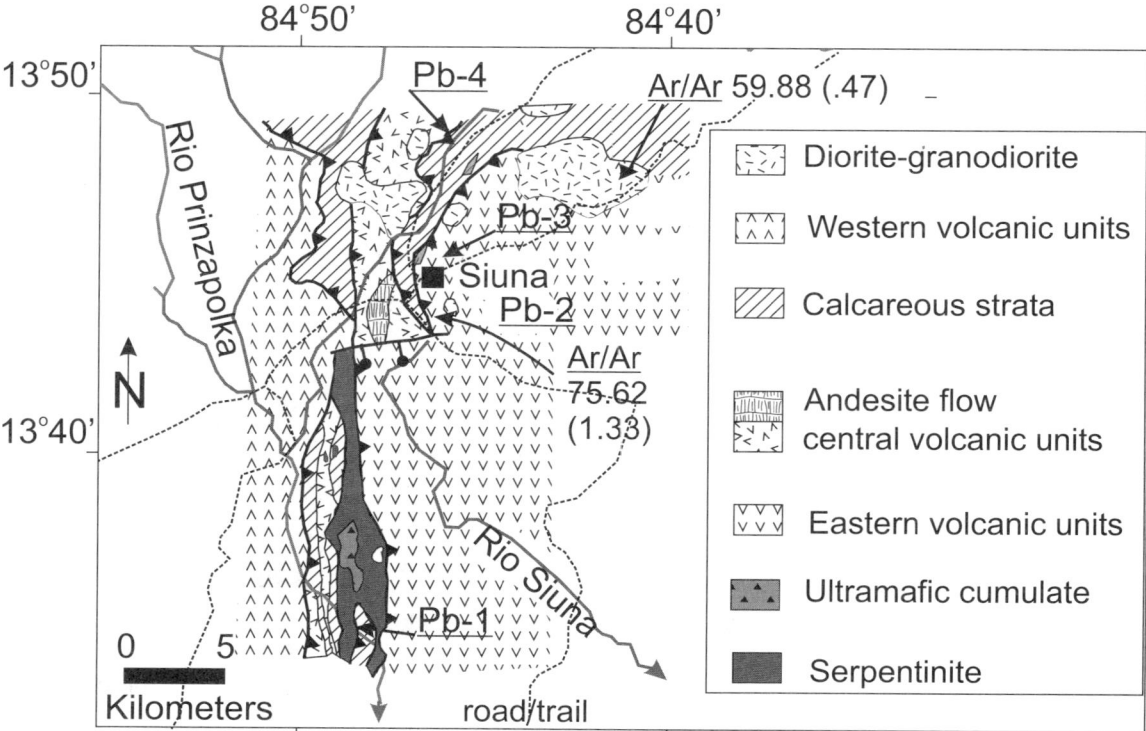

Figure 7. Geology of Siuna mining district, Nicaragua, modified from Venable (1994) showing distribution of thrusted ultramafic and igneous rocks. Location of map is shown on Figure 2. Locations of samples analyzed for Pb isotopic ratios (see Fig. 10 and Table 2) and of sample dated by argon-argon (Ar/Ar) are indicated.

TABLE 1. STRATIGRAPHIC CORRELATION OF THIS STUDY AND MILLS AND BARTON, 1996

Mills and Barton (1996)	This study	Lithology described by Mills and Barton (1996)
Ahuas beds	Tabacon Formation	Soft to hard, medium- to fine-grained, reddish sandstone siltstone, and conglomerate; volcaniclastic
Río Riba Formation	Valle de Angeles Formation	Red, fine-grained argillaceous hard sandstone and shale
Not observed	Krausirpe Formation	
Not observed	Atima Formation	
Agua Fria Formation	Agua Fria Formation	Dark tightly cemented sandstone and shale
Paleozoic (?) metasedimentary	Agua Fria Formation metamorphosed	Dark gray phyllite

thrust front. This contrasts with the footwall and foreland of the Colon belt, which is dominated by high-amplitude (>200 gamma) broad wavelength (20–30 km peak to trough distance) anomalies that range in color from magenta to dark blue in the area to the north and northwest of the thrust front.

Superimposing the thrust front—known from mapping in the Colon Mountains and the Mosquitia subsurface study—onto the aeromagnetic map clearly shows a northeast-trending deep basement structure expressed on the aeromagnetic map that is consistent with the shallow structure observed on the seismic data and by field mapping (Fig. 8). This alignment implies involvement of the underlying magnetic basement in the shortening either as an inversion of the rifted Jurassic margin (Rogers et al., this volume, Chapter 4) or as a backstop preventing further advancement of the thin-skinned thrust sheets.

THE COLON FOLD-THRUST BELT BENEATH THE NICARAGUA RISE, CARIBBEAN SEA

The Nicaragua Rise is a shallow-water (<100 m), tectonically stable, Cenozoic carbonate platform overlying both continental and arc crust (Arden, 1975; Case et al., 1990). Areas along the Honduran coast are dominated by terrigenous sedimentation related to major deltas of the Patuca and Coco Rivers (Fig. 2).

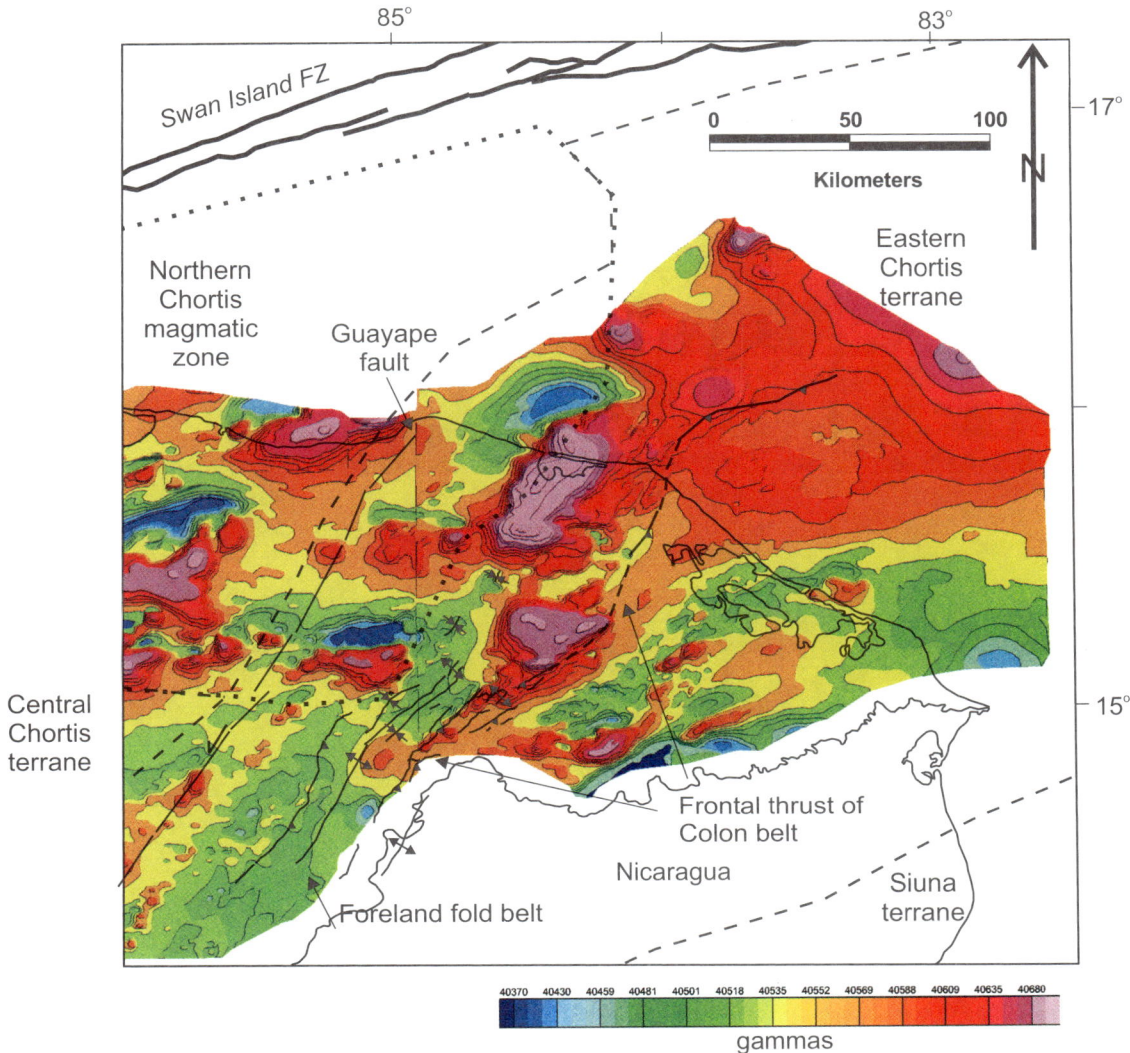

Figure 8. Aeromagnetic map of eastern Honduras modified from Dirección General de Energia (Honduras) (used with permission). Scale is in gammas. Thrust front from the geologic map in Figure 2 is superimposed to show the abrupt change in magnetic patterns that coincides with the leading edge of the Colon fold-thrust belt and the parallelism of surface thrust with magnetic trends in basement rocks. FZ—fault zone.

Rockwell (1985) reported findings based on 850 km of seismic reflection airgun data (48 fold) and 315 km of original Texaco data collected by surveys from 1977 to 1980 (offshore survey area shown in Fig. 2). This area of the Nicaragua Rise is known as the Gracias a Dios platform. The objective of Rockwell (1985) was to reevaluate petroleum prospects in the area and plan additional, non-exclusive seismic surveys in the area. Specific objectives included the enhancement of pre-Tertiary reflectors and clarification of shallow structural and stratigraphic relationships. Reprocessing substantially improved data down to the base of the Tertiary and marginally improved pre-Tertiary data.

Three horizons were mapped: top of Eocene, top of Cretaceous, and top of Atima Formation (mid-Cretaceous) (Figs. 6D and 6E). A seismic line tied to the Main Cape-1 well is located on Figure 2 and confirms the age of the first two horizons. No well tie in this vintage was able to substantiate the top of the Atima Formation although this has been demonstrated by other industry wells and seismic profiles in the vicinity of the Gracias a Dios platform. Data from other wells on the Nicaragua Rise compiled by Robertson Research (1984) support direct correlation between the Cretaceous stratigraphy in eastern Honduras and the Cretaceous stratigraphy underlying the Eocene-Recent carbonate platform. For example, the Caribe-1, Caribe-2, and Caribe-3 wells all encountered mid- to early Albian limestone below Late Cretaceous sandstone. This stratigraphic succession suggests the regional extent of Albian Atima Formation overlain by Late Cretaceous Valle de Angeles Formation. The Diamante-1 well penetrated Cretaceous limestone, no older than late Albian, overlain by dark shale with minor sandstone. This succession suggests Krausirpe Formation overlying Atima Formation (Fig. 3B).

The top of Eocene horizon represents an unconformity between Eocene and Oligocene carbonate lithologies and shows that Eocene rocks pinch out in a westward and northwestward direction on the platform by erosion (Fig. 6E). The wedge of Eocene rocks are cut by down-to-the-southeast normal faults. The relationship of the normal faults to the older shortening structures of the Colon belt is attributed to continued shortening thrusting and uplift along the thrust front and hanging wall of the Colon belt shown on Figure 2, or alternatively, to post-collisional collapse.

The top of Cretaceous unit dips east-southeast and is subparallel to an east-southeast–dipping frontal thrust. This fault aligns with the subsurface thrust mapped beneath the Mosquitia Plain and exposed in the Colon Mountains (Figs. 2 and 6D). The structure of this frontal thrust is similar in the Colon Mountains and in the Mosquitia plain: an anticlinal hanging wall block developed in mid-Cretaceous Atima limestone and overthrust clastic sedimentary units correlated to the Late Cretaceous Valle de Angeles Formation (Fig. 6D).

Offshore observations constrain the age of thrusting on the Nicaragua Rise as post-dating the deposition of the Valle de Angeles Formation, or Late Cretaceous–Tertiary (post-80 Ma). Tilting and normal faulting of the Eocene unit indicates that thrusting could have continued into the Eocene but ended by the beginning of the Oligocene.

Reconstructing the Southern Margin of North America in Latest Cretaceous Times

Restoring the Chortis block to its pre-translation (pre-Eocene) position adjacent to the truncated margin of southwestern Mexico realigns the Colon fold-thrust belt of the Eastern Honduras with the fold-thrust belts and ophiolites north of the Motagua suture in Guatemala (Burkart, 1994; Donnelly et al., 1990) (Fig. 9A). The present day configuration of these elements is shown in Figure 9B. This reconstruction also provides a best fit of the following elements common to the Chortis block and SW Mexico, which are discussed in detail in Rogers et al. (this volume, Chapter 4):

1. Influx of Late Cretaceous terrigenous sandstone and shale over Early Cretaceous shallow marine platform limestone of both southern Mexico and Chortis;
2. Grenville-age basement common to both areas;
3. Late Cretaceous shortening structures common to both areas; and
4. Mid-Cretaceous arc volcanism that is geochemically similar in both areas (Rogers, 2003; Rogers et al., this volume, Chapter 5).

These common features of the two regions are used to reconstruct the Late Cretaceous position of the Chortis block against southwestern Mexico in Figure 9A.

Three independent lines of evidence support our interpretation in Figure 9A that the eastern Chortis block records the collision of the Caribbean arc with the southern margin of North America in the Late Cretaceous. The first is the 350-km-long Colon fold-thrust belt with Late Cretaceous, northwest-directed shortening as described in this study. The second is the spatial association and inferred accretion of the intra-oceanic Siuna island arc complex on the southern margin of the Chortis block in the Late Cretaceous (Venable, 1994). The third is the Pacific origin of the Caribbean arc and its position at the leading edge of the Caribbean large igneous province (Pindell and Barrett, 1990; Sinton et al., 1997). The entry of this oceanic arc plateau into the proto-Caribbean Sea, as shown in Figure 9A, led to the subduction of proto-Caribbean oceanic crust and partial accretion of the Caribbean arc and the oceanic plateau at the "gateways" to the Caribbean in Colombia in South America (Kerr et al., 1998) and in southern Central America, where large areas of the crust appear to have been built on oceanic plateau material (Sinton et al., 1997).

In the reconstruction shown in Figure 9A, the Guayape fault system of eastern Honduras manifests collisional deformation by left-lateral strike-slip faulting and oroclinal bending of the inverted rift basins adjacent to the fault (Rogers, 2003; Rogers et al., this volume, Chapter 5).

Comparison of Lead Isotopic Composition of Siuna Terrane with Chortis Terranes and Maya Block

Kesler et al. (1990) showed that the continental blocks of Central America and Mexico display distinct clustering of lead isotopic ratios, which provides a useful basis for distinguishing among the complexly amalgamated terranes of the region. We follow their approach by utilizing lead data from Venable (1994) from the Siuna terrane (Table 2). As previously discussed, the crust of the Chortis block can be subdivided into (1) continental crust of the central and northern Chortis terranes with Precambrian-Paleozoic crystalline basement; (2) crust of the Eastern Chortis terrane, consisting of attenuated continental crust corresponding to exposed Jurassic metasedimentary basement (Agua Fria Formation); and (3) accreted oceanic island arc crust of the Siuna terrane (Rogers et al., Chapter 4, this volume) (Fig. 10A).

The plot in Figure 10B of $^{207/206}$Pb versus $^{206/204}$Pb data from volcanic host rock displays distinctive clustering of common lead isotopes grouped by terrane. Also displayed on the plot are lead isotopic ratios from the Caribbean large igneous province (Sinton et al., 1997; Hauff et al., 2000; Hoernle et al., 2002), the Maya block of the North America plate (Cumming and Kesler 1976; Cumming et al., 1981; Sunblad et al., 1991), and the Miocene volcanic cover of western Nicaragua (Cumming et al., 1981). The lead ratios of the Siuna terrane cluster outside of the Caribbean large igneous province cluster, indicating that the Siuna arc is not underlain by the Caribbean plateau material. Instead, the Siuna arc probably developed as an arc terrane at the leading edge of the Caribbean oceanic plateau province (Fig. 9A).

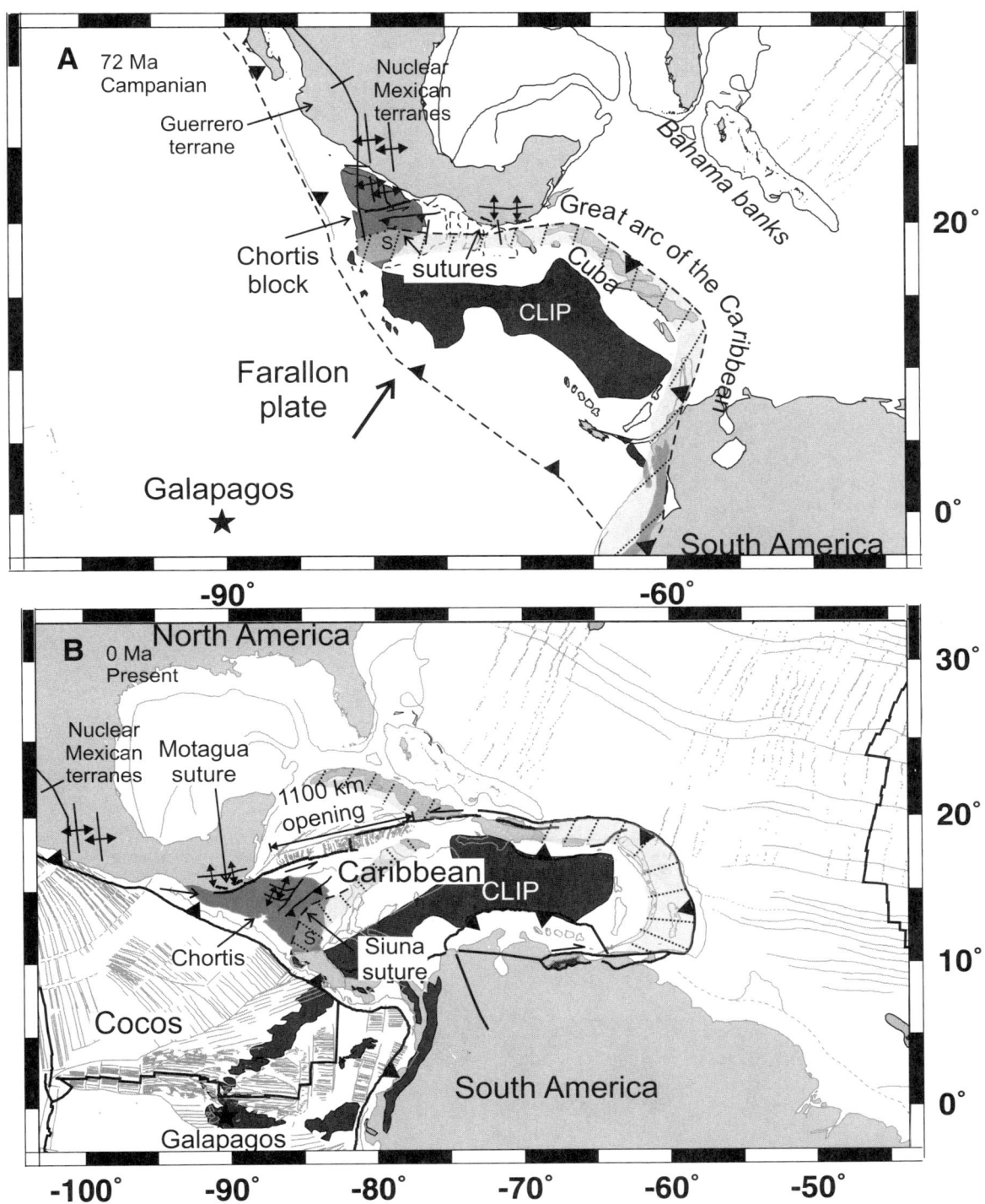

Figure 9. (A) Plate reconstruction during Campanian time (ca. 72 Ma). The reconstruction uses a mantle reference frame, and the Galapagos hotspot is held fixed. At this time, the Chortis block (gray area) forms the southwest corner of the North America plate. The Caribbean large igneous province (CLIP) formed ca. 88 Ma in the eastern Pacific Ocean as part of the Farallon plate and moved northeastward with the Caribbean arc at its leading edge. The Caribbean arc and trailing Caribbean large igneous province (dark area on figure) collided with the edges of North and South America on either side of the gap (proto-Caribbean Sea) formed by Late Jurassic–Early Cretaceous rifting and oceanic spreading between the two Americas. Accreted parts of the Caribbean large igneous province in South America are shown in medium-gray shading. Note the alignment of the Colon fold-thrust belt in the Chortis block with ophiolite belts to the east along the southern margin of North America (Motagua Valley, Guatemala). (B) Present-day position of the same tectonic elements shown in A. Note how eastward translation and rotation of the Chortis block has produced anomalous northeast trends in the Colon fold-thrust belt. S—Siuna.

TABLE 2. LEAD ISOTOPIC COMPOSITION OF THE SIUNA TERRANE, NICARAGUA

	Sample	Mineral	Host rock	$^{206}Pb/^{204}Pb$	$^{207}Pb/^{204}Pb$	$^{208}Pb/^{204}Pb$
Pb-1	417-231	galena	Cretaceous volcanic	18.613	15.591	38.399
Pb-2	403-196	galena	serpentinite	18.598	15.559	38.315
Pb-3	371-031	galena	Cretaceous volcanic	18.550	15.606	38.415
Pb-4	417-290	sphalerite	Cretaceous volcanic	18.583	15.575	38.341

Note: Sample locations shown in Figure 7; data from Venable, 1994. Analysis by A. Baadsgard, University of Alberta.

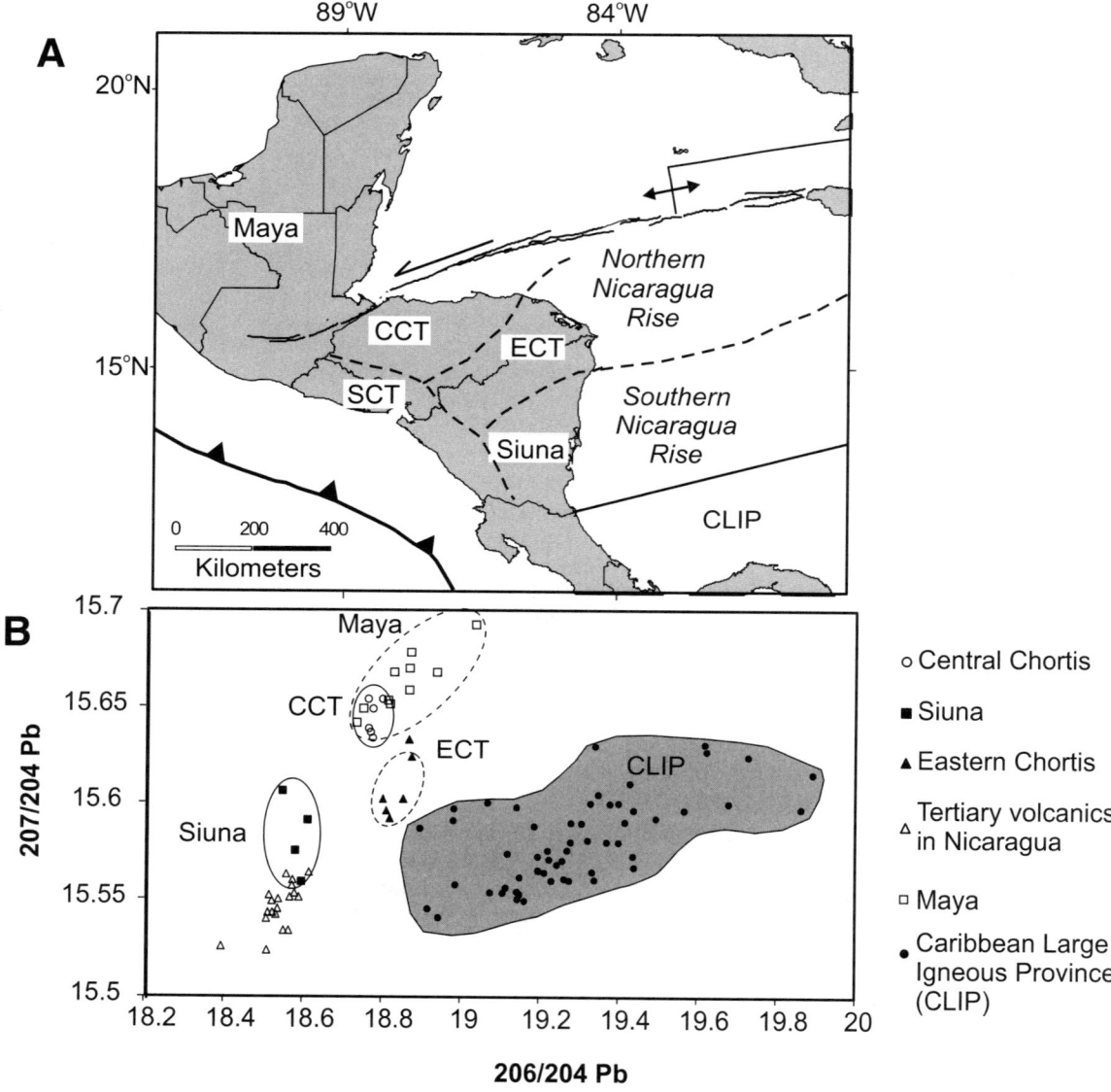

Figure 10. (A) Map showing Maya and Chortis blocks and proposed terranes of the Chortis block: CLIP—Caribbean large igneous province; CCT—Central Chortis terrane; ECT—Eastern Chortis terrane; SCT—Southern Chortis terrane; Siuna—Siuna terrane. (B) Comparison of common lead isotope data for Chortis block (including Siuna terrane) with surrounding terranes and Maya block. Note that lead data distinguishes the accreted Siuna terrane inferred to have originated in the Caribbean arc system from the Chortis block. (Lead data compiled from Cumming and Kesler, 1976; Cumming et al., 1981; Sunblad et al., 1991; Sinton et al., 1997; Hauff et al., 2000; Hoernle et al., 2002.)

The isotopic ratios from the Miocene volcanic cover of western Nicaragua plot close to the Siuna cluster, suggesting that the volcanic pyroclastic deposits may bury the Siuna arc terrane to the southwest where Garayar and Viramonte (1973) described peridotite-bearing ultramafic rocks locally exposed beneath the Nicaraguan volcanic cover. Walther et al.'s (2000) observation of high velocity mantle material in an upper crustal position beneath western Nicaragua may also represent a part of the oceanic Siuna terrane.

CONCLUSIONS

1. The Colon fold-thrust belt of eastern Honduras and the Nicaragua Rise is comprised of southeast-dipping imbricate thrusts and folds, which record a Late Cretaceous northwest-southeast tectonic shortening event (present geographic coordinates).
2. The Late Cretaceous Siuna oceanic island arc complex of northern Nicaragua consists of calc-alkaline volcanic strata, serpentinite, and ultramafic cumulates. The Siuna terrane is inferred to represent the leading edge of the Caribbean arc system that was accreted onto the southern edge of the Chortis block in latest Cretaceous time.
3. Coeval, Late Cretaceous deformation of the Colon belt and the accretion of the Siuna complex are attributed to collision between the intra-oceanic Caribbean arc and the continental Chortis block.
4. This collision is coeval or slightly older than the collision of the Caribbean arc with the Maya block in Guatemala. This event emplaced the northern subduction complex of the Motagua valley and produced widespread shortening deformation to the north of the Motagua valley in Guatemala.
5. We propose that the collision belts of Guatemala and the Colon belt were originally collinear features that have now been offset by the Cenozoic rotation and translation of the Chortis block by 1100 km to the east relative to the deformed Maya block in Guatemala, providing the first pre-Tertiary piercing line oriented east to west across this margin.

ACKNOWLEDGMENTS

Funding for R. Rogers and P. Mann was provided by the Petroleum Research Fund of the American Chemical Society (grant 33935-AC2 to P. Mann). We thank Dirección de Energia of Honduras and JAPEX Geoscience Institute for releasing data for this study. Roger Barton of True Oil Company provided seismic data. Lisa Gahagan at Institute for Geophysics provided plate reconstructions. Special thanks to Lisa Bingham for help with figures and text. This is University of Texas Institute for Geophysics contribution no. 1869. The authors acknowledge the financial support for publication costs provided by the University of Texas at Austin's Geology Foundation and the Jackson School of Geosciences.

REFERENCES CITED

Arden, D., 1975, Geology of Jamaica and Nicaraguan Rise, in Nairn, A.E.M., and Stehli, F.G., eds., The ocean basins and margins: New York, Plenum Press, v. 3, p. 616–661.

Azéma, J., Biju-Duval, B., Bizon, J.J., Carfantan, J.C., Mascle, A., and Tardy, M., 1985, Le Honduras (Amerique Centrale Nucleaire) et le bloc d'Oaxaca (Sud du Mexique): Deux ensembles comparables du continent Nord-Americain separés par le jeu decrochant senestre des failles du systeme Polochic-Motagua, in Mascle, A., ed., Géodynamique des Caraibes: Paris, Editions Technip, p. 427–438.

Bullard, E.C., Everett, J.E., and Smith, A.G., 1965, The fit of the continents around the Atlantic: Philosophical Transactions of the Royal Society of London, v. A258, p. 41–51.

Burkart, B., 1994, Northern Central America, in Donovan, S., and Jackson, T., eds., Caribbean geology: An introduction: Jamaica, University of the West Indies Publisher's Association, p. 265–284.

Burkart, B., and Self, S., 1985, Extension and rotation of crustal blocks in northern Central America and effect on the volcanic arc: Geology, v. 13, p. 22–26, doi: 10.1130/0091-7613(1985)13<22:EAROCB>2.0.CO;2.

Carpenter, R.H., 1954, Geology and ore deposits of the Rosario Mining District and the San Juancito Mountains, Honduras, Central America: Geological Society of America Bulletin, v. 65, p. 23–38, doi: 10.1130/0016-7606(1954)65[23:GAODOT]2.0.CO;2.

Case, J., MacDonald, W., and Fox, P., 1990, Caribbean crustal provinces; seismic and gravity evidence, in Dengo, G., and Case, J.E., eds., The Caribbean Region: Boulder, Colorado, Geological Society of America, The Geology of North America, v. H, p. 15–36.

Cumming, G., and Kesler, S., 1976, Source of lead in Central America and Caribbean mineralization: Earth and Planetary Science Letters, v. 31, p. 262–268, doi: 10.1016/0012-821X(76)90218-1.

Cumming, G., Kesler, S., and Krstic, D., 1981, Source of lead in Central America and Caribbean mineralization: II. Lead isotope provinces: Earth and Planetary Science Letters, v. 56, p. 199–209, doi: 10.1016/0012-821X(81)90127-8.

DeMets, C., Jansma, P., Mattioli, G., Dixon, T., Farina, F., Bilham, R., Calais, E., and Mann, P., 2000, GPS geodetic constraints on Caribbean–North American plate motion: Geophysical Research Letters, v. 27, p. 437–440, doi: 10.1029/1999GL005436.

Dengo, G., 1973, Estructura geológica, historia tectonica y morfologia de America Central (Geologic structure, tectonic history and morphology of Central America): Guatemala City, Instituto Centroamericano de Investigacion Tecnologia Industrial, p. 1–52.

Dengo, G., 1985, Mid America: Tectonic setting for the Pacific margin from southern Mexico to northwestern Columbia, in Nairn, A., and Stehli, F., eds., The Ocean Basins and Margins, vol. 7: New York, Plenum Press, p. 123–180.

Donnelly, T., Horne, G., Finch, R., and López-Ramos, E., 1990, Northern Central America: The Maya and Chortis blocks, in Dengo, G., and Case, J.E., eds., The Caribbean Region: Boulder, Colorado, The Geological Society of America, The Geology of North America, v. H, p. 37–76.

Finch, R., 1981, Mesozoic stratigraphy of central Honduras: AAPG Bulletin, v. 65, p. 1320–1333.

Finch, R.C., and Ritchie, A.W., 1991, The Guayape fault system, Honduras, Central America: Journal of South American Earth Sciences, v. 4, p. 43–60, doi: 10.1016/0895-9811(91)90017-F.

Garayar, J., and Viramonte, J., 1973, Hallazgo de peridotitas en Nicaragua, Informe y trabajos técnicos presentados en la IV Reunión de Geólogos de América Central: Publicaciones Geológicas del ICAITI, p. 31–42.

Gordon, M.B., 1990, Strike-slip faulting and basin formation at the Guayape Fault–Valle de Catacamas Intersection, Honduras, Central America [Ph.D. dissertation]: Austin, Texas, University of Texas at Austin, 259 p.

Gordon, M.B., 1993a, Revised Jurassic and Early Cretaceous (Pre-Yojoa Group) stratigraphy of the Chortis block: Paleogeographic and tectonic implications, in Pindell, J., and Perkins, R., eds., Mesozoic and Early Cenozoic Development of the Gulf of Mexico and Caribbean Region: A Context for

Hydrocarbon Exploration: Austin, Texas, Gulf Coast Section Society of Economic Paleontologists and Mineralogists Foundation, p. 143–154.

Gordon, M., 1993b, Mapa Geológica de Honduras, Hoja de Santa Maria del Real: Tegucigalpa, Honduras, Instituto Geográfico Nacional, escala 1:50,000, 1 sheet.

Gordon, M., and Young, K., 1993, Bathonian and Valanginian fossils from Honduras: Géobios, Mémoire Spécial, n. 15, p. 175–179.

Gordon, M., and Muehlberger, W., 1994, Rotation of the Chortís block causes dextral slip on the Guayape fault: Tectonics, v. 13, p. 858–872, doi: 10.1029/94TC00923.

Harlow, G., Hemming, S., Avé Lallemant, H., Sisson, V., and Sorensen, S., 2004, Two high-pressure-low-temperature serpentinite-matrix mélange belts, Motagua fault zone, Guatemala; a record of Aptian and Maastrichtian collisions: Geology, v. 32, p. 17–20, doi: 10.1130/G19990.1.

Hauff, F., Hoernle, K., Tilton, G., Graham, D., and Kerr, A., 2000, Large volume recycling of oceanic lithosphere over short time scales: Geochemical constraints from the Caribbean Large Igneous Province: Earth and Planetary Science Letters, v. 174, p. 247–263, doi: 10.1016/S0012-821X(99)00272-1.

Horne, G., Atwood, M., and King, A., 1974, Stratigraphy, sedimentology, and paleoenvironment of Esquias Formation of Honduras: AAPG Bulletin, v. 58, p. 176–188.

Hoernle, K., van den Bogaard, P., Werner, R., Lissinna, B., Hauff, F., Alvarado, G., and Garbe-Schoenberg, D., 2002, Missing history (16–71 Ma) of the Galapagos hotspot: Implications for the tectonic and biological evolution of the Americas: Geology, v. 30, p. 795–798, doi: 10.1130/0091-7613(2002)030<0795:MHMOTG>2.0.CO;2.

INETER (Instituto Nicaragüense de Estudios Territoriales), 1995, Mapa Geológica Minero de la Republica de Nicaragua: Managua, Nicaragua, escala 1:500,000, 5 sheets.

Jordan, B.R., Sigurdsson, H., Carey, S., Lundin, S., Rogers, R., Singer, B., and Barquero-Molina, M., 2007, this volume, Petrogenesis of Central American Tertiary ignimbrites and associated Caribbean Sea tephra, in Mann, P., ed., Geologic and tectonic development of the Caribbean plate boundary in northern Central America: Geological Society of America Special Paper 428, doi: 10.1130/2007.2428(07).

Kerr, A., Tarney, J., Nivia, A., Marriner, G., and Saunders, A., 1998, The internal structure of oceanic plateaus: Inferences from obducted Cretaceous terranes in western Colombia and the Caribbean: Tectonophysics, v. 292, p. 173–188, doi: 10.1016/S0040-1951(98)00067-5.

Kesler, S., Levy, E., and Martín, F., 1990, Metallogenic evolution of the Caribbean region, in Dengo, G., and Case, J.E., eds., The Caribbean Region: Boulder, Colorado, The Geological Society of America, The Geology of North America, v. H, p. 459–482.

Kozuch, M., 1991, Mapa Geológica de Honduras: Tegucigalpa, Honduras, Instituto Geográfico Nacional, 1:500,000, 3 sheets.

Leroy, S., Mauffret, A., Patriat, P., and Mercier de Lepinay, B., 2000, An alternative interpretation of the Cayman Trough evolution from a reidentification of magnetic anomalies: Geophysical Journal International, v. 141, p. 539–557, doi: 10.1046/j.1365-246x.2000.00059.x.

Mann, P., 1999, Caribbean sedimentary basins: Classification and tectonic setting from Jurassic to Present, in Mann, P., ed., The Caribbean Basin: Sedimentary Basins of the World series: Amsterdam, Elsevier, v. 4, p. 3–31.

Manton, W., 1996, The Grenville of Honduras: Geological Society of America Abstracts with Programs, v. 28, no. 7, p. 493.

Manton, W., and Manton, R., 1999, The southern flank of the Tela basin, Republic of Honduras, in Mann, P., ed., Caribbean Basins, Sedimentary Basins of the World series: Amsterdam, Elsevier, v. 4, p. 219–236.

Mills, R., and Barton, R., 1996, Geology of the Ahuas area in the Mosquitia Basin of Honduras: Preliminary report: AAPG Bulletin, v. 80, p. 1627–1640.

Mills, R., Hugh, K., Feray, D., and Swolfs, H., 1967, Mesozoic stratigraphy of Honduras: AAPG Bulletin, v. 51, p. 1711–1786.

Mills, R., and Hugh, K., 1974, Reconnaissance geologic map of Mosquitia region, Honduras and Nicaragua Caribbean Coast: AAPG Bulletin, v. 58, p. 189–207.

Nelson, B., Herrmann, U., Gordon, M., and Ratschbacher, L., 1997, Sm-Nd and U-Pb evidence for Proterozoic crust in the Chortis block, Central America: Comparison with the crustal history of southern Mexico: Terra Nova, v. 9, p. 496.

Paz-Rivera, N., 1963, Reconocimiento geológico de la costa Pacífico de Nicaragua: Boletín Servicio Geológico Nacional (Nicaragua), v. 8, p. 69–87.

Pindell, J., and Dewey, J., 1982, Permo-Triassic reconstruction of western Pangea and the evolution of the Gulf of Mexico/Caribbean region: Tectonics, v. 1, p. 179–211.

Pindell, J., and Barrett, S., 1990, Geological evolution of the Caribbean region; a plate-tectonic perspective, in Dengo, G., and Case, J.E., eds., The Caribbean Region: Boulder, Colorado, The Geological Society of America, The Geology of North America, v. H, p. 405–432.

Riller, U., Ratschbacher, L., and Frisch, W., 1992, Left-lateral transtension along the Tierra Colorada deformation zone, northern margin of the Xolapa magmatic arc of southern Mexico: Journal of South American Earth Sciences, v. 5, p. 237–249, doi: 10.1016/0895-9811(92)90023-R.

Ritchie, A.W., and Finch, R.C., 1985, Widespread Jurassic strata on the Chortis block of the Caribbean plate: Geological Society of America Abstracts with Programs, v. 17, no. 7, p. 700–701.

Robertson Research, 1984, A biostratigraphic study of wells in Honduras: Tegucigalpa, Honduras, unpublished report to Dirección General de Minas e Hidrocarburos, 112 p.

Rockwell, D., 1985, Interpretation of marine seismic data, Gracias a Dios platform, offshore northeast Honduras: Tegucigalpa, Honduras, unpublished report to Direccion General de Minas e Hidrocarburos, 13 p.

Rogers, R., 1995, Mapa geológica de Honduras, hoja de Valle de Jamastran: Tegucigalpa, Honduras, Instituto Geográfico Nacional, escala 1:50,000, 1 sheet.

Rogers, R., 1996, Geología a lo largo del Río Patuca y Wampu, La Mosquitia, Honduras: Tegucigalpa, Honduras, Instituto Geográfico Nacional, Revista, v. 4, p. 86–106.

Rogers, R., 2003, Jurassic–Recent tectonic and stratigraphic history of the Chortis block of Honduras and Nicaragua (northern Central America) [Ph.D. dissertation]: Austin, Texas, University of Texas at Austin, 264 p.

Rogers, R., and O'Conner, E., 1993, Mapa Geológica de Honduras: Hoja de Tegucigalpa (segunda edición): Tegucigalpa, Honduras, Instituto Geográfico Nacional, escala 1:50,000, 1 sheet.

Rogers, R., Kárason, H., and van der Hilst, R., 2002, Epeirogenic uplift above a detached slab in northern Central America: Geology, v. 30, p. 1031–1034, doi: 10.1130/0091-7613(2002)030<1031:EUAADS>2.0.CO;2.

Rogers, R., and Mann, P., 2007, this volume, Transtensional deformation of the western Caribbean-North America plate boundary zone, in Mann, P., ed., Geologic and tectonic development of the Caribbean plate boundary in northern Central America: Geological Society of America Special Paper 428, doi: 10.1130/2007.2428(03).

Rogers, R., Mann, P., and Emmet, P., 2007, this volume, Tectonic terranes of the Chortis block based on integration of regional aeromagnetic and geologic data, in Mann, P., ed., Geologic and tectonic development of the Caribbean plate boundary in northern Central America: Geological Society of America Special Paper 428, doi: 10.1130/2007.2428(04).

Rogers, R.D., Mann, P., Scott, R., and Patino, L., 2007, this volume, Cretaceous intra-arc rifting, sedimentation and basin inversion in east-central Honduras, in Mann, P., ed., Geologic and tectonic development of the Caribbean plate boundary in northern Central America: Geological Society of America Special Paper 428, doi: 10.1130/2007.2428(05).

Rosencrantz, E., Ross, M., and Sclater, J., 1988, Age and spreading history of the Cayman Trough as determined from depth, heat flow, and magnetic anomalies: Journal of Geophysical Research, v. 93, p. 2141–2157.

Rosenfeld, J.H., 1981, Geology of the Western Sierra de Santa Cruz, Guatemala, Central America: An ophiolite sequence [Ph.D. dissertation]: Binghamton, New York, State University of New York at Binghamton, 313 p.

Schaaf, P., Morán-Zenteno, D., Hernández-Bernal, M., Solís-Pichardo, G., Tolson, G., and Kohler, H., 1995, Paleogene continental margin truncation in southwestern Mexico: Geochronological evidence: Tectonics, v. 14, no. 6, p. 1339–1350, doi: 10.1029/95TC01928.

Scott, R., and Finch, R., 1999, Cretaceous carbonate biostratigraphy and environments in Honduras, in Mann, P., ed., Caribbean Basins, Sedimentary Basins of the World series: Amsterdam, Elsevier, v. 4, p 151–166.

Simonson, B., 1977, Mapa Geológico de Honduras, El Porvenir: Tegucigalpa, Honduras, Instituto Geográfico Nacional, escala 1:50,000, 1 sheet.

Sinton, C., Duncan, R., and Denyer, P., 1997, Nicoya Peninsula, Costa Rica: A single suite of Caribbean oceanic plateau magmas: Journal of Geophysical Research, v. 102, p. 15,507–15,520, doi: 10.1029/97JB00681.

Southernwood, R., 1986, Late Cretaceous limestone clast conglomerates of Honduras [M.S. thesis]: Dallas, Texas, University of Texas at Dallas, 300 p.

Sundblad, K., Cumming, G., and Krstic, D., 1991, Lead isotope evidence for the formation of epithermal gold quartz veins in the Chortis block, Nicaragua:

Economic Geology and the Bulletin of the Society of Economic Geologists, v. 86, p. 944–959.

Tardy, M., Lapierre, H., Freydier, C., Coulon, C., Gill, J., Mercier de Lepinay, B., and Beck, C., Martinez-R., Talavera-M., Ortiz-H., E., Stein, G., Bourdier, J., and Yta, M., 1994, The Guerrero suspect terrane (western Mexico) and coeval arc terranes (the Greater Antilles and the Western Cordillera of Colombia): A late Mesozoic intra-oceanic arc accreted to cratonal America during the Cretaceous: Tectonophysics, v. 230, p. 49–73.

Venable, M., 1994, A geological, tectonic, and metallogenetic evaluation of the Siuna terrane (Nicaragua) [Ph.D. dissertation]: Tucson, Arizona, University of Arizona, 154 p.

Viland, J., Henry, B., Calix, R., and Diaz, C., 1996, Late Jurassic deformation in Honduras: Proposals for a revised regional stratigraphy: Journal of South American Earth Sciences, v. 9, p. 153–160, doi: 10.1016/0895-9811(96)00002-8.

Walther, C., Flueh, E., Ranero, C., von Huene, R., and Strauch, W., 2000, Crustal structure across the Pacific margin of Nicaragua; evidence for ophiolitic basement and a shallow mantle sliver: Geophysical Journal International, v. 141, p. 759–777, doi: 10.1046/j.1365-246x.2000.00134.x.

Weiland, T., Suayah, I., and Finch, R., 1992, Petrologic and tectonic significance of Mesozoic volcanic rocks in the Río Wampú area, eastern Honduras: Journal of South American Earth Sciences, v. 6, p. 309–325, doi: 10.1016/0895-9811(92)90049-5.

Zoppis Bracci, L., and del Giudice, 1958, Un reconocimiento geológico del Río Bocay y parte del Río Coco: Boletín del Servicio Geológico Nacional de Nicaragua, v. 2, p. 81–102.

MANUSCRIPT ACCEPTED BY THE SOCIETY 22 DECEMBER 2006

Petrogenesis of Central American Tertiary ignimbrites and associated Caribbean Sea tephra

B.R. Jordan*
Department of Geology, Brigham Young University–Idaho, Rexburg, Idaho 83460, USA

H. Sigurdsson
S. Carey
Graduate School of Oceanography, University of Rhode Island, Narragansett, Rhode Island 02882, USA

S. Lundin
Department of Earth Atmospheric and Planetary Science, Massachusetts Institute of Technology, 77 Massachusetts Ave., Cambridge, Massachusetts 02139, USA

R.D. Rogers
Department of Geology, California State University, 801 W. Monte Vista Ave., Turlock, California 95382, USA

B. Singer
Department of Geology and Geophysics, University of Wisconsin, 1215 W. Dayton St., Madison, Wisconsin 53706, USA

M. Barquero-Molina
Department of Geosciences, Jackson School of Geosciences, University of Texas at Austin, 1 University Station C1100, Austin, Texas 78712, USA

ABSTRACT

Ignimbrites, as widespread sheets tens of meters thick, form the Central American Tertiary Ignimbrite Province. Geochemical data were collected from 99 Cenozoic marine ash layers within Caribbean Sea sediments, 76 vitrophyres, and 21 mafic lavas from Nicaragua and Honduras. Two major eruptive periods, one in the Eocene and one in the Miocene, have been broadly identified. $^{40}Ar/^{39}Ar$ laser fusion ages, determined from sanidine or plagioclase in 10 of the vitrophyre samples, have been interpreted to indicate that the bulk of the younger group of ignimbrites formed largely in the middle Miocene during a 3.5-m.y. period between 16.9 and 13.4 Ma. Modeling indicates that initial melts were from a normal mid-oceanic-ridge basalt (N-MORB)–type source, rather than the enriched mid-oceanic-ridge basalt (E-MORB)–type source postulated for the modern arc. All of the ignimbrites analyzed have $^{87}Sr/^{86}Sr$ isotope

---

*oceanographer@byu.net

Jordan, B.R., Sigurdsson, H., Carey, S., Lundin, S., Rogers, R.D., Singer, B., and Barquero-Molina, M., 2007, Petrogenesis of Central American Tertiary ignimbrites and associated Caribbean Sea tephra, in Mann, P., ed., Geologic and tectonic development of the Caribbean plate boundary in northern Central America: Geological Society of America Special Paper 428, p. 151–179, doi: 10.1130/2007.2428(07). For permission to copy, contact editing@geosociety.org. ©2007 The Geological Society of America. All rights reserved.

values ($^{87}Sr/^{86}Sr$ = 0.7040–0.7069) within the range of continental crust. Trace element trends are similar to those estimated for lower continental crust. Assimilation–fractional crystallization and melt mixing models produce trends that are consistent with ignimbrite compositions. This evidence is consistent with a large influence of continental crust in the ignimbrite formation. In addition, the ignimbrite magmas, like those of the modern arc, have also been determined to have been contaminated by sediment-derived fluids. Abnormally rapid subduction of the Farallon-Cocos plate, which coincides with the formation of the ignimbrites, may have resulted in their generation. A slab gap that currently exists beneath the modern arc may be the cause of a change in source from N-MORB to E-MORB by allowing rising asthenospheric material to "recharge" the mantle wedge in trace elements that had been depleted prior to the gap formation.

Keywords: Caribbean Sea, Central America, ignimbrites, petrogenesis, assimilation, slab detachment.

INTRODUCTION

The Central American Tertiary Ignimbrite Province

Ignimbrites are deposits of volcanic rocks produced by pyroclastic flows that occur during explosive eruptions of volatile-rich, siliceous magmas (Jackson, 1997). Large-scale explosive eruptions of silicic magma can erupt volumes on the order of tens to thousands cubic kilometers and form large geographic provinces of ignimbrites and associated volcanic deposits.

In Central America, ignimbrites are found as sheets tens of meters in thickness, which form the Central American Tertiary Ignimbrite Province (Fig. 1). The Central American Tertiary Ignimbrite Province extends from the -Mexican-Guatemalan border to southern Nicaragua and forms the Central American highlands (Williams et al., 1964; McBirney and Williams, 1965; Williams et al., 1969; Weyl, 1980). The entire area of the province is 170,000 km^2, covering up to 22% of Central America. In Honduras alone, the volume of the ignimbrite formation is estimated as 10,000 km^3 (Pushkar, 1972; Weyl, 1980). Two eruptive episodes have been recognized, one in the Eocene and one in the Miocene (Sigurdsson et al., 1997; Jordan et al., 2006).

The purpose of this paper is to discuss the petrogenetic and possible tectonic processes that have led to the production of the Miocene magmas, with specific examples modeling the formation of Jordan et al.'s (2006) geochemical ignimbrite groups. In particular, the role of mantle melts, fractional crystallization, and continental crustal melting and assimilation is evaluated. A working model for the Miocene ignimbrite petrogenesis is developed, which suggests that the ignimbrites were generated during a period of rapid subduction in which melts from a depleted normal mid-oceanic-ridge basalt (N-MORB)–like source assimilated continental crust and fractionally crystallized prior to eruption. The enhanced mid-oceanic-ridge basalt (E-MORB)–like source of the modern arc is a result of upper mantle contamination that occurred after subduction slowed.

GEOLOGIC SETTING AND BACKGROUND

The Modern Arc

The present Central American Volcanic Arc lies along the western edges of Guatemala, El Salvador, Honduras, Nicaragua, and Costa Rica (Fig. 1). The chain of volcanoes that make up the arc extends over 1,100 km from the Mexican-Guatemalan border to Panama (Reynolds, 1980; Prosser and Carr, 1987; Elming, 1998). The presence of the arc is associated with the subduction of the Cocos plate northeastward beneath the North American and Caribbean Plates at ~7 cm/yr.

There is evidence for a detached segment of slab beneath Central America (Tatsumi and Eggins, 1995; Rogers et al., 2002). The upper part of the detached portion lies at a depth of 500 km and the gap between this and the currently subducting slab extends from ~300 km to this 500 km depth. The detachment has been estimated to have taken place between 10 and 4 Ma (Rogers et al., 2002).

Tectonics and Paleoarc History

The terrestrial volcanic rocks analyzed in this study come from one of the three major geologic provinces of Central America. This province is the Chortis block (underlying southern Guatemala, El Salvador, Honduras, northern Nicaragua, and the western Nicaraguan Rise) (Donnelly et al., 1990). The Chortis block has a Paleozoic, and possibly older, continental basement made up of igneous and metamorphic rock (Gose, 1983; Case et al., 1984; Donnelly et al., 1990; Rogers, Mann, and Emmet, this volume).

Previous work on the Central American arc has shown that tephra layers in sediments recovered in drill cores during the Ocean Drilling Program (ODP) Leg 165 in the Caribbean Sea can be geochemically correlated with the ignimbrites of Nicaragua and Honduras. This work classified the marine and land glasses into 14 distinctive geochemical groups based on their rare earth element (REE) signatures (Jordan et al., 2006).

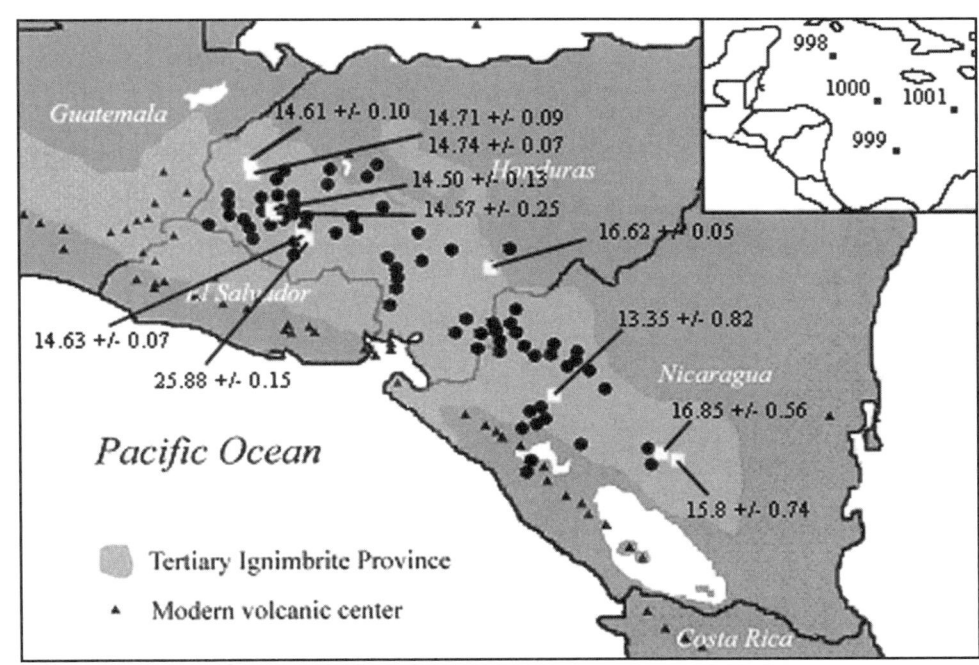

Figure 1. The Tertiary ignimbrite province in Central America. The modern arc is denoted by the linear array of triangles, representing active volcanic centers. Closed circles represent geochemical sample locations. Open squares represent dated sample localities. All dates are Ma. The light gray area is the Tertiary Ignimbrite Province.

As recorded in the ODP cores, two major eruptive episodes occurred in Central America from the Eocene to the late Miocene (Sigurdsson et al., 1997). The first of these major volcanic episodes began in the early Eocene, ca. 47 Ma, and then ended in the early Oligocene, ca. 30 Ma. The second episode began at the end of the Oligocene, ca. 24 Ma, and then ended in the mid- to late Miocene, ca. 12 Ma (Fig. 2) (Williams and McBirney, 1969; Sigurdsson et al., 1997).

The initiation of the second ignimbrite flare-up in the late Oligocene–early Miocene seems to coincide with a change in the direction of motion of the Farallon plate subducting beneath Central America, when the Farallon-Pacific spreading center was subducted. At this time, a new spreading center was formed that divided the Farallon plate into the Nazca and Cocos plates ca. 23–25 Ma (Meschede and Barckhausen, 2001). Just afterward, ca. 18–11 Ma, and possibly earlier, there was a pulse of extremely rapid seafloor spreading between the Cocos, Nazca, and Pacific plates. Wilson (1996) found that the half-spreading rate during this period was between 9 and 10.5 cm/yr compared to a modern (0–3 Ma) subduction rate of ~7 cm/yr. The exact start of this rapid spreading is difficult to determine, but an upper bound may have been as early as 25 Ma, which would coincide with the reorganization of the plates (Fig. 2) (Wilson, 1996; Sigurdsson et al., 2000).

Field and Petrographic Observations

The ignimbrites are primarily weakly to moderately welded, but are often indurated, or hardened by post-emplacement processes. The deposits vary in color, and it is thought that the reddish tints are due to fumarolic activity, while the greens and yellows are a result of alteration to clay minerals (Weyl, 1980). Most deposits are poorly sorted and massive, but bedding and surge deposits with fine-scale stratification and cross bedding are not uncommon within individual ignimbrite layers. Hydrated, pumice-rich fall layers are also common. Some of the vitrophyres at the base of individual layers of ignimbrites are columnar jointed. Individual ignimbrite layers vary from less than a meter to >15 m in thickness, and it is quite common to find multiple distinct ignimbrite layers overlying each other, with good examples near La Entrada and Gracias, Honduras.

Basalts and andesites are common in the modern Central American Volcanic Arc (Williams et al., 1969), with rhyolites and dacites occurring rarely and only at some localities in Costa Rica and Panama (Johnston and Thorkelson, 1997). However, the Miocene ignimbrite vitrophyres of Honduras and Nicaragua are predominantly alkali rhyolite, rhyolite, and rhyodacite in composition. Basaltic andesite and andesitic lavas are found interbedded with the ignimbrite layers, but are not extensive. A wide variety of breccias and lahars are common (McBirney and Williams, 1965). The layers are also cut by basaltic andesite, andesitic, and rhyolitic dikes, which are often columnar jointed—perpendicular to the contact between the dikes and the ignimbrite layers. The ignimbrites are also extensively faulted.

The vast majority of both the marine ash layers and the terrestrial ignimbrite samples contain plagioclase (both zoned and unzoned), clinopyroxene and/or orthopyroxene, and Fe-Ti oxides. Some samples also include K-feldspar, biotite, amphibole, and quartz. Accessory minerals are present in some ash layers and ignimbrites, but other than apatite, which is found in a few of the ignimbrite samples, they were not identified.

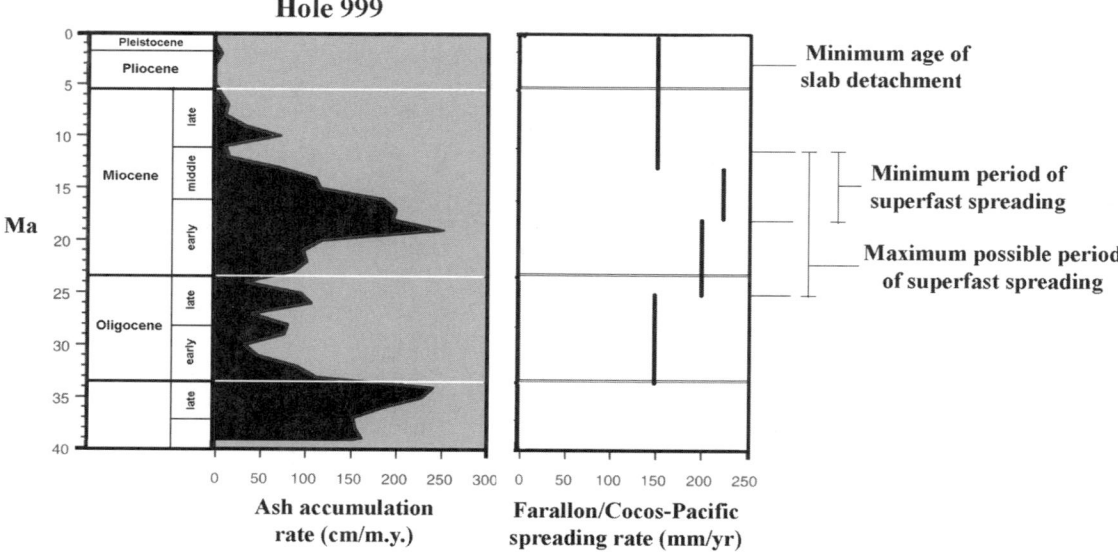

Figure 2. Age estimates for slab detachment (Rogers et al., 2002) and period of super-fast spreading at the Miocene Cocos–Pacific plate boundary (Wilson, 1996) in relation to volcanic ash layer accumulation rates in Ocean Drilling Program Leg 165 (Hole 999) Caribbean sediments (Sigurdsson et al., 1997) and the Central American Tertiary ignimbrite flare-up.

All of the basaltic andesites, andesites, and dacites of this study also contain plagioclase phenocrysts. Most of these lavas also contain phenocrysts of orthopyroxene and clinopyroxene. Olivine is also present in some. All but two of the basaltic andesites, andesites, and dacites have a dark, aphanitic matrix. Sample Hon-56, an andesite, also has an aphanitic matrix, but is purplish in color and has very large Fe-Ti oxide phenocrysts. Some of its plagioclase is chloritized, which is true in a few of the other samples as well (Hon-42, Hon-58, Hon-76, and Hon-107). Sample Hon-82 is a light gray dacite. In addition to the chloritization of some samples, several also contain zeolites (Hon-5, Hon 59, Hon-76, and Hon-140). Sample Hon-35 is an unusual basaltic andesite, containing very large plagioclase phenocrysts, possibly labradorite, some of which are up to 2 cm in length.

GEOCHEMISTRY

Samples

Geochemical data were collected for a total of 99 marine volcanic ash layers taken from within ODP Leg 165 Caribbean sediment cores recovered from sites 998a, 999a, 999b, 1000a, and 1000b. Seventy-six terrestrial samples, including whole-rock and glass from vitrophyres, glassy or vitric clasts, and fiamme were collected, predominately from outcrops, as well as from a few detached blocks of ignimbrites in Nicaragua and Honduras. Whole rock analysis of these glasses was used to determine magma compositions because few unaltered pumice samples were found.

In addition, six Honduran continental crustal rocks were obtained. They are likely Paleozoic (orthogneisses and gneiss) and Late Cretaceous or early Tertiary (granites) in age (Table 1). Granite sample 2-22-00A and orthogneiss sample 3-31-00F were collected from the western edge of the Chindona batholith, a nonmetamorphosed granitoid. It is the largest (1200 km^2) of several batholiths in north-central Honduras. The batholith contains metamorphic xenoliths, but data were not obtained for these. Orthogneiss sample 3-09-00A is granitic in appearance. Sample 3-31-00A is a high-grade banded gneiss common to the north coast of Honduras and may reflect deeper crustal levels than the other basement samples, including 1-26-01C.

Major Oxides

Whole-rock analysis was performed on 52 of the vitrophyric samples and 21 basaltic andesite, andesite, and dacite samples using X-ray fluorescence spectrometry (XRF) (Table 2), revealing a calc-alkaline trend. The whole rock compositions range from basalt to rhyolite, but show a somewhat bipolar distribution with the majority of samples clustering either in the basaltic-andesite field or the dacite to rhyolite (mostly rhyolite) fields. Only four samples classify as andesite (Fig. 3).

Plots of oxide data from both glass and bulk rock show typical trends of decreasing wt% values of most oxides with increasing SiO_2 wt% and increasing K_2O wt% with increasing SiO_2 wt%. The only oxide that does not show the same trend is Na_2O, which shows a flat trend in the bulk rock (Fig. 4). This is very similar to southern Sierra Madre ignimbrites (Webber et al., 1994).

TABLE 1. WHOLE-ROCK OXIDE AND TRACE ELEMENT DATA FOR AVERAGE CONTINENTAL CRUST AND HONDURAS CONTINENTAL CRUST ROCKS USED IN THIS STUDY*

Sample	Lower continental arc crust[†]	Upper continental crust[§]	Lower continental crust[§]	3-09-00A (orthogneiss)	3-31-00A (gneiss)	3-31-00F (orthogneiss)	1-26-01C (granite)	2-22-00A (granite)
SiO_2 (wt%)	50.60	61.5		78.11	69.71	66.49	75.63	68.98
TiO_2	0.90	0.68		0.55	0.64	0.64	0.36	0.46
Al_2O_3	16.80	15.1		10.94	15.13	16.39	13.73	16.29
Fe_2O_3	8.80*	6.28		3.37	3.45	4.09	2.41	3.37
MnO	0.10	0.1		0.04	0.15	0.06	0.03	0.07
MgO	7.70	3.7		0.92	0.93	2.00	0.63	1.23
CaO	10.10	5.5		0.46	3.14	4.50	0.17	3.16
Na_2O	2.50	3.2		2.21	3.10	3.90	4.12	3.87
K_2O	0.40	2.4		3.25	3.65	1.78	2.83	2.42
P_2O_5	0.10	0.18		0.15	0.10	0.16	0.08	0.14
Rb (ppm)	7.00	110	41	125.00	132.80	39.10	59.20	99.30
Ba	228.00	668	568	229.54	485.91	453.32	1098.27	727.24
K	3320.60	28650	13140	26564.80	29968.42	14610.64	22995.16	19757.57
Th	0.70	10.3	6.6	19.93	7.08	5.66	8.57	6.82
U	0.10	2.5	0.93	5.11	2.40	2.21	2.18	4.76
Nb	5.00	26	11.3	22.98	17.81	13.65	9.78	14.48
Ta	0.60	1.5	0.84	2.04	1.05	0.96	2.88	1.38
La	7.00	32.3	26.8	29.28	27.20	24.48	18.20	17.86
Ce	16.00	65.7	53.1	71.03	72.16	52.99	47.21	42.50
Pb	2.80	17	12.5	28.62	30.41	7.76	4.17	30.79
Sr	357.00	316	352	54.90	421.77	428.58	55.19	208.98
Nd	9.20	25.9	28.1	30.30	37.17	18.91	16.02	17.55
Hf	1.50	5.8	4	6.24	9.10	3.93	5.17	3.11
Zr	57.00	237	165	246.00	399.00	171.00	191.00	126.00
Ti	5395.63	3117	5010	3237.38	3776.94	3776.94	2098.30	2697.81
Yb	1.40	1.5	2.5	4.56	5.71	1.80	2.95	2.30
Y	15.00	20.7	27.2	44.68	53.09	12.84	20.15	20.80

Note: See Jordan et al. (2006) for sample locations.
*Oxide values for upper and lower continental crust were unavailable, so total continental crustal values were used.
[†]Data from Rudnick and Fountain (1995).
[§]Data from Wedepohl (1995).

Trace Element Data

Of the intermediate and mafic samples, six basaltic andesites, two andesites, and two dacites were also analyzed by solution inductively coupled plasma–mass spectrometry (ICP-MS) (Table 3). The trace element concentrations for the ignimbrite samples used in this study were obtained using laser ablation (LA)–ICP-MS (see Jordan et al., 2006) for data tables). All ICP-MS work was done at the Boston University Department of Earth Sciences. The spot-to-spot precision for the LA-ICP-MS mineral analysis was better than 3% relative standard deviation (RSD) for Sc, Cr, Co, Zn, Sr, Y, Zr, Nb, Hf, Ba, and the REE. RSD spot-to-spot precision for Ta, Th, Pb, and U was 5%–10%. A complete description of sample preparation and analytical procedures can be found in Jordan et al. (2006).

Modeling and Partition Coefficient Selection

Since the ignimbrite province is dominated by evolved silicic magmas, modeling focused primarily on fractional crystallization (FC) and assimilation-fractional crystallization (AFC). However, the large volume of ignimbrites in Central America, complex zoning of minerals, and xenocrysts commonly found in the ignimbrites, including the vitrophyres, also indicate magma mixing, so AFC and continental crustal melting and mixing models were also used. FC, AFC, and mixing modeling were performed using Igpet for Windows© version 6 (Carr, 2003). Values for r varied from 0.3 to 1, similar to results in Reiners et al. (1995) for late stage AFC models.

The greatest difficulty in modeling the Central American ignimbrite petrogenesis was choosing proper bulk partition

TABLE 2. X-RAY FLUORESCENCE SPECTROMETRY DATA FOR CENTRAL AMERICAN SILICIC IGNIMBRITES AND MAFIC LAVAS USED IN THIS STUDY

Sample	Decimal Latitude	Decimal Longitude	SiO_2	TiO_2	Al_2O_3	FeO*	MnO	MgO	CaO	Na_2O	K_2O	P_2O_5
Ignimbrites												
HON 1	14.8591	−87.7996	70.54	0.44	15.35	2.82	0.07	0.66	2.20	2.84	4.94	0.13
HON 2	14.8528	−87.8154	70.37	0.42	15.34	2.97	0.08	0.56	2.21	4.15	3.77	0.12
HON 24	14.5489	−88.5259	76.85	0.15	12.56	1.67	0.08	0.05	0.31	2.67	5.66	0.00
HON 25	14.5134	−88.5159	76.59	0.14	12.47	1.68	0.08	0.04	0.31	3.14	5.55	0.00
HON 29	14.3967	−88.4181	74.38	0.34	14.46	1.39	0.06	0.18	0.80	2.15	6.20	0.04
HON 32	14.2488	−88.5102	73.80	0.35	14.25	1.71	0.09	0.30	1.15	3.14	5.14	0.07
HON 34	14.1178	−88.5389	73.41	0.38	14.04	1.86	0.09	0.42	1.32	3.43	4.97	0.08
HON 55	14.4417	−88.6911	66.11	0.77	15.48	6.00	0.13	0.97	4.30	3.90	2.13	0.21
HON 57	14.4653	−88.6746	76.46	0.15	13.17	1.68	0.10	0.06	1.27	3.57	3.52	0.03
HON 59	14.4625	−88.6732	61.77	1.25	14.80	9.20	0.50	1.13	4.03	3.50	3.34	0.47
HON 76	14.5778	−89.0702	62.88	1.32	17.15	5.43	0.10	1.14	4.98	3.69	2.96	0.35
HON 77	14.5778	−89.0702	74.93	0.22	13.71	1.26	0.06	0.33	1.19	3.44	4.82	0.05
HON 82	14.2563	−88.8575	65.84	0.98	16.94	4.92	0.13	1.03	2.32	6.96	0.44	0.43
HON 83	14.2929	−88.8672	73.28	0.30	14.42	2.64	0.26	0.32	2.46	3.61	2.64	0.07
HON 87	14.3775	−88.9184	75.57	0.16	13.32	1.74	0.05	0.05	0.83	3.15	5.12	0.01
HON 91	14.4192	−88.4247	74.57	0.34	14.36	1.43	0.06	0.16	0.74	3.12	5.20	0.02
HON 92	14.3972	−88.4152	74.36	0.33	14.60	1.42	0.05	0.14	0.77	2.10	6.18	0.05
HON 98	14.2269	−88.0415	75.11	0.29	13.90	1.35	0.09	0.29	0.99	3.38	4.54	0.06
HON 100	14.3767	−88.0833	72.17	0.51	14.50	2.48	0.09	0.41	1.63	2.16	5.96	0.10
HON 102	14.3621	−88.0095	71.76	0.51	14.58	2.55	0.12	0.42	1.65	2.90	5.40	0.10
HON 104	14.5296	−87.7907	74.31	0.35	14.28	1.54	0.07	0.18	1.22	3.03	4.99	0.02
HON 106	14.2869	−87.4904	75.71	0.18	12.94	2.03	0.06	0.12	1.38	2.97	4.60	0.02
HON 134	13.6821	−87.7274	71.81	0.44	14.72	2.00	0.07	0.55	1.48	2.26	6.59	0.07
HON 140	13.5057	−87.1212	69.39	0.63	15.57	3.00	0.12	0.75	2.19	3.73	4.49	0.13
NIC 39	12.3052	−85.5072	68.97	0.87	15.44	3.88	0.14	1.05	4.40	3.45	1.59	0.21
NIC 41	12.3297	−85.5167	68.99	0.84	15.51	3.84	0.15	1.26	3.18	4.42	1.59	0.20
NIC 69	12.5212	−86.0521	61.69	0.78	16.45	6.31	0.16	2.04	4.93	3.53	3.42	0.69
NIC 72	12.9106	−85.9189	74.15	0.27	14.04	1.71	0.07	0.50	1.88	3.96	3.35	0.07
NIC 80	13.2416	−86.1457	75.44	0.12	13.97	1.01	0.08	0.19	1.51	3.31	4.35	0.03
NIC 91	12.6824	−86.4233	72.98	0.58	13.69	2.82	0.11	0.46	1.70	2.53	5.03	0.09
NIC 93	12.6784	−86.4268	66.38	1.03	15.46	5.67	0.19	1.48	3.80	3.54	2.09	0.35
NIC 98	12.5014	−86.7007	49.70	0.75	18.03	10.09	0.20	6.69	11.90	2.13	0.41	0.10
NIC 111	11.9892	−86.1667	51.69	1.05	15.63	11.50	0.23	5.29	10.28	2.96	1.11	0.25
NIC 118	13.4750	−86.6961	66.18	0.90	15.84	4.68	0.11	1.41	3.26	2.50	4.85	0.27
NIC 119	13.4290	−86.7137	67.54	0.78	15.94	3.50	0.13	1.18	2.88	2.39	5.44	0.22
NIC 120	13.4539	−86.6741	68.77	0.44	15.67	3.28	0.10	0.97	3.15	3.21	4.24	0.18
Basaltic andesites												
HON 10	15.1266	−88.2100	52.71	1.67	16.99	8.42	0.16	4.24	9.15	3.37	1.61	0.75
HON 35	14.1069	−88.5384	52.96	1.86	17.08	8.98	0.21	3.70	8.38	3.48	1.59	0.75
HON 50	14.4050	−88.7287	53.03	1.57	17.56	8.90	0.13	4.19	8.38	3.27	1.26	0.71
HON 51	14.4066	−88.7285	52.33	1.52	17.13	8.78	0.16	5.16	8.70	3.29	1.26	0.69
HON 78	14.1949	−88.7977	52.68	1.57	16.76	8.76	0.19	5.16	8.58	3.25	1.26	0.82
HON 79	14.2174	−88.8303	53.23	1.58	16.70	8.79	0.18	4.95	8.31	3.15	1.30	0.84
HON 88	14.8614	−89.1208	52.57	1.30	18.30	9.03	0.16	4.30	8.47	3.81	0.84	0.21
HON 96	14.2538	−88.0511	52.61	1.37	17.60	8.42	0.16	5.65	8.56	3.19	1.10	0.39
HON 103	14.5587	−87.9691	52.25	1.67	16.87	8.92	0.20	5.62	8.45	2.96	1.36	0.71
HON 107	14.2648	−87.3961	52.44	1.16	18.01	8.20	0.17	5.37	9.09	3.05	1.16	0.42
HON 123	14.0262	−87.1979	52.41	1.05	18.14	8.32	0.18	4.94	9.09	2.80	1.75	0.39
NIC 26	12.1921	−85.0417	54.28	1.03	18.52	8.07	0.20	3.94	8.51	3.18	1.09	0.26
Andesites												
HON 5	14.7563	−88.1856	58.76	1.31	16.83	5.53	0.26	4.10	6.88	3.28	1.96	0.48
HON 42	14.7276	−88.4469	58.27	1.26	18.14	6.06	0.10	2.37	6.89	3.53	2.35	0.36
HON 56	14.4417	−88.6911	59.90	1.31	16.80	6.44	0.14	1.78	5.80	4.03	2.67	0.40
HON 58	14.4653	−88.6746	60.26	1.22	18.03	5.95	0.18	1.13	6.11	3.57	2.51	0.39

Note: See Jordan et al. (2006) for sample locations.
*Total iron calculated as FeO from original Fe_2O_3 values using wt% Fe_2O_3 × 0.8998122.

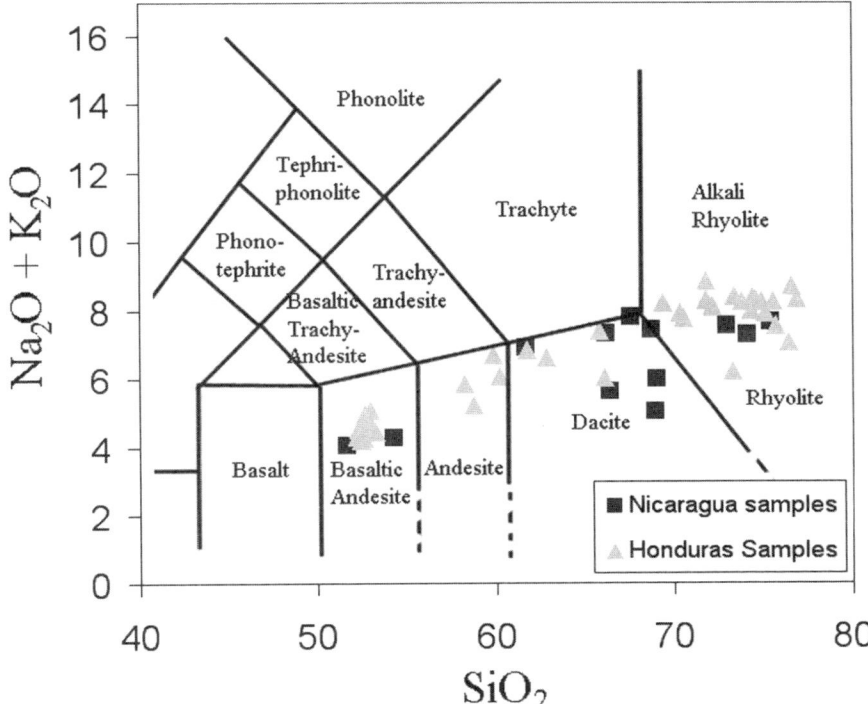

Figure 3. International Union of Geological Sciences classification of whole rock samples collected from Central American ignimbrites.

coefficients. Values for individual partition coefficients were calculated between each mineral and matrix. These values were, in turn, used to calculate bulk partition coefficients based on the modal proportion of each phase in the sample. For most samples, the modal proportions were suspect because of sizable proportions of possible xenocrysts within the vitrophyres. Fortunately, the sample Nic-120, a fresh-looking, glassy vitrophyre, had mineral-melt partition coefficients close to other accepted values for plagioclase and clinopyroxene in silicic rocks. In the models of this study, partition coefficient values for clinopyroxene and plagioclase were thus taken from Nic-120. However, since partition coefficients were not obtainable for all of the mineral phases present in Nic-120, some values were substituted from the Bishop Tuff (Michael, 1983) and the Sierra Madre Occidental of Mexico (Cameron and Cameron, 1986) (Table 4).

Because bulk partition coefficients were not obtained for the Central American basaltic andesites, appropriate bulk partition coefficients were taken from Verma (2001) (Table 4), which are representative of a basalt from Los Humeros, Mexico. The Los Humeros basalt has a mineralogy of ~35% olivine, 30% plagioclase, 30% clinopyroxene, and 5% opaque minerals.

In the modeling, E-MORB and N-MORB were used as primary magmas, and their oxide values were taken from Schilling et al. (1983). E-MORB and N-MORB trace element values were taken from Sun and McDonough (1989). Average upper and lower continental crust values were taken from Wedepohl (1995), additional lower continental arc crust values were taken from Rudnick and Fountain (1995), and carbonate and hemipelagic sediment compositions were taken from Patino et al. (2000) (Table 5).

Sr-Isotope Analysis

Powders of 6 glass samples (Hon-102, Hon-106, Hon-134, Hon-140, Nic-72, and Nic-120) were analyzed for Sr-isotope data at the Isotope Laboratory of the Scripps Institution of Oceanography. $^{87}Sr/^{86}Sr$ values for the ignimbrites are between mantle and continental crustal values (0.7030–0.7100). They have a high value of 0.7069 in Hon 106 and a low value of 0.7040 in Nic-72 (Table 6).

The earliest isotopic study on the Central American Tertiary Ignimbrite Province was by Pushkar et al. (1972), who found that $^{87}Sr/^{86}Sr$ isotope ratios in the rhyolites ranged from 0.7035 to 0.7145. They concluded that these values are much higher than the other calc-alkaline rocks of the ignimbrite province. Although these Sr isotope ratio values overlap continental crustal values of rocks underlying the ignimbrites, it was concluded that no one model of partial or total melting, or crustal assimilation, completely explained the entire Central American ignimbrite petrogenesis.

The $^{87}Sr/^{86}Sr$ values of all but one of the Honduras samples in this study are higher than those of Nicaragua (0.7041–0.7069 vs. 0.7040–0.7041). Honduras is underlain by continental crust that is ~45 km thick, similar to that underlying Guatemala, while

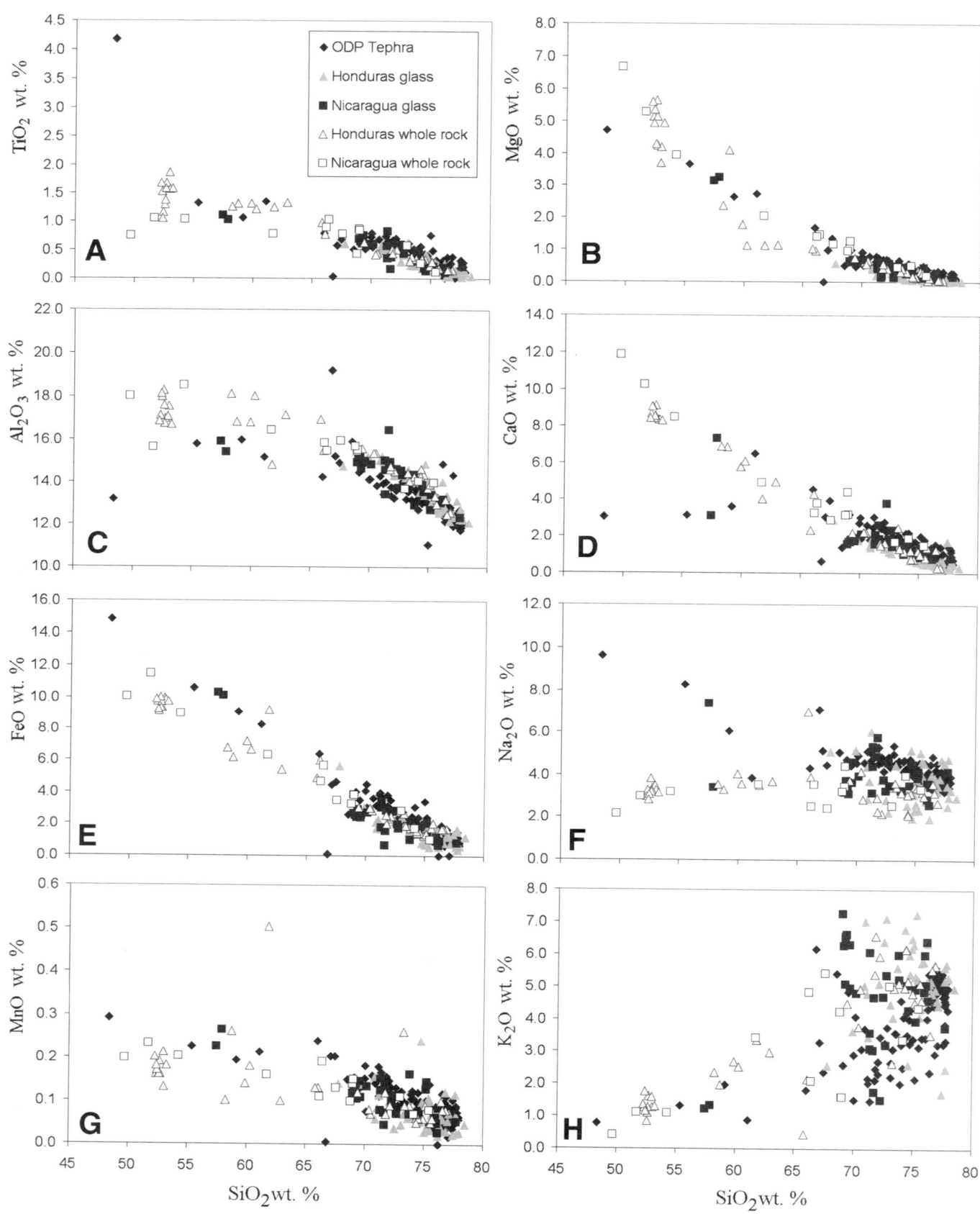

Figure 4. Oxide trends versus SiO_2 of glass and whole rock analyses for Miocene ignimbrites and Caribbean Sea tephra. The symbols for each plot are the same as those given in A. (A) TiO_2 versus SiO_2. (B) MgO versus SiO_2. (C) Al_2O_3 versus SiO_2. (D) CaO versus SiO_2. (E) FeO versus SiO_2. (F) Na_2O versus SiO_2. (G) MnO versus SiO_2. (H) K_2O versus SiO_2.

TABLE 3. INDUCTIVELY-COUPLED PLASMA–MASS SPECTROMETRY DATA USED IN THIS STUDY

Sample	La	Ce	Pr	Nd	Sm	Eu	Gd	Tb	Ho	Yb	Lu
Basement rocks											
3-09-00A	29.28	71.03	8.1	30.3	6.92	0.52	7.23	1.19	1.52	4.56	0.62
3-31-00A	27.2	72.16	9.24	37.17	8.54	2.43	8.48	1.45	1.79	5.71	0.8
3-31-00F	24.48	52.99	5.58	18.91	3.67	1.15	3.3	0.48	0.45	1.8	0.2
1-26-01C	18.2	47.21	4.44	16.02	3.39	0.97	3.78	0.57	0.74	2.95	0.42
2-22-00A	17.86	42.5	4.72	17.55	4.2	1.07	3.95	0.62	0.73	2.3	0.31
Basaltic andesites											
HON 10	33.33	73.97	9.89	40.92	8.58	2.39	7.92	1.22	1.29	3.11	0.49
HON 35	22.76	55.81	8.13	36.89	8.95	2.14	9.27	1.51	1.82	4.74	0.75
HON 51	25.87	54.46	8.34	36.39	7.92	2.43	7.45	1.11	1.14	2.68	0.41
HON 88	8.15	19.3	2.87	13.44	3.51	1.32	3.84	0.63	0.77	2.03	0.32
HON 123	16.8	35.85	5.61	24.47	5.68	1.79	5.69	0.88	0.96	2.41	0.38
NIC 26	10.46	23.83	3.58	15.94	4.02	1.27	4.33	0.71	0.91	2.47	0.39
Andesites											
HON 5	31.74	68.07	8.57	33.28	6.31	1.79	5.52	0.81	0.76	1.74	0.27
HON 42	20.1	44.18	5.83	23.77	5	1.61	4.33	0.65	0.58	1.28	0.19
Dacites											
HON 59	29.08	66.03	8.96	37.26	8.68	2.05	8.85	1.45	1.74	4.42	0.69
HON 140	28.44	58.39	8.45	35.02	7.84	2.03	7.85	1.28	1.58	4.54	0.73

Note: See Jordan et al. (2006) for sample locations.

Nicaragua is underlain by ~35-km-thick continental crust. The greater values of $^{87}Sr/^{86}Sr$ in Honduras are likely a result of greater continental crustal contamination due to a longer melt column through the crust.

All of the geochemical data points to a relationship between the Miocene ignimbrites of Central America and the Caribbean Sea tephras with the more mafic rocks of the Central American Tertiary Ignimbrite Province, as well as the strong influence of crustal melting upon the compositions of both the mafic and silicic magmas.

$^{40}Ar/^{39}Ar$ GEOCHRONOLOGY

Sanidine or plagioclase phenocrysts were separated from 11 ignimbrite-vitrophyre samples, eight from Honduras and three from Nicaragua, by standard methods and irradiated for 40 or 50 h at Oregon State University along with sanidine crystals of the 28.34 Ma Taylor Creek Rhyolite as a neutron flux monitor (Renne et al., 1998). Using the 25 W CO_2 laser at the University of Wisconsin–Madison Rare Gas Geochronology Laboratory, single crystals of sanidine or plagioclase were fused, isotopic compositions measured, and apparent $^{40}Ar/^{39}Ar$ ages calculated following procedures in Smith et al. (2003).

Uncertainties are at the two sigma level and include analytical contributions only. Between 6 and 17 single crystal laser fusion measurements were obtained from each ignimbrite. For most samples, all of the measured crystals yielded a common apparent age that is indistinguishable from the inverse isochron age calculated from these data. In some samples, a few crystals gave apparent ages slightly younger or older than the majority of the crystals, and these were omitted from the final age calculations (see Smith et al., 2003, for an example of how data were treated). The latter crystals most probably reflect low-temperature alteration effects, and in one case may record minor xenocrystic contamination of the ignimbrite. Inverse-variance weighted mean apparent ages of the crystals are compared to the inverse isochron ages for these same crystals in Table 7. Because the isochrons make no assumption regarding the initial composition of argon in the sample, these give the preferred age for each ignimbrite.

All but one sample age fall between 13 and 17 Ma, clearly within the Miocene eruptive period determined in the Caribbean Sea tephra. One sample, Hon-38, has an Oligocene age of 25.88 ± 0.15 Ma (Table 7).

DISCUSSION

Evidence of Continental Crust Assimilation

One of the main lines of evidence for the involvement of continental crust in the formation of the Central American Tertiary Ignimbrite Province is the very large volume of the rhyolite ignimbrites in comparison to the rare occurrences of mafic magmas. The quantity of silicic magmas produced by melting and/or assimilation of continental crust by intrusion of a given volume of mafic magmas is much larger than those produced by strictly fractional crystallization of the same volume of mafic magma. For example, it has been calculated that in order to produce, by FC alone, the 1 km average thickness of the Sierra Madre ignimbrites, 4 km of mafic material must have fractionally crystallized, or ~75% mafic cumulate to ~25% silicic differentiate melt

TABLE 4. Nic 120 TRACE ELEMENT BULK Ds USED IN MODELING, COMPARED WITH BISHOP TUFF (MICHAEL, 1983), MOUNT MAZAMA (BACON AND DRUITT, 1988), AND SMO (CAMERON AND CAMERON, 1986)

Samples	Nic 120 glass	Nic 120 plag	Nic 120 amp*	Nic 120 bio†	Nic 120 px	Nic 120 FeTi†	Nic 120 bulk D	Bishop Tuff bulk D	Mount Mazama bulk D	Batopilas, México bulk D	Basaltic andesite bulk D§
Observed mode	77.68	15.2	2.6	1.9	1.3	1.3					
Experimental mode	68.00	23.61			7.03	1.43					
SiO$_2$		0.71			0.69		0.70				0.532
TiO$_2$		0.00			0.96		0.02				0.553
Al$_2$O$_3$		2.16			0.04		0.00				
FeO		0.25			18.35		0.04				
MnO		0.00			22.55		13.32				
MgO		0.00			103.29		74.96				
CaO		9.87			0.73		2.42				
Na$_2$O		1.24			0.00		0.39				
K$_2$O		0.03			0.00		0.00				
La		0.47		3.6	0.05	14.5	1.51	1.95			0.062
Ce		0.45	1.8	2.95	0.11	12.5	1.54	1.53	0.78	1	0.065
Nd		0.56	4.6	2.65	0.31	9.25	1.75	1.04		1.3	0.12
Sm		0.42	7.7	1.85	0.52	6.15	1.78	0.75	1.27	1.4	0.15
Eu		1.44	8.5	1.35	0.30	2.2	2.27	3.08	2.12	4.1	0.31
Gd		0.45	10.2		0.69		1.57			1.6	0.15
Tb		0.33		1.15	0.78	4.2	0.63	0.28			0.16
Dy		0.30	10.5	0.95	0.61	2.25	1.72	0.25		1.5	
Ho		0.28			0.63		0.23				0.15
Er		0.22	9.2		0.58		1.29			1.4	0.16
Yb		0.17	6.7	0.75	0.44		1.02	0.27	0.7	1.4	0.18
Lu		0.12			0.37		0.10	0.27			0.208
Cs		0.05		6.45	0.01		0.62	0.2	0.02		
Rb		0.06		7.3	0.00		0.70	0.4	0.22		0.021
Ba		0.13		6.65	0.02		0.69	3.22	0.37		0.257
K		1.43					1.05	1.04			0.056
Th		0.10		1.4	0.00	5.85	0.55	0.43	0.08		
U		0.06		0.23	0.00		0.06	0.19			
Nb		0.10			0.16		0.08				0.15
Ta		0.05		1.55	0.09	0.8	0.23	0.1	0.38		0.088
Pb		0.51			0.01		0.35				
Sr		1.07			0.09		0.73	2.98	3.23		0.036
Hf		0.12		0.625	0.09		0.14		0.08		
Zr		0.12			0.06		0.08		0.2		0.09
Y		0.28			0.71		0.23				0.205

Note: amp—amphibolite; bio—biotite; plag—plagioclase; px—pyroxene; bulk D—bulk partition coefficient. See Jordan et al. (2006) for sample locations.
*Batopilas, México (Sierra Madres Oriental) values.
†Bishop Tuff values.
§Los Humeros basalt values.

TABLE 5. WHOLE-ROCK OXIDE AND TRACE ELEMENT DATA FOR E-MORB, N-MORB, AND OCEAN SEDIMENTS

Sample	E-MORB	N-MORB	Carbonate sediments	Hemipelagic sediments
SiO_2 (wt%)	49.72	50.55	7.39	55.78
TiO_2	1.46	1.31	0.02	0.58
Al_2O_3	15.81	16.38	0.37	11.82
FeO			1.97	6.25
Fe_2O_3	10.04	9.17		
MnO	0.16	0.16	0.29	0.14
MgO	7.9	7.8	0.73	2.20
CaO	11.84	11.62	49.90	2.70
Na_2O	2.35	2.79	0.24	2.08
K_2O	0.5	0.09	0.18	1.84
P_2O_5	0.22	0.13	0.13	0.12
Cs (ppm)	0.06	0.01	0.15	2.17
Rb	5.04	0.56	4.28	40.78
Ba	57.00	6.30	2145.48	3941.49
K	2100.00	600.00	1494.27	15274.76
Th	0.60	0.12	0.16	3.00
U	0.18	0.05	0.15	4.89
Nb	8.30	2.33	0.44	5.03
Ta	0.47	0.13		
La	6.30	2.50	8.78	17.96
Ce	15.00	7.50	2.40	28.05
Pb	0.60	0.30	3.70	9.59
Sr	155.00	90.00	1504.12	336.16
Nd	9.00	7.30	6.79	17.77
Hf	2.03	2.05		
Zr	73.00	74.00		
Ti	6000.00	7600.00	119.90	3477.18
Yb	2.37	3.05	1.18	2.78
Y	22.00	28.00	17.69	31.06

Note: Enriched mid-oceanic-basalt (E-MORB) and normal mid-oceanic-ridge basalt (N-MORB) values taken from Sun and McDonough (1989). Sediment values taken from Patino et al. (2000). See Jordan et al. (2006) for sample locations.

TABLE 6. Sr ISOTOPIC VALUES OF CENTRAL AMERICAN IGNIMBRITES

Sample	$^{87}Sr/^{86}Sr$	2 sigma	IUGS rock class
Hon-102	0.704659	0.000010	rhyolite
Hon-106	0.706914	0.000010	rhyolite
duplicate	0.706899	0.000010	
Hon-134	0.704444	0.000012	rhyolite
Hon-140	0.704061	0.000010	rhyolite
duplicate	0.704073	0.000014	
Nic-72	0.704019	0.000010	rhyolite
Nic-120	0.70411	0.000010	trachybasalt
duplicate	0.704129	0.000010	

Note: IUGS—International Union of Geological Sciences. See Jordan et al. (2006) for sample locations.

(Cameron et al., 1980). Geophysical data does not support the presence of such a thick, dense lower crust beneath the Sierra Madre (Ruiz et al., 1988).

Another possible example of qualitative evidence for contamination of the ignimbrite magmas by continental crust is the apparent abundance of xenocrysts within the ignimbrites. This suggests the incorporation of mineral phases from continental crustal sources.

Petrogenetic Sources and Processes

Basaltic andesite, andesite, dacite, and ignimbrite group compositions fall on or close to AFC trend lines extending from N-MORB (Fig. 5B), but not extending from E-MORB (Fig. 5A). The same is true for FC. An N-MORB source melt can produce nearly all of the magmas by fractional crystallization or by assimilation of continental crust similar in composition to the Honduras crustal rocks, as well as average upper and lower continental crust.

Plots of other trace elements and trace element ratios, such as Figure 6 and 7, are also consistent with AFC involving N-MORB as a parental magma source and either lower continental crust or Honduras crustal basement rocks as assimilants.

Figure 6 illustrates that although AFC trends from E-MORB–parallel, the basaltic andesite trend, it is N-MORB AFC trends involve assimilants of similar composition to lower continental arc crust or the Honduras granite 1-26-01C that are closest to the basaltic andesite trend. The basaltic andesites appear to follow an AFC trend from N-MORB, with the more evolved magmas, such as the andesites, dacites, and even the rhyolitic ignimbrite groups, also following in semi-progressive order, after a change in fractionating mineralogy on this trend.

Figure 7A indicates similar trends. Although Figure 7B allows for E-MORB to be the parental magma source for the basaltic andesites, the earlier plots suggest that N-MORB is the stronger candidate. FC trends are also illustrated in Figures 7A and 7B, showing that it is unlikely that the basaltic andesites are evolved from either E-MORB or N-MORB by purely FC processes. Simple FC fails to explain the trends in magma compositions found in either the whole-rock or glass geochemical groups. These results support the view that the silicic ignimbrite magmas and the basaltic andesites are genetically related. Ayalew et al. (2002) found similar trends in Ethiopian ignimbrites and flood basalts.

Assimilation-Fractional Crystallization and Magma Mixing

The Central American Tertiary basaltic andesites, like the ignimbrites, have trace element trends typical of subduction zone environments except that they lack a pronounced negative Sr- or Ti-anomaly (Figs. 8A and 8D). Trace element spider plots comparing the basaltic andesites with average upper continental crust (UC) and lower continental crust (LC) (Wedepohl, 1995; Rudnick and Fountain, 1995) (Fig. 8B) and with Honduras basement rocks (Fig. 8C) show that the Central American Tertiary basaltic andesites have trace element compositions similar to both lower continental crust and some of the Honduran continental basement rocks.

Because of the stability of the REE, they are strong indicators of the source and evolution of magmas (Wilson, 1989; Rollinson, 1993). The basaltic andesites are most similar to LC rocks, but dissimilar from UC, as are the Central American Tertiary dacitic rocks (Figs. 9A, 9B, and 9D). This implies that

TABLE 7. SUMMARY OF $^{40}Ar/^{39}Ar$ LASER FUSION ANALYSES OF SINGLE FELDSPARS FROM 11 CENTRAL AMERICAN IGNIMBRITE SAMPLES

Sample ID	Latitude	Longitude	Mineral*	mean K/Ca	$N^†$	Age (ka)§ ± 2s	MSWD#	MSWD#	$^{40}Ar/^{36}Ar_i$ ± 2s	Age (ka)§ ± 2s
Honduras										
HON-020	14.77	271.32	san	59	7 of 7	14.59 ± 0.04	0.8	0.4	386.3 ± 136.5	14.50 ± 0.13
HON-021	14.74	271.34	san	44	5 of 6	14.64 ± 0.06	1.2	1.4	314.8 ± 69.3	14.57 ± 0.25
HON-014	15.04	271.84	san	53	7 of 7	14.58 ± 0.07	0.4	0.4	278.5 ± 45.5	14.61 ± 0.10
HON-043	14.84	271.39	san	3200	5 of 9	14.60 ± 0.04	0.5	0.4	263.6 ± 55.4	14.63 ± 0.07
HON-015	14.81	271.22	san	60	5 of 9	14.67 ± 0.08	1.2	1.0	204.5 ± 86.2	14.71 ± 0.09
HON-016	14.81	271.22	san	57	9 of 7	14.65 ± 0.07	1.6	0.4	174.4 ± 64.4	14.74 ± 0.07
HON-112	14.08	273.18	san	1363	9 of 9	16.61 ± 0.04	0.7	0.8	294.2 ± 20.9	16.62 ± 0.05
HON-038	14.69	271.48	san	2326	6 of 15	25.96 ± 0.09	1.6	1.4	314.8 ± 31.1	25.88 ± 0.15
Nicaragua										
NIC-063	12.75	273.88	plag	0.01	14 of 14	13.53 ± 0.56	0.5	0.5	297.0 ± 4.9	13.35 ± 0.82
NIC-026	12.19	274.96	plag	0.10	14 of 14	16.32 ± 0.28	1.4	1.2	320.6 ± 33.2	15.80 ± 0.74
NIC-021	12.19	274.69	plag	0.01	14 of 14	16.89 ± 0.20	1.2	1.3	296.1 ± 9.2	16.85 ± 0.56

Note: See Jordan et al. (2006) for sample locations.
*san—sanidine; plag—plagioclase.
†N = number if single crystal fusion analyses included in weighted mean apparent age and inverse isochron calculations.
§ages calculated relative to 28.34 Ma Taylor Creek rhyolite sanidine (Renne et al., 1998).
#MSWD—mean square of weighted deviates.

assimilation of the lower crust was likely dominant. The Central American Tertiary basaltic andesites are enriched in the light rare earth elements (LREE) relative to N-MORB, with compositions of the heavy rare earth elements (HREE) close to those of N-MORB. With the one exception of Hon-35, they have no Eu-anomaly. This is also true of the Central American Tertiary andesites, which exhibit a greater depletion in the HREEs relative to N-MORB (Fig. 9C). The dacites are very close in composition to the basaltic andesites, but they exhibit a clear negative Eu-anomaly (Fig. 9A). The basaltic andesites and dacites have REE trends very similar to average LC (Wedepohl, 1995) (Fig. 9B), while the andesites have trends closer to those of average UC (Wedepohl, 1995) (Fig. 9D).

Wark (1991) noted that due to their relative incompatibilities, the K/Rb ratio is greatly affected by continental crustal assimilation and can thus be used as a potential tracer of wall rock assimilation AFC. Increasing AFC will result in decreasing K/Rb ratios, with an increase in Rb. The K/Rb ratios for the Central American Tertiary volcanic rocks appear to indicate evidence of continental crustal AFC for the generation of the ignimbrite groups. AFC modeling on plots of K/Rb versus Rb using an N-MORB as a source and Honduras continental crust samples as assimilants generates trends that are very close to those of the Central American Tertiary geochemical groups (Figs. 10A and 10B). Subsequent mixing of the residual magmas produces even closer trends (Fig. 10C). Unfortunately, no specific examples of this are found in the glasses, but in several cases evidence for mixing in the form of two distinct glass types was found in the Miocene ignimbrites and in the Caribbean Sea tephra.

Although the AFC models given above indicate that it is possible to generate nearly all of the silicic ignimbrite groups by two-stage AFC processes, continued AFC, without magma mixing, can also account for the origin of a number of the geochemical groups in the likely event of assimilation of a heterogeneous continental crust. Assimilation of a granitic-type crust of the composition of Honduras continental crustal sample 1-26-01C, while fractionally crystallizing from N-MORB melt, will produce the basaltic andesite magmas. These may, in turn, continue to fractionate, but with different mineral modes and also assimilating different continental crustal material, such as samples 3-09-00A and 3-31-00A, which will evolve to compositions of the rhyolitic ignimbrites (Fig. 11).

Another model involving the 10% fractional crystallization of an N-MORB–like melt, which is a reasonable assumption for producing basaltic andesite major oxide compositions, assimilating rock with an average LC REE composition in the proportions 4:1, 3:2, 2:3, and 1:4 produces magmas with REE geochemical characteristics that encompass the range of the Central American basaltic andesite magmas (Fig. 12).

Calculations of reasonable estimates of the major oxide concentrations and proportions of the mafic melt and the assimilated continental crustal material indicate that SiO_2 can also be explained by AFC, but not necessarily the high K_2O values. The K_2O may be the result of some other process, such as alteration. For example: 25% of a mafic melt at 1.5% K_2O and 55% SiO_2 combined with 75% of a continental crustal melt at 3.6% K_2O and 75% SiO_2 should produce a magma with a composition of 3% K_2O and 70% SiO_2.

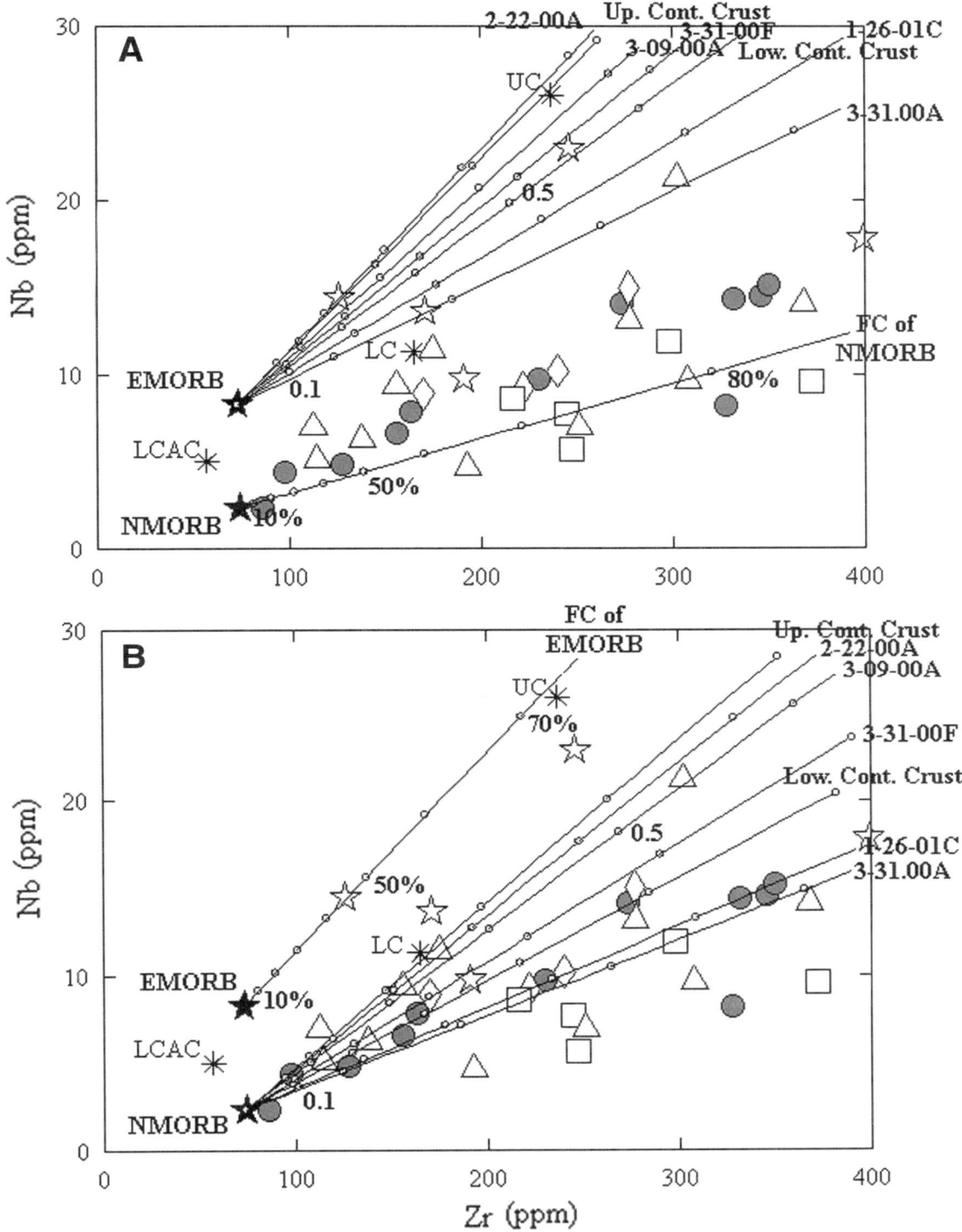

Figure 5. Plot of Nb versus Zr in Central American Tertiary volcanic rocks illustrating trends extending from normal mid-oceanic-ridge basalt (N-MORB), rather than enriched (E)-MORB. Closed circles are basaltic andesites; open diamonds are andesites; open squares are dacites; open triangles are ignimbrite geochemical groups; open stars are Honduras whole-rock continental basement rocks (Rogers and Patino, unpublished data); asterisks are average continental crust (Rudnick and Fountain, 1995; Wedepohl, 1995); filled stars are E-MORB and N-MORB (Sun and McDonough, 1989). Points on assimilation-fractional crystallization (AFC) lines indicate melt fraction increments of 0.1, and r for all trends is 0.5. Points on fractional crystallization (FC) lines indicate percent crystallization. UC—upper continental crust; LC—lower continental crust; LCAC—lower continental arc crust. (A) AFC trends for E-MORB and FC for N-MORB. (B) AFC trends for N-MORB and FC for E-MORB.

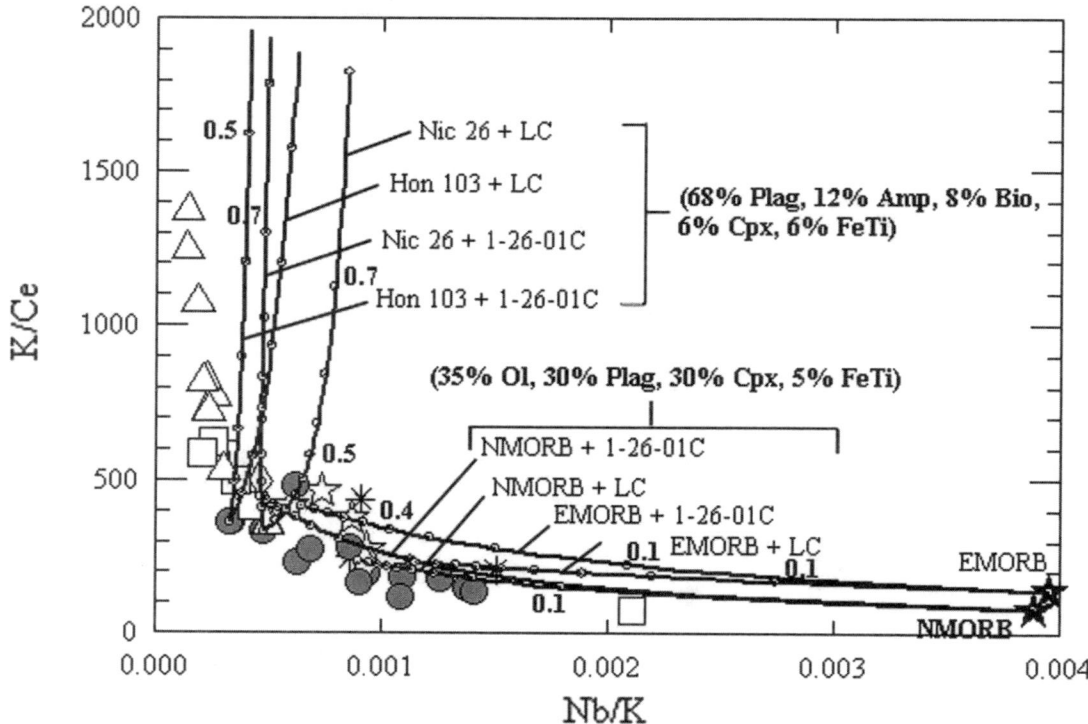

Figure 6. The variation in K/Ce versus Nb/K in Central American Tertiary volcanic rocks. The basaltic andesites fall along the assimilation-fractional crystallization (AFC) trend extending from normal mid-oceanic-ridge basalt (N-MORB). AFC trends extending from the basaltic andesites using a more silicic modal mineralogy are close to the observed trends for the ignimbrite geochemical groups. Points on lines indicate melt fraction increments of 0.1, and $r = 0.5$ for all trends. E-MORB—enriched mid-oceanic-ridge basalt; Amp—amphibolite; Bio—biotite; Cpx—clinopyroxene; Ol—olivine; Plag—plagioclase. Symbols are the same as in Figure 5.

Evolution and Formation of Ignimbrite Geochemical Groups

As mentioned, the same REE groups from Jordan et al. (2006) were used for this study. However, three of the groups were not considered. Two were not used because they consisted only of marine tephra (Groups 1c and 3b). The third was not considered because it consists of only two samples whose average and standard deviation produces a doubtful and likely altered REE pattern with upper and lower trendlines not parallel to each other (Group 2b). For full descriptions of each group, see Jordan et al. (2006).

All of the groups determined by Jordan et al. (2006) have significant negative Nb and Ti anomalies, indicative of a subduction zone environment (Wilson, 1989). It has been suggested that the negative Nb anomaly in subduction zone magmas may be due to the retention of Nb within titaniferous mineral phases, such as ilmenite, in the downgoing slab (Wilson, 1989; Rogers and Hawkesworth, 2000). This may also account for the negative Ti anomaly. However, it is also possible to obtain such anomalies by remelting of earlier subduction-related magmatic rocks.

Evidence for such remelting has been found in other Central America rocks, specifically in Costa Rica (Vogel et al., 2004).

Most groups also are enriched nearly 1000 times in Cs relative to N-MORB. The notable exceptions to this are Group 1a, Group 5d, and Group 6. Group 1a is enriched almost 10,000 times that of N-MORB, Group 5d is enriched between 100 and 600 times N-MORB, and Group 6 is enriched 1000–3000 times N-MORB. Most of the groups, with the exceptions of Group 4b, Group 5a, and Group 5d, have negative Sr anomalies. Group 1b has a very large negative Sr anomaly and also is the only group to have a negative Ba anomaly as well. These fit with the large negative Eu anomaly also found in Group 1b and probably indicate that the melts were in equilibrium with fractionating plagioclase. The rest of the groups either have no Ba anomaly or a small, positive one.

Each of these geochemical groups in the Central American Tertiary Ignimbrite Province likely has its own petrogenetic history reflecting some variability in the overall evolutionary processes of the province. The overall REE trends for each group, with the exception of Eu-anomalies, are very similar to the REE trends of the basaltic andesites, which indicates that they had the

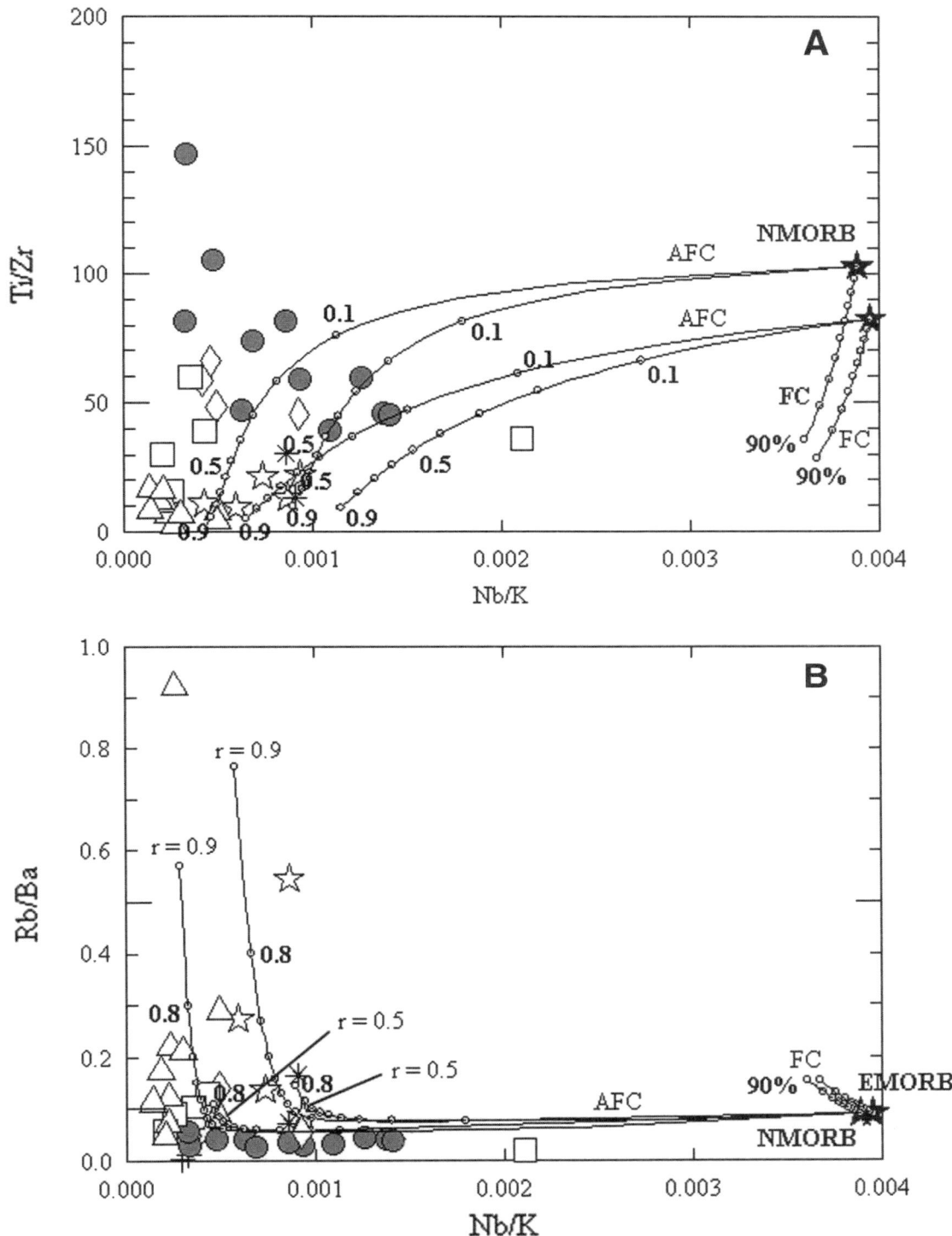

Figure 7. Ti/Zr and Rb/Ba versus Nb/K trends for Central American Tertiary volcanic rocks and assimilation-fractional crystallization (AFC) versus fractional crystallization (FC) trends for normal mid-oceanic-ridge basalt (N-MORB) and enriched (E)-MORB. The suite of Tertiary volcanic rocks found in the ignimbrite province clearly follows the AFC trends. Points on lines indicate melt fraction increments of 0.1. Symbols are the same as in Figure 5. (A) Ti/Zr versus Nb/K. Upper line for both E-MORB and N-MORB is AFC with Honduras crustal rock 1-26-01C. Lower line is AFC with lower continental crust. (B) Rb/Ba versus Nb/K. AFC trends for $r = 0.5$ and $r = 0.9$. Upper line is AFC involving N-MORB and lower continental crust. Lower line is AFC involving N-MORB and Honduras crustal rock 1-26-01C. E-MORB AFC trends are nearly identical to the N-MORB AFC trends.

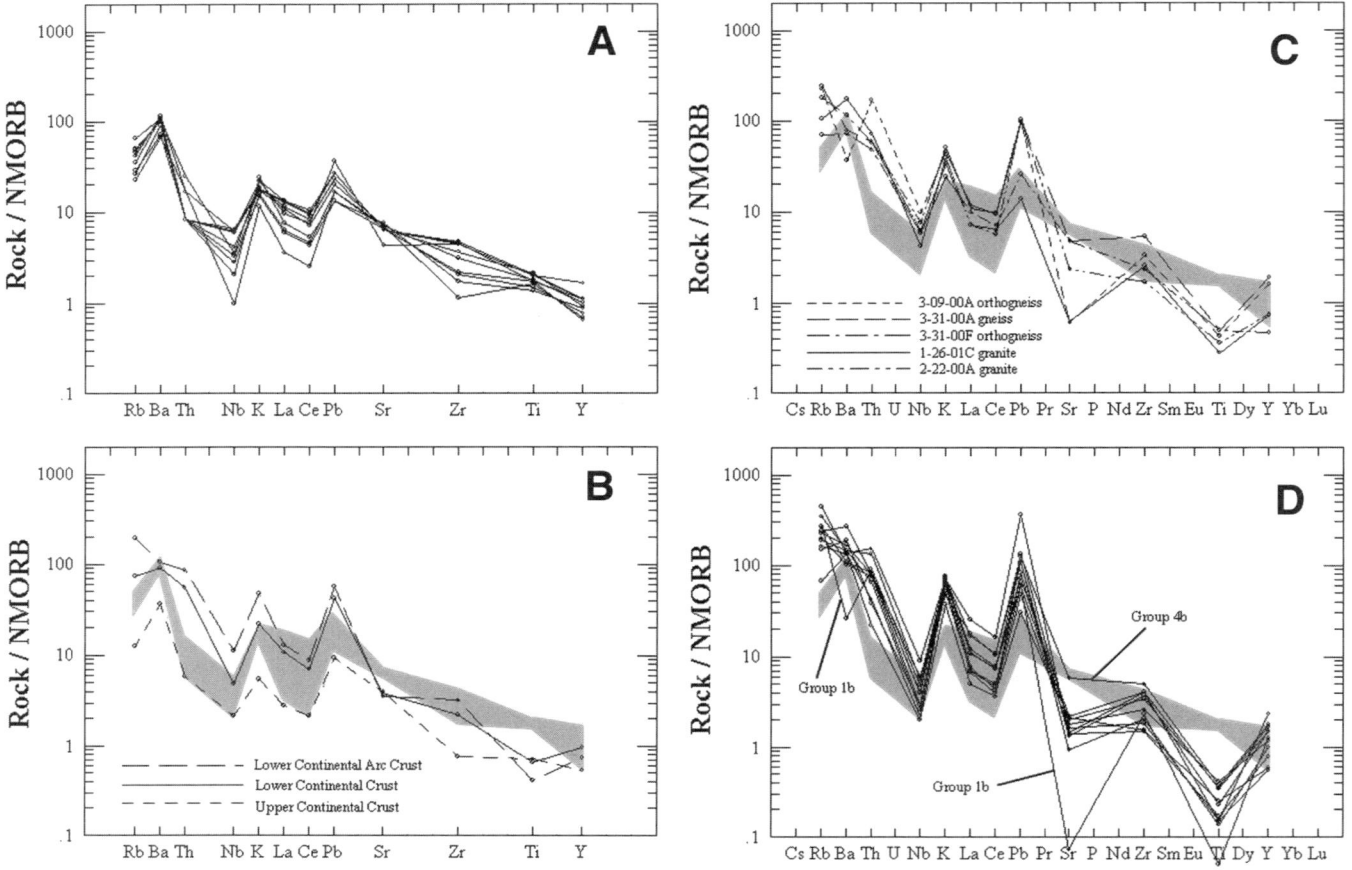

Figure 8. (A) Trace element compositions of Central American basaltic andesites normalized to normal mid-oceanic-ridge basalt (N-MORB). (B) Comparison of one sigma range of basaltic andesites with average continental crustal compositions (Wedepohl, 1995; Rudnick and Fountain, 1995). Basaltic andesites (shaded region) are most similar to lower continental crust. (C) Trace element compositions of Central American continental basement rocks compared to one sigma range of Central American Tertiary basaltic andesites (shaded region). (D) Comparison of one sigma range of basaltic andesites (shaded region) with Central American Tertiary ignimbrite geochemical group averages.

same source, allowing for the involvement of the basaltic andesites in the formation of the silicic ignimbrite magmas. As mentioned, the binary plots indicate the possibility of a two-step AFC process involving N-MORB–type melt and lower continental crust to form the basaltic andesites. AFC of these basaltic andesites continues to form the silicic ignimbrites. A cogenetic relationship is apparent for Hon-34, a silicic vitrophyre, and Hon-35, a basaltic andesite dike, which are associated in the field. Hon-35 has a less evolved, but similar, subparallel, trace element trend to that of Hon-34 (Fig. 13).

It should be pointed out clearly that the models for each of the groups are based on the glass data, *NOT* the whole rock data. The reason for this is mainly because a substantial amount of the geochemical data was collected from the ODP cores, which consisted of glasses shards. It was thought that in order to used the data from the ODP cores, it was best to compare the glasses only. Secondly, few fresh pumice fragments were found. Using just the glass compositions is not as accurate as using other compositions (bulk pumice analysis, for example), but since the glass is considered to be the residual melt at the time of eruption, its composition does serve as a general record of the evolution of the ignimbrite magmas.

It is important to keep in mind that given the heterogeneous nature of the continental crust in Central America (Herrstrom, 1994) it would not be unusual for more than one type of continental crustal contaminant to be involved in the Central American Tertiary ignimbrite petrogenesis. The following is a discussion of models used to reproduce each geochemical group. The models involve mixing and assimilation of various continental crustal contaminants with mafic melts derived by FC of average Central American basaltic andesite compositions.

Group 1a

Volcanic glasses with geochemical characteristics falling into geochemical Group 1a are found in ignimbrites along the whole length of the Central American Tertiary Ignimbrite Province and

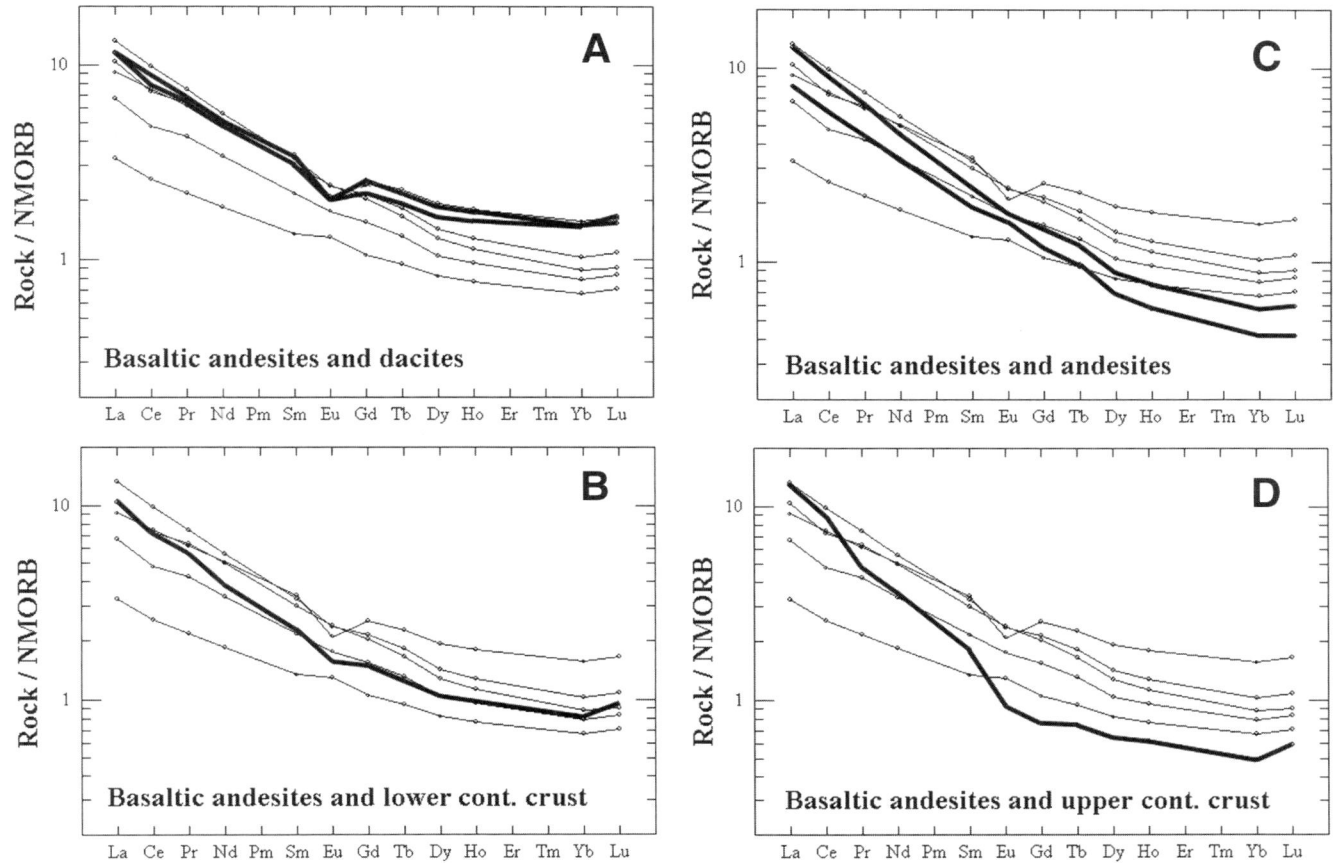

Figure 9. (A) Rare earth element (REE) compositions of Central American Tertiary basaltic andesites compared to Tertiary dacites (bold lines) from the same region. (B) Comparison of Central American Tertiary basaltic andesites with average lower continental crust (bold line) (Wedepohl, 1995). (C) REE compositions of Central American Tertiary basaltic andesites compared to Tertiary andesites (bold lines) from the same region. (D) Comparison of Central American Tertiary basaltic andesites with average upper continental crust (bold line) (Wedepohl, 1995). Note the dissimilarity between the Tertiary rocks and the average upper continental crust.

have been identified in three volcanic ash layers found in Caribbean Sea sediments (Jordan et al., 2006). Plagioclase, pyroxenes, and Fe-Ti oxides are common phenocrysts, with some samples containing biotite, amphibole, and quartz.

Figure 14 part 1A illustrates the results of an AFC mixing model involving a mafic melt formed by a 10% fractionally crystallized (~35% olivine, 30% plagioclase, 30% clinopyroxene, and 5% opaque minerals) basaltic andesite melt and a crustal assimilant with the composition of the Honduras gneissic rock 3-09-00A. Model lines represent the results of mixing. The best fit appears to be mixing proportions of mafic melt to continental crustal melt between 40%–75% and 60%–25%, respectively.

Group 1b

Group 1b is a small geochemical group of only three samples. Two are from Honduras and one is a Caribbean Sea volcanic ash layer. Each of the samples in this group contain prominent plagioclase phenocrysts. The evolution of this group is best explained by an AFC mixing model of 10% fractionally crystallized basaltic andesite and a melt of the Honduran orthogneiss continental crustal rock 3-09-00A (Fig. 14 part 1B). However, the very large negative Eu anomaly of this group is not fully accounted for in the model. It is possible that the pronounced Eu anomaly may be the result of late-stage plagioclase fractionation in the magma chamber preceding eruption. Additional evidence of this is that Group 1b also has very prominent negative Ba and Sr anomalies, which are also indicative of plagioclase-dominated fractionation.

Group 2a

Volcanic ash with Group 2a geochemistry was found in six layers in Caribbean Sea sediments ranging in age from 10.3 Ma to 19.2 Ma. Most of these samples contain plagioclase phenocrysts with some pyroxenes. Other minerals are rare. The model that produces the best-fit AFC mixing trends involves two Honduran continental crustal samples, the orthogneiss 3-09-00A and

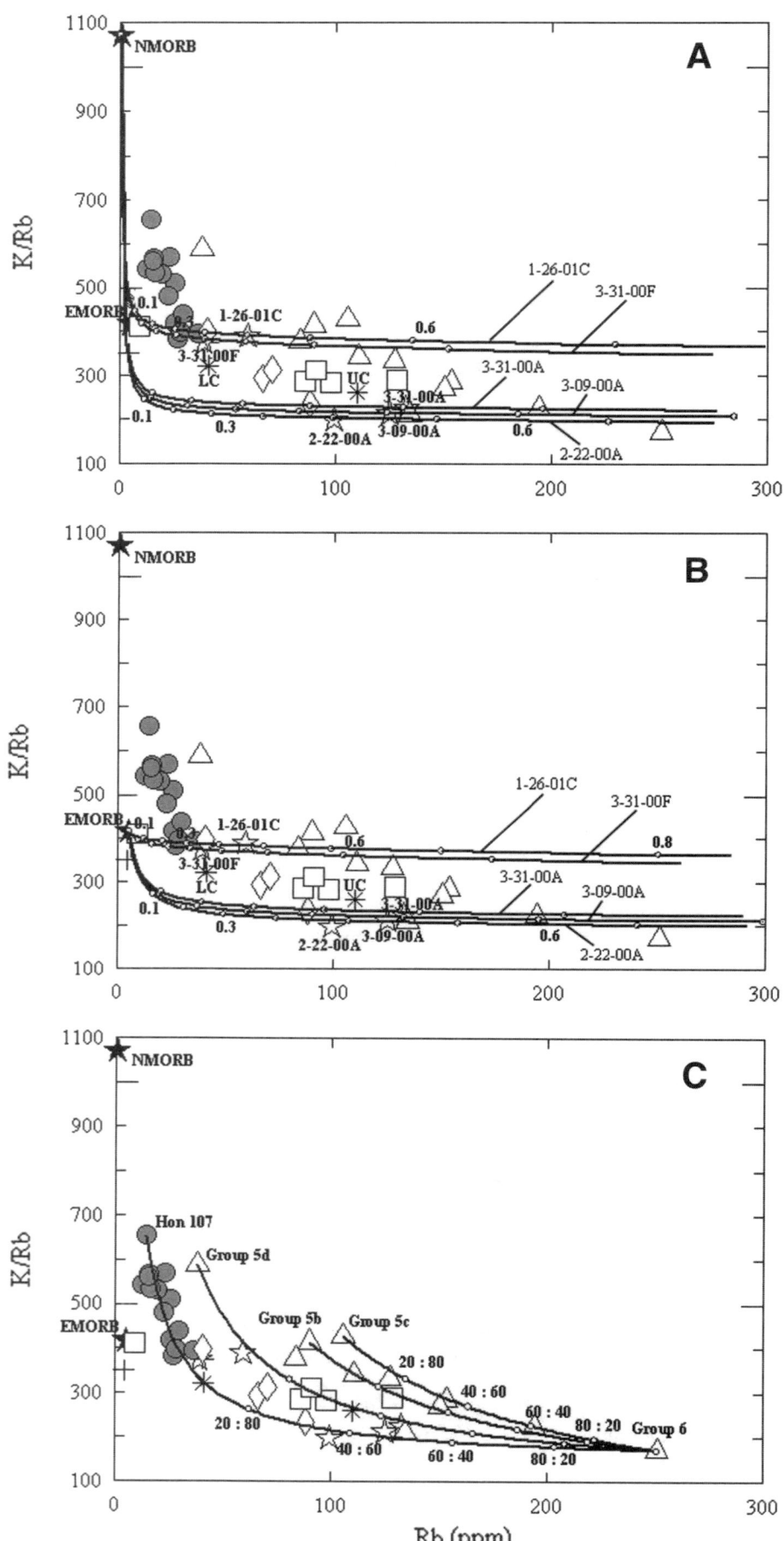

Figure 10. Assimilation-fractional crystallization (AFC) trends for (A normal mid-oceanic-ridge basalt (N-MORB) and (B) enriched (E-MORB) with Honduras continental crustal rocks as assimilants. Bold sample numbers are compositions. Normal sample numbers indicate the AFC trend for that assimilant. Points on lines indicate melt fraction of assimilant in increments of 0.1. (C) Result of mixing models for Central American Tertiary volcanic rocks. The endpoints are geochemical groups 5b, 5c, and 6, along with the basaltic andesite Hon-107, which can be formed by AFC (A and B) and subsequently mix between themselves to produce most of the other geochemical groups and lavas (C). It should be noted that Group 6 appears to be a crustal melt (see text) and does not need to be produced by AFC. Mixing between the Group 6-type melt and other geochemical groups produced by AFC can generate the whole spectrum of magmas present in the Central American Tertiary ignimbrite province. Points on lines indicate the mixing percentage for each endpoint. Symbols are the same as in Figure 5. LC—lower continental crust; UC—upper continental crust; E-MORB—enriched mid-oceanic-ridge basalt; N-MORB—normal mid-oceanic-ridge basalt.

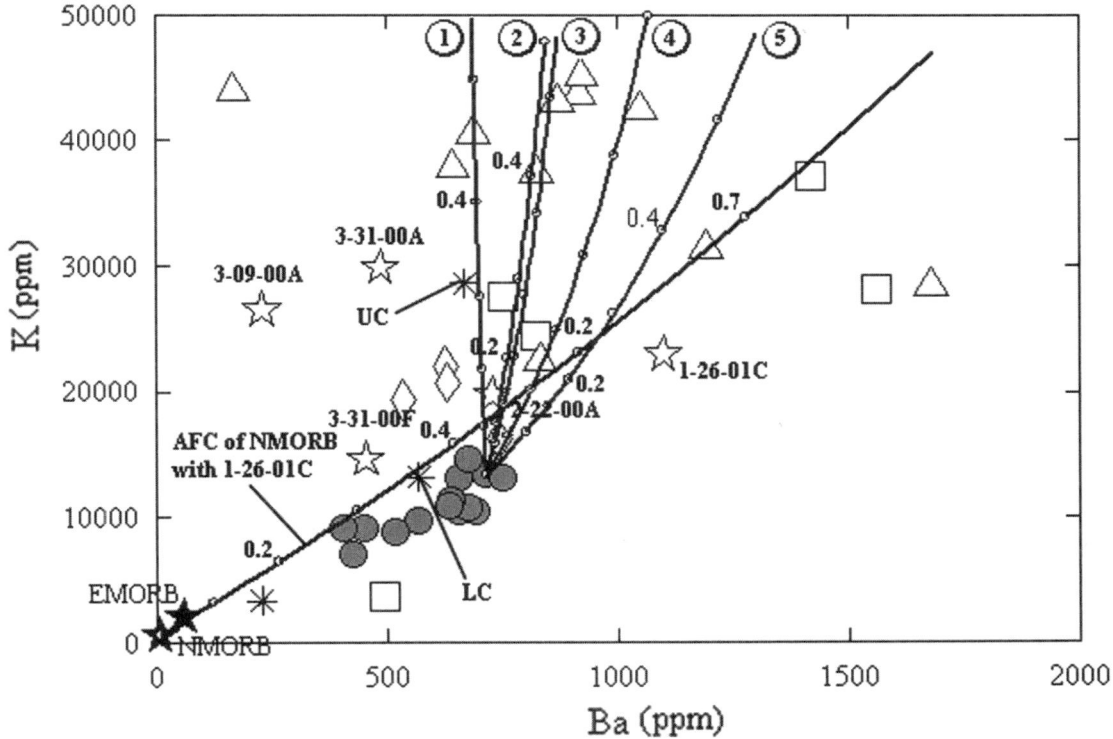

Figure 11. Model for two-step process of formation for the silicic ignimbrite groups. Initial assimilation-fractional crystallization (AFC) takes place with a normal mid-oceanic-ridge basalt (N-MORB) source and a continental crustal assimilant (1-26-01C—granite) to form the basaltic andesites. The basaltic andesite magma then assimilates additional continental crust rock in the following trends: (1) 3.09.00A; (2) 3-31-00A; (3) 3-31-00F; (4) 2-22-00A; (5) 1-26-01C. This two-step AFC process can give rise to most of the silicic ignimbrite magmatic groups, other basaltic andesites, and the andesites and dacites. Points on lines indicate melt fraction of assimilant in increments of 0.1, and $r = 0.5$ for all trends. LC—lower continental crust; UC—upper continental crust. Symbols are the same as in Figure 5.

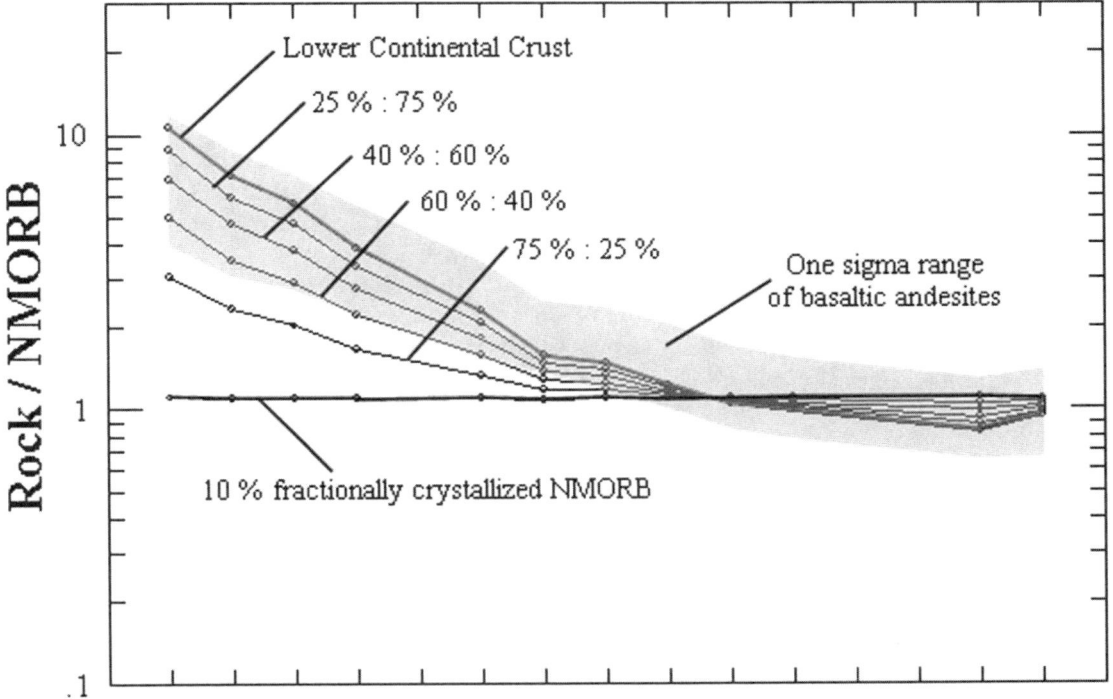

Figure 12. Rare earth element mixing model of a lower continental crustal melt with a mafic melt derived by 10% fractional crystallization of normal mid-oceanic-ridge basalt (N-MORB). Percentage ratios are proportions of N-MORB residual melt to lower continental crustal melt. Shaded region is one sigma range of Central American basaltic andesite lavas. Bold line is lower continental crust values from Wedepohl (1995).

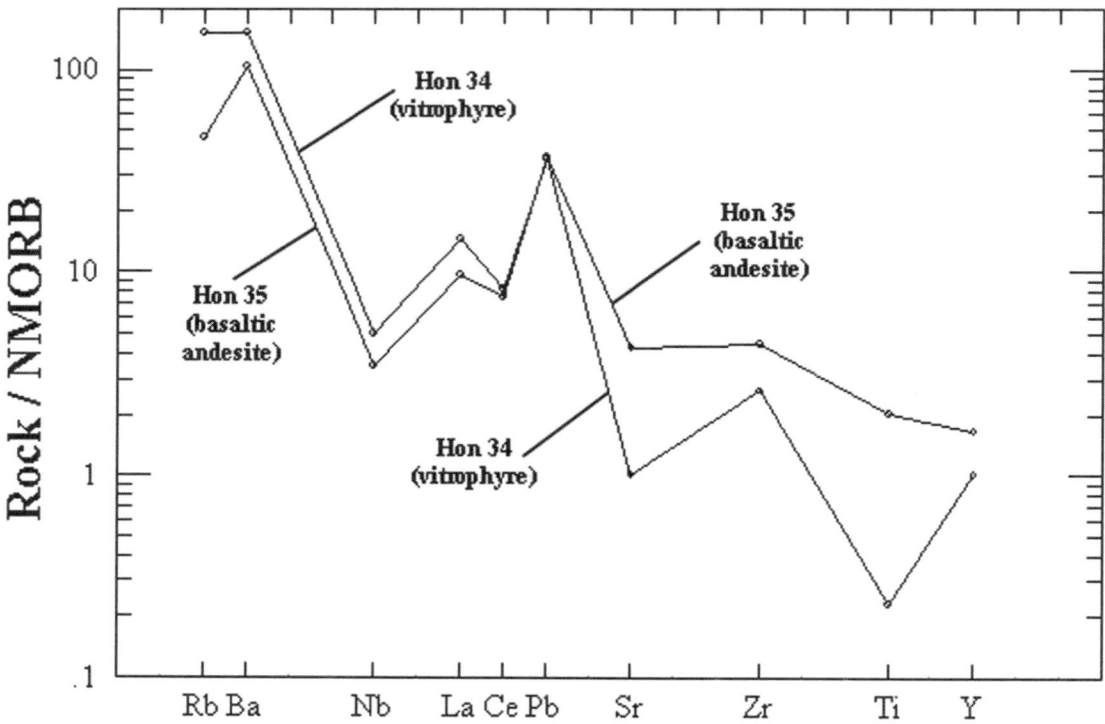

Figure 13. Trace element trends for a field associated Central American silicic vitrophyre and basaltic andesite dike. Their trends are subparallel to each other with the basaltic andesite (Hon 35) having a less evolved trend than that of the vitrophyre (Hon 34). N-MORB—normal mid-oceanic-ridge basalt

the granite 1-26-01C, mixing with a residual melt from 10% fractionally crystallized basaltic andesite (Fig. 14 part 1C).

Group 3a

Group 3a glasses were found in four Caribbean Sea marine tephra samples and seven Central American Tertiary ignimbrite samples. Plagioclase, pyroxenes, and Fe-Ti oxides are common phenocrysts in most of the samples. Group 3a magmas are already close in composition to 3-31-00A Honduran continental crustal rocks but are not as enriched in the LREEs. This depletion in the LREEs relative to the continental crustal rocks and even the fractionated basaltic andesite is difficult to explain unless allanite is present in the magmas of this group. If it is, it was not identified. A model involving AFC mixing of continental crustal melts derived from the Honduran rocks 3-09-00A and 3-31-00A and a mafic melt produced by 10% fractionation of Central American Tertiary basaltic andesites appear to be the best fit for this group (Fig. 14 part 1D).

Group 4a

Group 4a is geographically extensive, found as ignimbrites throughout Honduras and Nicaragua, as well as eight volcanic ash layers identified in the ODP Leg 165 Caribbean Sea sediments. Plagioclase, pyroxenes, Fe-Ti oxides, and biotite are common phenocrysts, with some samples containing amphibole and quartz as well. Some unidentified accessory minerals are also present. The REE compositions of this group are best accounted for by the AFC mixing of a magma derived from 10% fractional crystallization of a Central American Tertiary basaltic andesite and an assimilant similar in composition to the Honduran orthogneiss 3-09-00A (Fig. 14 part 1E).

Group 4b

Group 4b is made up of four samples—one marine volcanic ash layer, two Honduran ignimbrites, and one Nicaraguan ignimbrite. Group 4b is most similar to the Honduran continental crustal rock 3-31-00A. However, unlike the granite 3-31-00A, it has a positive Eu-anomaly (Fig. 14 part 1F). High temperature, hydrothermal fluids typically have a positive Eu-anomaly (Cocherie et al., 1994). If the ignimbrites were altered by hydrothermal fluids after deposition then they could have been imprinted with a positive Eu anomaly, but that would not explain the volcanic ash from the Caribbean Sea also having a positive Eu anomaly. However, it may be possible that fluids derived from a hydrothermally altered subducting oceanic crust may have contaminated the Group 4b melts. Another possibility related to subduction may be the influence of sediment-derived fluids from the downgoing slab (see Sediment Influence section). Positive Eu anomalies can develop during diagenesis of oceanic sediments in areas of high oxidation (MacRae et al., 1992). It has also been suggested that crustal melts dominated by plagioclase can also produce positive Eu anomalies. Also, some hybrid melts in the lower crust may have a positive Eu anomaly (Mazzucchelli et al., 1992). Given the likelihood of crustal contamination in the Central American Tertiary ignimbrites, this last possibility involving crustal anatexis may be the best one. The lack of a positive Eu anomaly in the gneiss 3-31-00A likely implies that the Group 4b positive anomaly may be the result of its source magma being a crustal melt and may be only similar to 3-31-00A in its overall REE trend, but not in terms of Eu.

Group 5a

Group 5a ignimbrites are found throughout Honduras and Nicaragua, along with fourteen Caribbean Sea volcanic ash layers. Group 5a ignimbrites contain phenocrysts of plagioclase and pyroxenes, and Fe-Ti oxides are evident in some of the samples. The best-fit mixing model that reproduces the Group 5a REE trend involves mixing of two continental crustal melts similar in compositions to the Honduran orthogneiss 3-09-00A and granite 1-26-01C (Fig. 14 part 2G). This possible mixing of two continental crustal melts suggests processes similar to those postulated by Huppert and Sparks (1988) in which density and buoyancy contrasts prevent mixing between silicic magmas and the mafic melts, with the silicic melt overlying the mafic one. This would not prevent mixing between two similar silicic-density melts, however. Mixing proportions of 25% 1-26-01C to 75% 3-09-00A give the best fit.

Group 5b

Group 5b ignimbrites are found throughout the ignimbrite province in Honduras and Nicaragua. It is made up of seven ignimbrites and two volcanic ash layers. Mineralogically, most samples in this group contain plagioclase, pyroxene, Fe-Ti oxides, and biotite phenocrysts, but some also contain amphibole. Its overall evolution is best fit by a crustal melt similar in composition to Honduran granite 1-26-01C (Fig. 14 part 2H). Once again, the positive Eu anomaly of this group might be evidence for a crustal melting or fluid contamination derived from subducting sediments. The more depleted HREE in Group 5b relative to the Honduran continental crustal rock may be related to the presence of amphibole, which tends to deplete melts in the HREEs.

Group 5c

Group 5c consists of two marine volcanic ash and three ignimbrites from Nicaragua. However one of the ignimbrite contains two slightly different silicic glasses (glass 1 SiO_2 = 76.06 wt%; glass 2 SiO_2 = 75.97 wt%), which suggests some magma mixing. These rocks contain plagioclase, pyroxene, and biotite phenocrysts. Group 5c magmas are very similar to the Honduran granite 1-26-01C (Fig. 14 part 2I). This group is slightly more depleted in middle rare earth elements (MREE) and HREE than 1-26-01C. Amphibole and apatite fractionation can cause this type of depletion trend (Hanson, 1978). Apatite may be present as an accessory mineral in Group 5c ignimbrites. Accessory minerals were present in at least three of the five Group 5c samples, but were not identified.

Group 5d

Group 5d consists of two marine volcanic ash layers and one vitrophyre from Nicaragua. Plagioclase and pyroxene phenocrysts are present, but other phases are not readily apparent. This group has rather low LREE concentrations for a silicic ignimbrite (Fig. 14 part 2J). The geochemical evolution of HREEs and the negative Eu anomaly of this group are well explained by AFC mixing between a melt derived by 50% fractionally crystallized basaltic andesite and a melt of continental crust similar to the Honduran granite 1-26-01C. The LREE compositions are harder to explain. Given their low concentrations compared to the other ignimbrite geochemical groups, it is possible that allanite was fractionated during Group 5d's magmatic evolution. Allanite readily takes up the LREEs and thus suppresses their compositions in the residual glass compositions. However, allanite was not identified in any of the glasses.

Group 6

Group 6 consists of eight samples—four volcanic ash layers in Caribbean Sea sediments and four ignimbrites in Honduras. The samples have a fairly limited geographic distribution in western Honduras in the region between Santa Barbara and Comayagua. Plagioclase and Fe-Ti oxide phenocrysts are common, with half of the samples clearly containing quartz and biotite.

Generation of Group 6 can be modeled by FC of a melt similar in composition to the Honduran orthogneiss 3-09-00A. Simple FC using a modal mineralogy of 68% plagioclase, 12% amphibole, 9% biotite, 6% pyroxene, and 6% Fe-Ti oxides, provides enough enrichment to place the evolved melt within the Group 6 magma field (Fig. 14 part 2K). The problem with the initial generation of a pure crustal melt for Group 6 is similar to that encountered for groups 4b, 5a, 5b, and 5d. The most likely possibility is continental crust melt generated and then fractionally crystallized, but not mixed or assimilated with, a more mafic magma, similar to the process proposed by Huppert and Sparks (1988).

Sediment Influence

The failure of AFC, FC, and AFC mixing models to completely explain some of the groups is most likely due the lack of characterization of all possible assimilants. The models used here are meant to explain the groups generally. However, there are some other identified sources that may have contributed to the geochemical variation of the ignimbrite magmas.

Relatively small contributions from fluids derived from sediments that are part of the downgoing slab can significantly modify arc magmas and make it difficult to infer mantle source compositions in arc settings (Patino et al., 2000). Geochemical components derived from these fluids released during sediment

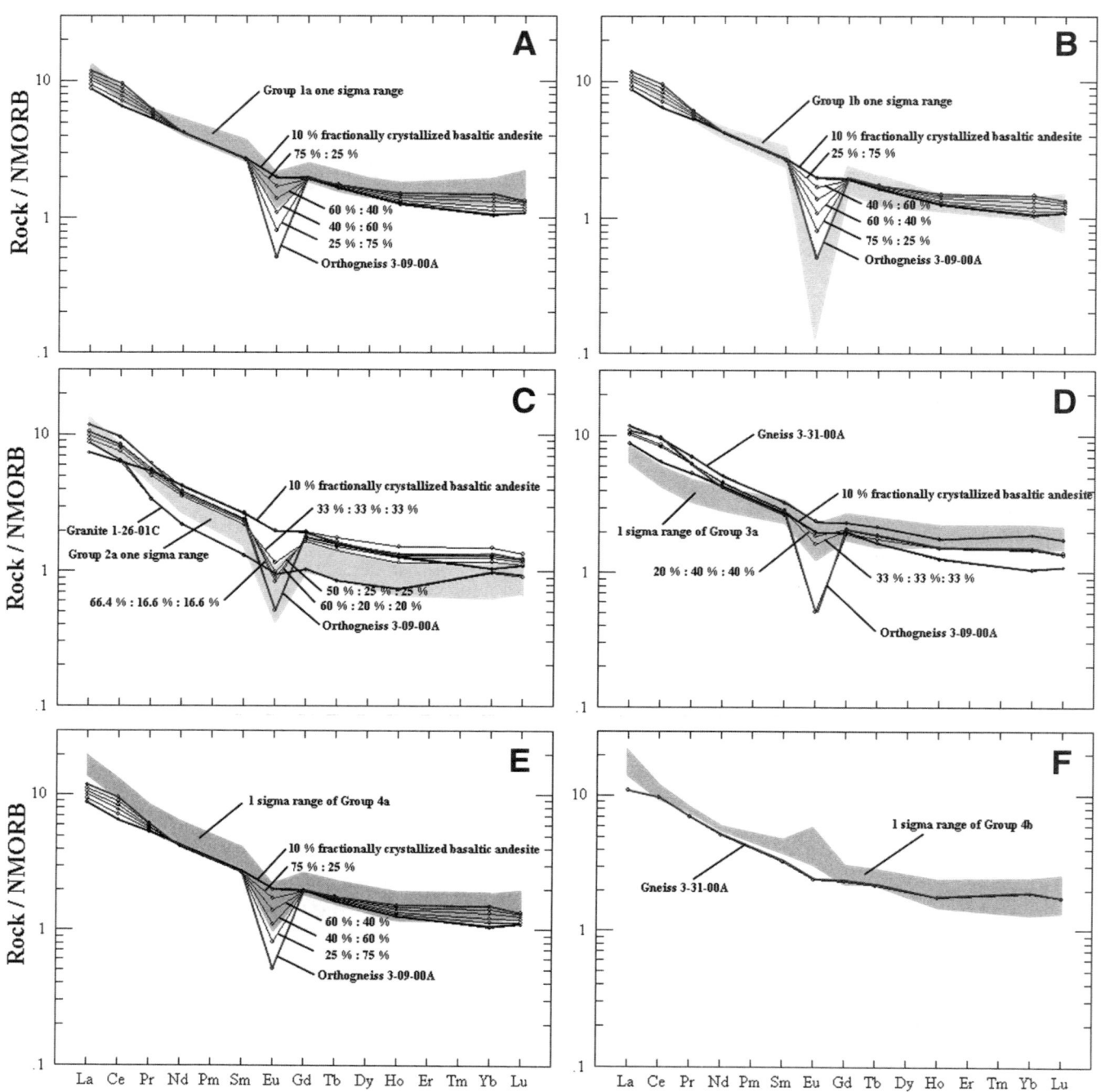

Figure 14. Assimilation-fractional crystallization (AFC) mixing models of Central American ignimbrite groups. Shaded areas represent 1 sigma range of each group. (A) Group 1a: Percentage ratios are proportions of a 10% fractionally crystallized basaltic andesite residual melt to a melt of Honduras continental crustal rock 3-09-00A composition. (B) Group 1b: Percentage ratios are proportions of a melt of Honduras continental crustal rock 3-09-00A composition to a 10% fractionally crystallized basaltic andesite residual melt. (C) Group 2a: Percentage ratios are proportions of melts of Honduras continental crustal rocks 3-09-00A, 1-26-01C, and a 10% fractionally crystallized basaltic andesite residual melt. (D) Group 3a: Percentage ratios are proportions of melts of Honduras continental crustal rocks 3-09-00A, 3-31-00A, and a 10% fractionally crystallized basaltic andesite residual melt. (E) Group 4a: Percentage ratios are proportions of a 10% fractionally crystallized basaltic andesite residual melt to a melt of Honduras continental crustal rock 3-09-00A composition. (F) Rare earth element (REE) trends of the Honduran gneissic rock 3-31-00A and the one sigma range of Central American Tertiary ignimbrite geochemical Group 4b. Group 4b has an unusual positive Eu anomaly, which may be related to subduction-derived fluids (see main text).

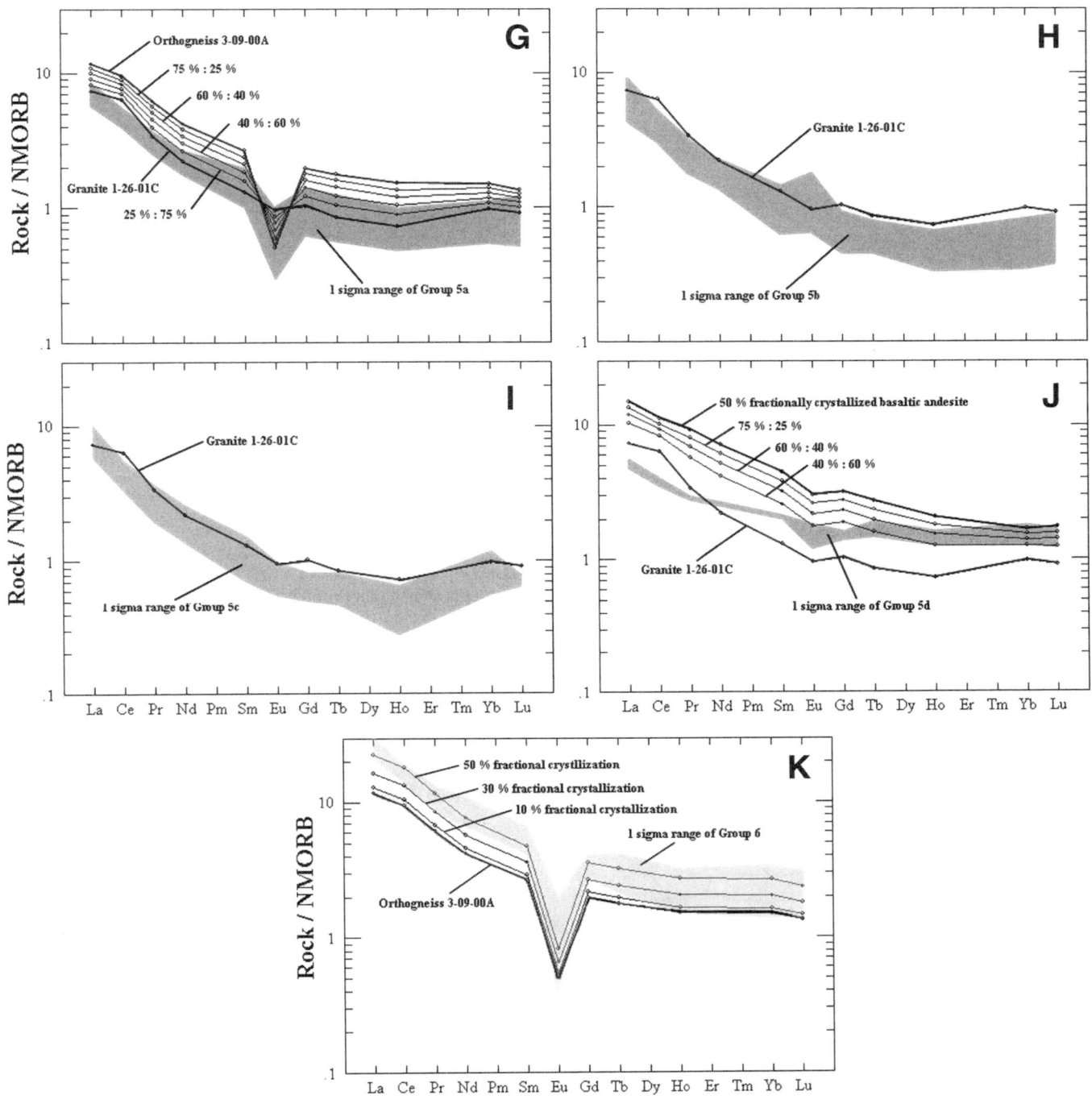

Figure 14 (*continued*). (G) Group 5a: Percentage ratios are proportions of melts of Honduras continental crustal rocks 3-09-00A and 1-26-01C. (H) REE trends of the Honduran granite 1-26-01C and the one sigma range of Central American Tertiary ignimbrite geochemical Group 5b. Just like Group 4b, Group 5b has an unusual positive Eu anomaly which may be related to subduction derived fluids (see main text). (I) REE trends of the Honduran granite 1-26-01C and the one sigma range of Central American Tertiary ignimbrite geochemical Group 5c. The depletion in the middle REE of Group 5c relative to 1-26-01C may be due to apatite fractionation. (J) REE mixing model for ignimbrite geochemical Group 5d. Percentage ratios are proportions of a 50% fractionally crystallized basaltic andesite residual melt to a melt of Honduras granite 1-26-01C composition. (K) REE fractional crystallization model for ignimbrite geochemical Group 6 involving 10, 30, and 50% crystallization a melt with the composition of the Honduran orthogneiss 3-09-00A. N-MORB—normal mid-oceanic-ridge basalt.

subduction may also have influenced the geochemistry of the Central American ignimbrites.

The Ba/La trend of the modern arc magmas has a maximum value in Nicaragua and shows minima in Guatemala and Costa Rica (Balzer and Planck, 1996; Patino et al., 2000). One suggestion for the variation in Ba/La has been the subduction of sediments, with greater sediment subduction occurring beneath Nicaragua (Carr et al., 1990; Walker et al., 1990; Balzer and Planck, 1996; Patino et al., 2000). Ba is enriched in marine sediments and is fluid-mobile during subduction (Carr et al., 1990; Plank and Langmuir, 1993).

Ba/La versus U/Th plots of mixing between an N-MORB magma with carbonate and hemipelagic sediments also indicate that the silicic ignimbrite magmas may have been contaminated by sediments (Fig. 15A). The groups seem to follow a mixing line between N-MORB and carbonate sediments, as do the basaltic andesites. However, Figure 15B illustrates that variation in the ignimbrite magmas and basaltic andesites is not completely the result of such mixing, likely evidence for the influence of the other factors described earlier, like continental crustal contamination.

Origin of the Central American Tertiary Ignimbrite Province Ignimbrites

Arguments for continental crustal contamination and AFC in the petrogenesis of the Central American Tertiary Ignimbrite Province are based on four criteria:
1. The volumetric predominance of rhyolite over other magma types. Although basaltic andesite, andesite, and dacites, occur in the Tertiary volcanics, they are rare;
2. Complex zoning in ignimbrite phenocrysts and the common occurrence of xenocrysts in the ignimbrite magmas;
3. $^{87}Sr/^{86}Sr$ isotopic values within the range of a continental crustal component; and
4. Geochemical trace element modeling that indicates that all of the ignimbrite geochemical groups of Jordan et al. (2006) can be produced by AFC of continental crust by N-MORB and/or basaltic andesites, mixing between two continental crustal melts, or by fractional crystallization of continental crustal melts. This is most likely a two-stage process.

A Working Petrogenetic Model

In the modern Central American Arc, Sr and Nd isotopic ratios have been used to indicate that the modern mafic magmas are the products of mixing of at least four different components: marine sediments, N-MORB, E-MORB, and continental crust (Patino et al., 2000). Carr et al. (1990) used Sr and Nd isotopic ratios to illustrate that most of the modern calc-alkaline magmas are dominated by mixing between subduction-modified mantle and E-MORB. Because these young arc volcanic rocks fall on a mixing trend between these two MORB sources, Carr et al. reason that mixing occurs between melt generated by fluid release into the mantle overlying the subducting slab and melt that is generated by decompression melting of asthenosphere (E-MORB) as it rises upward in the mantle wedge. On the other hand, as has been shown in this study, the most likely original source magma for the Tertiary magmas of this arc is N-MORB, rather than E-MORB.

As early as 25 Ma, a period of super-fast seafloor spreading occurred at the Farallon-Pacific plate boundary (Fig. 2) (Wilson, 1996). Shortly after (ca. 23 Ma), the Cocos-Nazca plate spreading center was formed, with the breakup of the Farallon plate (Meschede and Barckhausen, 2001). During the period of 25–12 Ma, rates of spreading are estimated to have been 180–210 mm/yr along the Farallon-Pacific and subsequent Cocos-Pacific plate boundaries. Modern spreading rate values along the Cocos-Pacific plate boundary are estimated to be 152 mm/yr (Wilson, 1996). The rapid spreading rate would correspond with an increase in subduction rate at the Central American trench. A likely result of this increase in subduction rate would have been the flux of much larger amounts of water into the mantle wedge, leading to generation larger quantities of melt. This melt generation would be due to a lowering of the melting temperature in a larger volume of the mantle wedge than in the present arc system at its lower subduction rate.

Mantle-derived basaltic melts (Tatsumi, 2005) rise due to density differences, would likely pond at the mantle-crust boundary or within the LC, where densities are equal, until it cooled and fractionated enough mafic minerals or assimilated enough crustal material (or both), producing the basaltic andesites and lowering its density enough to continue to rise (Fig. 16).

In more recent work than that of Huppert and Sparks (1988), Annen and Sparks (2002) modeled the intrusion of 50 m sills of dry mantle melts into a wet lower crust every 10,000 yr. They showed that this flux will form a partial melt zone 4 km thick in 0.8 million years or 8 km thick in just 1.6 million years in the lower crust near the Moho. In the context of Central America, given that the Tertiary ignimbrite eruption events occurred over a period of ~10 m.y., regular intrusions of 50 m sills of mantle melts would result, if no melt was removed, in a partial melt zone 50 km thick. A zone of such thickness seems unlikely, but extrapolation from the model suggests that regular intrusions of mantle melts into the lower crust, over the period of the Central American Tertiary Ignimbrite Province formation, would produce a partial melt zone of silicic magma that would be sufficient in size to serve as a source for the large volume of ignimbrite erupted at the surface.

This would also suggest that even with a fairly constant rate of mantle melt intrusion into the LC, a greater quantity of silicic magma would be produced with time due to the lower temperature of fusion in the LC. Thus, the evidence of the flare-up in the marine accumulation rate record (Fig. 2) showing an increase and decrease over time rather than an abrupt start and stop may be the result of fusion temperatures in the continental crust during constant intrusion and then cessation of mantle melts into the continental crust, even if those intrusions are a result of a

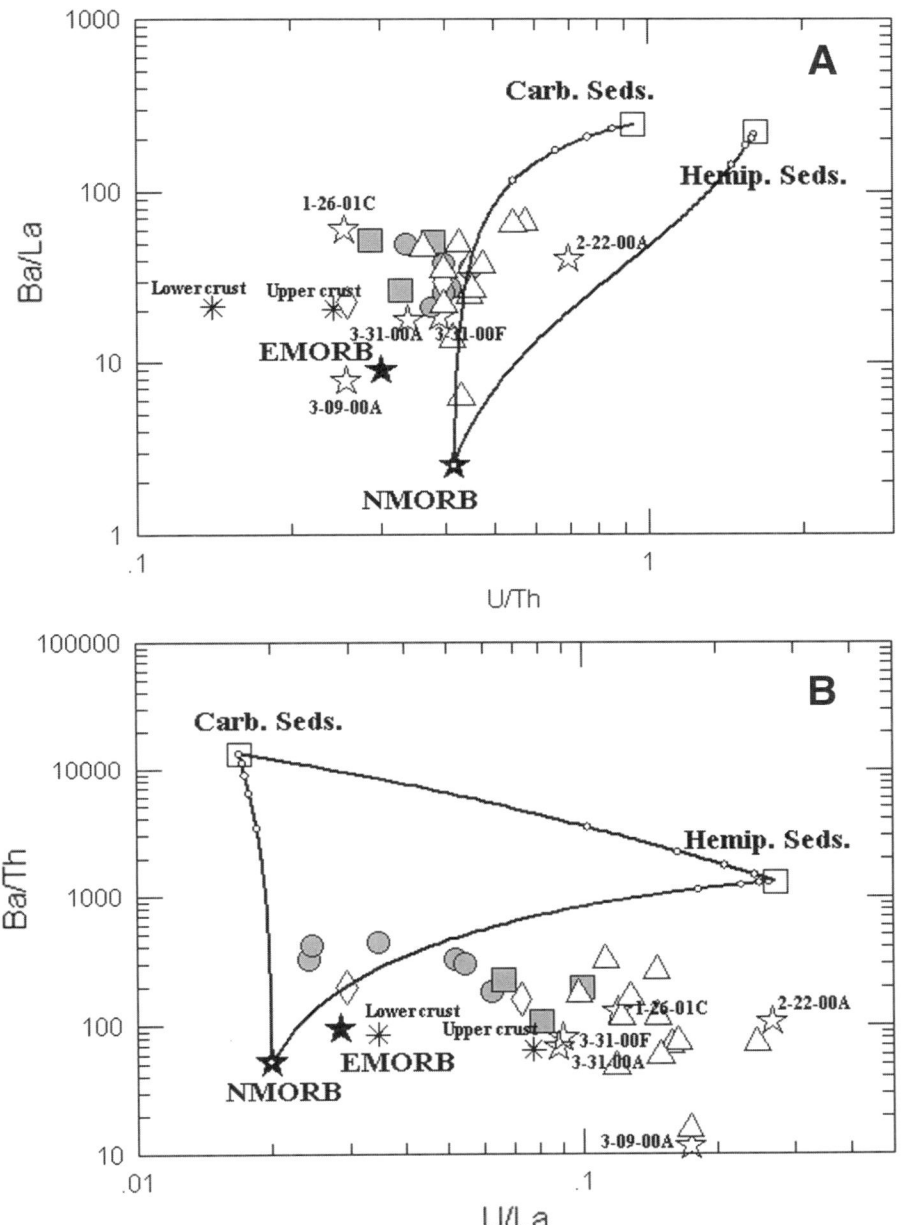

Figure 15. Trace element ratios used to evaluate sediment influence in the generation of the Central American Tertiary silicic ignimbrite magmas. Lines are binary mixing trends between the endpoints in 20% increments. For similar plots of the modern Central American arc magmas, see Patino et al. (2000). (A) Ba/La versus U/Th seems to suggest sediment contamination. (B) Ba/Th versus U/La trends indicate that although sediment contamination is a possibility, other processes are more likely the dominant cause of ignimbrite element trends. Open squares are sediments (Patino et al., 2000); closed squares are dacites; open triangles are ignimbrite groups; open stars are Honduras basement rocks; asterisks are average upper and lower continental crust (Wedepohl, 1995). E-MORB—enriched mid-oceanic-ridge basalt; N-MORB—normal mid-oceanic-ridge basalt.

relatively sudden increases or decreases of subduction. It may also be notable that the continental crustal thickness of Annen and Sparks' (2002) models (38 km) closely approximates that of Central American (45–35 km). Given this modeling and the length of the flare-up, assimilation and mixing of continental crustal melts is very likely.

The basaltic andesites appear to be products of AFC involving lesser amounts of N-MORB derived magmas (10% fractionally crystallized) relative to continental crustal material. In turn, six of the ignimbrite magmatic groups appear to be products of AFC involving lesser amounts of basaltic andesite derived magmas (~25%–35%) relative to continental crustal material, or even almost pure crustal melts (Groups 1b, 4b, 5a, 5b, 5c, 6), which may be explained by a crustal intrusion scenario similar to that described by Huppert and Sparks (1988). The other groups (1a, 2a, 3a, 4a, 5d), which involve AFC of up to 50% fractional crystallization of a basaltic andesite melt and the same relative amount of N-MORB (25%–35%), fit the melting, assimilation, storage, and homogenization (MASH) model better (Hildreth and Moorbath, 1988). The MASH model involves rising, staging, and mixing of the magmas within the crust rather than just ponding in the lower crust.

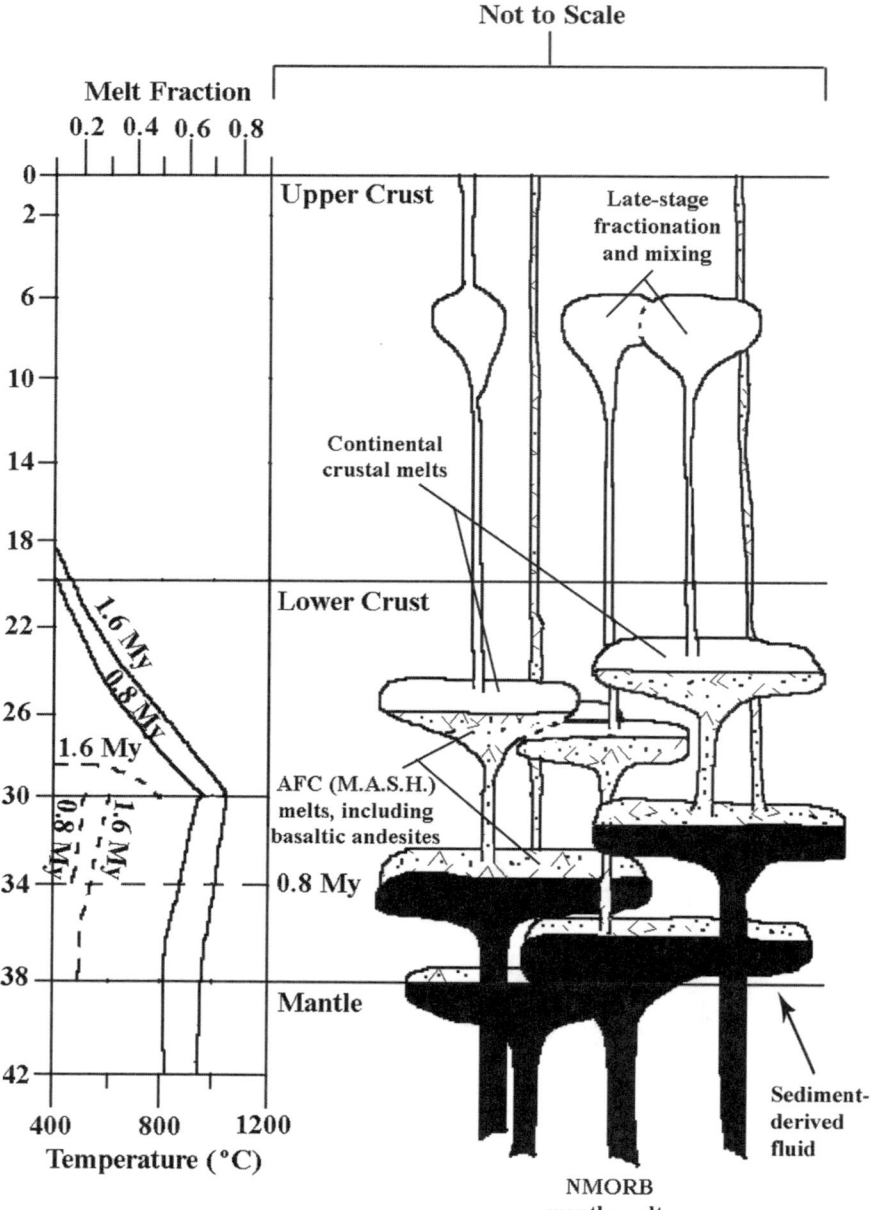

Figure 16. Working model for the generation of the Central American Tertiary ignimbrites. Initial normal mid-oceanic-ridge basalt (N-MORB) melts stage, fractionate, and assimilate continental crust. The new melt rises, stages, and continues to fractionate and assimilate, with some mixing occurring. Buoyancy contrasts prevent most mafic melts from reaching the surface. Many melts are continental crustal melts. Temperature (solid lines) and melt fraction profile (dashed lines) from Annen and Sparks (2002). AFC—assimilation-fractional crystallization; M.A.S.H.—melting, assimilation, storage, and homogenization.

The wide variation in trace element compositions in the Central American Tertiary silicic magmas could be the result of staging and assimilation at various levels within the continental crust. In some cases the mafic magmas would not reach the surface at all, as indicated by the lack of any known Tertiary lavas more mafic than basaltic andesite in Honduras. In addition, the heterogeneous nature of the continental crust of Central America would also tend to lead to generation of a wide variety of silicic magmas. These could rise to the surface independent of each other or mix in magma reservoirs prior to eruption, leading to the variety of the silicic ignimbrites now present at the surface (Fig. 16).

The slab gap that exists beneath Central America has been estimated to have formed between 3.8 Ma and 10 Ma (Rogers et al., 2002), after the major Miocene eruptions. Resulting asthenospheric flow through the break could explain why most of the modern lavas appear to have an E-MORB source, whereas the Miocene lavas have an N-MORB one. During the Miocene, the mantle wedge was made up predominantly of depleted, upper mantle material due to a long period of melt extraction. After the break-off, lower, enriched mantle material rose through the gap (Fig. 17). In this scenario, the mechanism for "shutting off" the large volumes of ignimbrite magma may have been the slowing

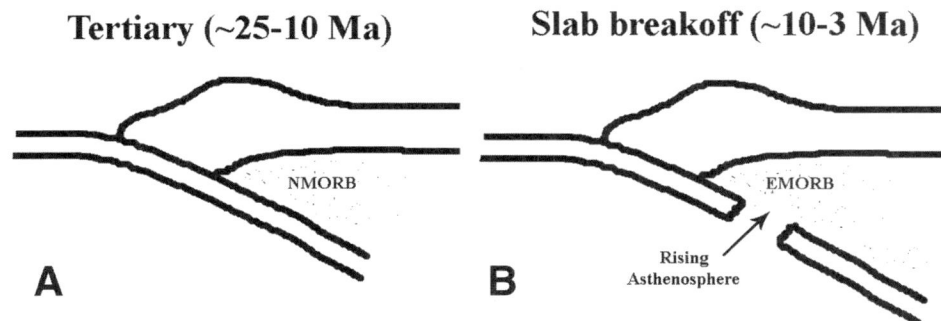

Figure 17. Explanation for normal mid-oceanic-ridge basalt (N-MORB)–type melt generation during the Tertiary in Central America volcanic arc (A) followed by enriched (E)-MORB–type melt generation in the modern Central American volcanic arc (B) due to breaking of the downgoing slab and rising asthenospheric material.

of the slab before break-off due to the subduction of younger, less dense, oceanic crust produced during the period of rapid spreading. Once the slab detached, spreading rates were no longer as high, and less water was fluxing into the wedge, preventing the large scale melting that was occurring previously.

CONCLUSIONS

The petrogenetic processes and tectonic mechanisms that produce ignimbrite flare-ups are not known, and various arguments have been made for their generation, either by fractional crystallization or assimilation-fractional crystallization.

The conclusions of this study are that the influence of the continental crust on the formation of the Central American Tertiary ignimbrites is substantial. Several reasons for this are as follows:

1. The very large volume of the rhyolite ignimbrites and the Caribbean Sea tephra layers in comparison to the rare occurrences of mafic magmas. The quantity of silicic magmas produced by melting and/or assimilation of continental crust by underplating and/or injection into the continental crust of a given volume of mafic magmas is much larger than those produced by strictly fractional crystallization of the same volume of mafic magma. It would require between 70% and 90% fractional crystallization of the Central American Tertiary basaltic andesites to evolve the compositions of the Tertiary rhyolitic ignimbrites.
2. All of the ignimbrites analyzed for $^{87}Sr/^{86}Sr$ isotopes have values within the range of continental crustal rocks.
3. REE and other trace element trends are similar to those of average lower continental crust.
4. In models, fractional crystallization fails to explain the major and trace element compositions of the ignimbrites.
5. Assimilation-fractional crystallization and continental crustal melt mixing models produce evolutionary trends that are consistent with the compositions of the ignimbrites. Six of the Central American ignimbrite geochemical groups appear to be products of AFC involving proportions of ~25%–50% basaltic andesite derived magmas relative to continental crustal material (Groups 1a, 1b, 2a, 3a, 4a, 5a, 5d), which may be explained by the MASH model of Hildreth and Moorbath (1988). A crustal melting scenario similar to that described by Huppert and Sparks (1988) may explain the formation of the other geochemical groups (4b, 5b, 5c, 6).

In addition to these conclusions that the petrogenesis of the Central American Tertiary Ignimbrite Province involved contamination by continental crust, there are three other important results of this study:

1. The dates obtained in this study are a significant contribution to the slowly growing geochronology of the province.
2. Geochemical modeling using assimilation-fractional crystallization models indicates that initial melts beneath the Central American Tertiary Ignimbrite Province were from an N-MORB–type source, rather than the E-MORB–type source postulated for the modern Central American arc.
3. The Central American Tertiary ignimbrite magmas were very likely contaminated by fluids derived from ocean sediments. The evidence for this includes the positive Eu anomaly in two of the ignimbrite geochemical groups as well as the plotting of the ignimbrite magmas within mixing fields between N-MORB and oceanic sediments.

Abnormally rapid subduction of the Farallon-Cocos plate beneath Central America that appears to coincide with the formation of the Central American Tertiary Ignimbrite Province in Central America may have caused a large influx of water into the mantle wedge beneath Central America, resulting in the generation of large volumes of mantle melt. This melt rose until, due to density contrasts it ponded at the crust-mantle boundary or within the lower crust, it formed a zone of melt similar to that modeled by Annen and Sparks (2002).

These N-MORB melts then fractionated and assimilated lower continental crust to produce basaltic andesites, which could have continued to rise and be contaminated by crust to produce the silicic ignimbrite magmas. It is also possible that the mantle melt could have melted enough crustal material to produce the silicic ignimbrite magmas without significant intermediate melts. Some of the silicic magmas also mixed with each other before reaching the surface. The final results were highly evolved, rhyolitic melts that produced the Central American Tertiary Ignimbrite Province and the deep-sea tephra layers within Caribbean Sea sediments during ODP Leg 165.

ACKNOWLEDGMENTS

Gratitude is expressed to the U.S. National Science Foundation, and thus American taxpayers, by whom this research was funded under NSF Project EAR-9909769. A special thanks also to Lina Patino at Michigan State University for her input and constructive comments, and to Tom Vogel at Michigan State and Shan de Silva of the University of North Dakota for helpful reviews of the manuscript. The authors are also very grateful to Joe Devine of Brown University, Terry Planck, Linda Farr, Katie Kelley, and Joel Sparks of Boston University, and Don Hermes at the University of Rhode Island for the use of their laboratory facilities. Richard Kingsley of the University of Rhode Island's Graduate School of Oceanography and Nick Pierce of the University of Wales (UK) were also most helpful in the obtaining and discussion of the ICP-MS data.

REFERENCES CITED

Annen, C., and Sparks, C.S.J., 2002, Effects of repetitive emplacement of basaltic intrusions on thermal evolution and melt generation in the crust: Earth and Planetary Science Letters, v. 203, p. 937–955, doi: 10.1016/S0012-821X(02)00929-9.

Ayalew, D., Barbey, P., Marty, B., Reisberg, L., Yirgu, G., and Pik, R., 2002, Source, genesis, and timing of giant ignimbrite deposits associated with Ethiopian continental flood basalts: Geochimica et Cosmochimica Acta, v. 66, p. 1429–1448, doi: 10.1016/S0016-7037(01)00834-1.

Bacon, C.R., and Druitt, T.H., 1988, Compositional evolution of the zoned calc-alkaline magma chamber of Mount Mazama, Crater Lake, Oregon: Contributions to Mineralogy and Petrology, v. 98, p. 224–256, doi: 10.1007/BF00402114.

Balzer, V., and Planck, T., 1996, Sediment recycling through time in Nicaragua: Eos (Transactions, American Geophysical Union), v. 77, no. 46, suppl., p. 789.

Cameron, K.L., and Cameron, M., 1986, Whole-rock/groundmass differentiation trends of rare earth elements in high-silica rhyolites: Geochimica et Cosmochimica Acta, v. 50, p. 759–769, doi: 10.1016/0016-7037(86)90352-2.

Cameron, M., Bagby, W.C., and Cameron, K.L., 1980, Petrogenesis of voluminous Mid-Tertiary ignimbrites of the Sierra Madre Occidental, Chihuahua, Mexico: Contributions to Mineralogy and Petrology, v. 74, p. 271–284, doi: 10.1007/BF00371697.

Carr, M.J., 2003, Igpet for Windows: Version 6, July 11, 2003 (CD-ROM).

Carr, M.J., Feigenson, M.D., and Bennett, E.A., 1990, Incompatible element and isotopic evidence for tectonic control of source mixing and melt extraction along the Central American arc: Contributions to Mineralogy and Petrology, v. 105, p. 369–380, doi: 10.1007/BF00286825.

Case, J.E., Holcombe, T.L., and Martin, R.G., 1984, Map of geological provinces in the Caribbean region, in Bonini, W.E., Hargraves, R.B., and Shagam, R., eds., The Caribbean-South American plate boundary and regional tectonics: Geological Society of America Memoir 162, p. 1–30.

Cocherie, A., Calvez, J.Y., and Oudin-Dunlop, E., 1994, Hydrothermal activity as recorded by Red Sea sediments: Sr-Nd isotopes and REE signatures: Marine Geology, v. 118, p. 291–302, doi: 10.1016/0025-3227(94)90089-2.

Donnelly, T.W., Horne, G.S., Finch, R.C., and Lopez-Ramos, E., 1990, Northern Central America: The Maya and Chortis blocks, in Dengo, G., and Case, J.E., eds., The Caribbean Region: Boulder, Colorado, Geological Society of America, The Geology of North America, v. H, p. 37–76.

Elming, S., 1998, A palaeomagnetic study and K-Ar age determinations of tertiary rocks in Nicaragua, Central America, in Elming, S., Widenfalk, L., and Rodriquez, D., eds., Geoscientific Research in Nicaragua: Sweden, Luleå University of Technology, p. 1–19.

Gose, W.A., 1983, Late-Cretaceous–Early Tertiary tectonic history of southern Central America: Journal of Geophysical Research, v. 88, n. B12, p. 10,585–10,592.

Hanson, G.N., 1978, The application of trace elements to the petrogenesis of igneous rocks of granitic composition: Earth and Planetary Science Letters, v. 38, p. 26–43, doi: 10.1016/0012-821X(78)90124-3.

Herrstrom, E.A., Reagan, M.K., and Morris, J.D., 1994, Discontinuous variations in mantle composition related to flow of asthenosphere beneath southern Central America: Geological Society of America Abstracts with Programs, v. 26, no. 7, p. A-331.

Hildreth, W., and Moorbath, S., 1988, Crustal contributions to arc magmatism in the Andes of Central Chile: Contributions to Mineralogy and Petrology, v. 98, p. 455–489, doi: 10.1007/BF00372365.

Huppert, H.E., and Sparks, R.S.J., 1988, The generation of granitic magmas by intrusion of basalt into continental crust: Journal of Petrology, v. 29, p. 599–624.

Jackson, J.A., ed., 1997, Glossary of Geology, 4th Edition: Alexandria, Virginia, American Geological Institute, 769 p.

Johnston, S.T., and Thorkelson, D.J., 1997, Cocos-Nazca slab window beneath Central America: Earth and Planetary Science Letters, v. 146, p. 465–474, doi: 10.1016/S0012-821X(96)00242-7.

Jordan, B.R., Sigurdsson, H., Carey, S., Rogers, R., and Ehrenborg, J., 2006, Geochemical correlation of Caribbean Sea tephra layers with ignimbrites in Central America, in Siebe, C., Macías, J.L., and Aguirre-Díaz, G.J., eds., 2006, Neogene-Quaternary continental margin volcanism: A perspective from Mexico: Geological Society of America Special Paper 402, p. 161–194.

MacRae, N.D., Nesbitt, H.W., and Kronberg, B.I., 1992, Development of a positive Eu anomaly during diagenesis: Earth and Planetary Science Letters, v. 109, p. 585–591, doi: 10.1016/0012-821X(92)90116-D.

Mazzucchelli, M., Rivalenti, G., Vannucci, R., Bottazzi, P., Ottolini, L., Hofmann, A.W., and Parenti, M., 1992, Primary positive Eu anomaly in clinopyroxenes of low-crust gabbroic rocks: Geochimica et Cosmochimica Acta, v. 56, p. 2363–2370, doi: 10.1016/0016-7037(92)90194-N.

McBirney, A.R., and Williams, H., 1965, Volcanic history of Nicaragua: Berkeley, California, University of California Press, 73 p.

Meschede, M., and Barckhausen, U., 2001, Plate tectonic evolution of the Cocos-Nazca spreading center, in Silver, E.A., Kimura, G., and Shipley, T.H., eds., Proceedings of the Ocean Drilling Program, Scientific Results, v. 170, p. 1–10 (CD-ROM).

Michael, P.J., 1983, Chemical differentiation of the Bishop Tuff and other high-silica magmas through crystallization processes: Geology, v. 11, p. 31–34, doi: 10.1130/0091-7613(1983)11<31:CDOTBT>2.0.CO;2.

Patino, L.C., Carr, M.J., and Feigenson, M.D., 2000, Local and regional variations in Central American lavas controlled by variations in subducted sediment input: Contributions to Mineralogy and Petrology, v. 138, p. 265–283, doi: 10.1007/s004100050562.

Plank, T., and Langmuir, C.H., 1993, Tracing trace elements from sediment input to volcanic input at subduction zones: Nature, v. 362, p. 739–743.

Prosser, J.T., and Carr, M.J., 1987, Poás volcano, Costa Rica: geology of the summit region and spatial and temporal variations among the most recent lavas: Journal of Volcanology and Geothermal Research, v. 33, p. 131–146, doi: 10.1016/0377-0273(87)90057-6.

Pushkar, P., MicBirney, A.R., and Kudo, A.M., 1972, The isotopic composition of strontium in Central American ignimbrites: Bulletin Volcanologique, v. 35, p. 265–294.

Reiners, P.W., Nelson, B.K., and Ghiorso, M.S., 1995, Assimilation of felsic crust by basaltic magma: Thermal limit and extents of crustal contamination of mantle-derived magmas: Geology, v. 23, p. 563–566, doi: 10.1130/0091-7613(1995)023<0563:AOFCBB>2.3.CO;2.

Renne, P.R., Swisher, C.C., Deino, A.L., Karner, D.B., Owens, T.L., and DePaolo, D.L., 1998, Intercalibration of standards, absolute ages and uncertainties in $^{40}Ar/^{39}Ar$ dating: Chemical Geology, v. 145, p. 117–152, doi: 10.1016/S0009-2541(97)00159-9.

Reynolds, J.H., 1980, Late Tertiary volcanic stratigraphy of Northern Central America: Bulletin of Volcanology, v. 43-3, p. 601–607.

Rogers, N., and Hawkesworth, C., 2000, Composition of magmas, *in* Sigurdsson, H., Houghton, B.F., McNutt, S.R., Rymer, H., and Stix, J., Encyclopedia of Volcanoes: New York, Academic Press, p. 89–147.

Rogers, R.D., Kárason, H., and van der Hilst, R.D., 2002, Epeirogenic uplift above a detached slab in northern Central America: Geology, v. 30, p. 1031–1034, doi: 10.1130/0091-7613(2002)030<1031:EUAADS>2.0.CO;2.

Rollinson, H.R., 1993, Using Geochemical Data: Evaluation, Presentation, Interpretation: New York, Prentice Hall, 352 p.

Rudnick, R.L., and Fountain, D.M., 1995, Nature and composition of the continental crust: A lower crustal perspective: Reviews of Geophysics, v. 33, p. 267–309, doi: 10.1029/95RG01302.

Ruiz, J., Patchett, P.J., and Arculus, R.J., 1988, Nd-Sr isotope composition of lower crustal xenoliths—Evidence for the origin of mid-Tertiary felsic volcanics in Mexico: Contributions to Mineralogy and Petrology, v. 99, p. 36–43, doi: 10.1007/BF00399363.

Schilling, J.-G., Zajac, M., Evans, R., Johnston, T., White, W., Devine, J.D., and Kingsley, R., 1983, Petrologic and geochemical variations along the Mid-Atlantic Ridge from 27° N to 73° N: American Journal of Science, v. 283, p. 510–586.

Sigurdsson, H., Leckie, R.M., Acton, G.D., and Shipboard Scientific Party, 1997, Caribbean volcanism, Cretaceous/Tertiary impact, and ocean-climate history: Synthesis of Leg 165: Proceedings of the Ocean Drilling Program, Initial Reports, v. 165, p. 377–400.

Sigurdsson, H., Kelley, R.M., Carey, S., Bralower, T., and King, J., 2000, History of circum-Caribbean explosive volcanism: $^{40}Ar/^{39}Ar$ dating of tephra layers: Proceedings of the Ocean Drilling Program, Scientific Results, v. 165, p. 299–314.

Smith, M.E., Singer, B., and Carroll, A.R., 2003, $^{40}Ar/^{39}Ar$ geochronology of the Eocene Green River Formation, Wyoming: Geological Society of America Bulletin, v. 115, p. 549–565, doi: 10.1130/0016-7606(2003)115<0549:AGOTEG>2.0.CO;2.

Sun, S.-s., and McDonough, W.F., 1989, Chemical and isotopic systematics of oceanic basalts: implications for mantle composition and processes, *in* Saunders, A.D., and Norry, M.J., eds., 1989, Magmatism in the Ocean Basins: Geological Society [London] Special Publication 42, p. 313–345.

Tatsumi, Y., 2005, The subduction factory: How it operates in the evolving Earth: GSA Today, v. 15, p. 4–10, doi: 10.1130/1052-5173(2005)015[4:TSFHIO]2.0.CO;2.

Tatsumi, Y., and Eggins, S., 1995, Subduction zone magmatism: Cambridge, Massachusetts, Blackwell Science, 211 p.

Verma, S.P., 2001, Geochemical and Sr-Nd-Pb isotopic evidence for a combined assimilation and fractional crystallization process for volcanic rocks from the Huichapan caldera, Hidalgo, Mexico: Lithos, v. 56, p. 141–164, doi: 10.1016/S0024-4937(00)00062-1.

Vogel, T.A., Patino, L.C., Alvarado, G.E., and Gans, P.B., 2004, Silicic ignimbrites within the Costa Rican volcanic front: evidence for the formation of continental crust: Earth and Planetary Science Letters, v. 226, p. 149–159, doi: 10.1016/j.epsl.2004.07.013.

Walker, J.A., Carr, M.J., Feigenson, M.D., and Kalamarides, R.I., 1990, The petrogenetic significance of interstratified high- and low-Ti basalts in central Nicaragua: Journal of Petrology, v. 31, p. 1141–1164.

Wark, D.A., 1991, Oligocene ash flow volcanism, Northern Sierra Madre Occidental: Role of mafic and intermediate-composition magmas in rhyolite genesis: Journal of Geophysical Research, v. 96, p. 13,389–13,411.

Webber, K.L., Fernandez, L.A., and Simmons, Wm.B., 1994, Geochemistry and mineralogy of the Eocene-Oligocene volcanic sequence, Southern Sierra Madre Occidental, Juchipila, Zacatecas, Mexico: Geofísica Internacional, v. 33, p. 77–89.

Wedepohl, K.H., 1995, The composition of the continental crust: Geochimica et Cosmochimica Acta, v. 59, p. 1217–1232, doi: 10.1016/0016-7037(95)00038-2.

Weyl, R., 1980, Geology of Central America: Berlin, Gebruder Borntrager, 371 p.

Williams, H., McBirney, A.R., and Dengo, G., 1964, Geologic reconnaissance of southeastern Guatemala: University of California Press, Berkeley, California, University of California Publications in Geological Sciences, v. 50, 56 p. with plates.

Williams, H., and McBirney, A.R., with Aoki, K., 1969, Volcanic history of Honduras: University of California Press, Berkeley, California, University of California Publications in Geological Sciences, v. 85, 101 p.

Wilson, D.S., 1996, Fastest known spreading on the Miocene Cocos-Pacific plate boundary: Geophysical Research Letters, v. 23, p. 3003–3006, doi: 10.1029/96GL02893.

Wilson, M., 1989, Igneous Petrogenesis: Boston, Massachusetts, Kluwer Academic Publishers, 466 p.

Manuscript Accepted by the Society 22 December 2006

Regional Books of Interest from GSA

Superior books at sensational prices!

Volcanic Hazards in Central America
edited by William I. Rose, Gregg J.S. Bluth, Michael J. Carr, John W. Ewert, Lina C. Patino, and James W. Vallance, 2006

This volume is a sampling of current scientific work about volcanoes in Central America with specific application to hazards. The papers reflect a variety of international and interdisciplinary collaborations and employ new methods. The book will be of interest to a broad cross section of scientists, especially volcanologists. The volume also will interest students who aspire to work in the field of volcano hazards mitigation or who may want to work in one of Earth's most volcanically active areas.

SPE412, 276 p. plus index, ISBN-10 0-8137-2412-0; ISBN-13 978-0-8137-2412-6
$80.00, member price $56.00

Caribbean–South American Plate Interactions, Venezuela
edited by Hans G. Avé Lallemant and Virginia B. Sisson, 2005

Rocks in plate boundary zones are generally strongly deformed. Rocks in the Mesozoic–Cenozoic Caribbean–South American plate boundary zone in Venezuela are no exception. The first of four major deformation events occurred in Jurassic to Early Cretaceous time and is expressed by normal faults recognized in seismic reflection lines and by extensional mylonites in the Tinaquillo alpine-type peridotite. Subsequently, Early Cretaceous subduction created high-pressure–low-temperature mélanges that were exhumed in the Late Cretaceous to Eocene. Next, north-south contraction resulted in an Eocene fold and thrust belt. The final event from Eocene to Recent resulted in west to east diachronous, right-oblique convergence and collision of the Leeward Antilles arc. All of this is documented with new geochronology, geochemistry, petrology, sequence stratigraphy, structural geology, and reflection seismology.

SPE394, 331 p. plus index, plates, ISBN 0-8137-2394-9
$40.00, member price $28.00 REDUCED PRICE!

Active Tectonics and Seismic Hazards of Puerto Rico, the Virgin Islands, and Offshore Areas
edited by Paul Mann, 2005

Puerto Rico and the Virgin Islands occupy a 450-km-long and 300-km-wide segment of the seismically active North America–Caribbean plate boundary zone. Geologic and seismological information on both onland and offshore plate boundary faults are critical for understanding the earthquake and tsunami hazards that these structures pose to a rapidly urbanizing island population of about 4 million inhabitants. This volume presents an integrated set of 15 chapters on the geological, geophysical, and seismological nature of late Quaternary plate boundary zone faults revealed by both onland and offshore studies. The volume chapters are grouped into four sections: (1) three introductory chapters establishing the regional tectonic setting of Puerto Rico and the Virgin Islands and its offshore area using GPS-based geodesy and regional geologic information; (2) three chapters on the instrumental and historical seismicity of the region; (3) five chapters on the identification of late Quaternary faults in Puerto Rico and its shallow coastal areas using onland mapping, fault trenching, and offshore geophysical mapping; and (4) four chapters on seismic sources, ground amplification, and paleoliquefaction.

SPE385, 287 p. plus index, ISBN 0-8137-2385-X
$50.00, member price $35.00 REDUCED PRICE!

Natural Hazards in El Salvador
edited by William I. Rose, Julian J. Bommer, Dina L. López, Michael J. Carr, and Jon J. Major, 2004

Scientists, especially geoscientists, must increasingly focus research, training, and other work on the mitigation of natural hazards in less-developed countries. Each country is different; natural hazard mitigation occupies varying ranks in national priorities, and it is often difficult to complete work that has an impact on a country and builds its infrastructure. The requirements of such work may leave scientists feeling unprepared, and little comprehensive literature is available regarding these challenges. This volume meets a vital need: it focuses on a single country but provides information that will be useful for other countries as well. El Salvador is a small, third-world country with significant seismic, landslide, and volcanic hazards. Recent events in El Salvador include civil war, floods, drought, and major hurricanes, earthquakes, and landslides, and the country has forged a new plan to help it face these severe natural hazards. This plan includes the development of a new geological agency, which is seeking outside assistance. Scientists throughout the world are already doing significant work aimed at helping with this effort, work that is part of a worldwide endeavor to enhance the local infrastructure's ability to mitigate natural hazards. The effort in El Salvador is of interest to the entire community of natural hazards workers and particularly to geoscientists concentrating on hazard mitigation. This volume highlights volcanic, seismic, and landslide hazards, with contributions from Salvadorans as well as scientists from North America and Europe.

SPE375, 480 p. plus index, ISBN 0-8137-2375-2
$50.00, member price $35.00 REDUCED PRICE!

Petrologic and Structural History of Tobago, West Indies: A Fragment of the Accreted, Mesozoic Oceanic Arc of the Southern Caribbean
by Arthur W. Snoke, David W. Rowe, J. Douglas Yule, and Geoffrey Wadge, 2001

Petrologic, geochemical, and isotopic data for the igneous and metamorphic rocks exposed on Tobago, West Indies, indicate an origin in an intraoceanic-arc system. However, these Mesozoic rocks now form part of an allochthonous terrane (Tobago terrane) within the South American–Caribbean plate boundary zone. Tobago provides an exceptional opportunity to study compositional and structural variations of an ancient oceanic-arc sequence at contrasting structural levels. A greenschist- to amphibolite-facies metavolcanic and metasedimentary schist belt forms the footwall block, whereas an Albian ultramafic to tonalitic plutonic suite and coeval Tobago Volcanic Group form the hanging-wall block of the plastic-to-brittle Central Tobago normal-sense fault system. This book is a detailed petrologic and structural analysis of the Mesozoic oceanic-arc crust exposed on Tobago. Geologic maps, cross sections, structural stereograms, detailed field and petrographic descriptions, and numerous photographs provide an extensive database for comparison with other oceanic-arc rocks exposed throughout the Caribbean region as well as in other orogens.

SPE354, 54 p., ISBN 0-8137-2354-X, 4 plates
$12.00 REDUCED PRICE! (Sorry, no additional discounts)

GSA SALES AND SERVICE P.O. Box 9140 • Boulder, Colorado 80301-9140, USA
+1.303.357.1000, option 3 • Toll free: +1.888.443.4472 • Fax: +1.303.357.1071

www.geosociety.org

THE GEOLOGICAL SOCIETY OF AMERICA®